Lecture Notes in Mathematics

Edited by A. Dold and B. Eckmann

T0185167

1190

Optimization and Related Fields

Proceedings of the "G. Stampacchia International School of Mathematics" held at Erice, Sicily
September 17–30, 1984

Edited by R. Conti, E. De Giorgi and F. Giannessi

Springer-Verlag
Berlin Heidelberg New York Tokyo

Editors

Roberto Conti
Istituto Matematico "U.DINI", Università di Firenze
Viale Morgagni 67/A, 50134 Firenze, Italy

Ennio De Giorgi
Scuola Normale Superiore
Piazza dei Cavalieri 7, 56100 Pisa, Italy

Franco Giannessi
Dipartimento di Matematica, Università di Pisa
Via F. Buonarroti 2, 56100 Pisa, Italy

Mathematics Subject Classification (1980): 49-XX, 65KXX, 90CXX

ISBN 3-540-16476-6 Springer-Verlag Berlin Heidelberg New York Tokyo
ISBN 0-387-16476-6 Springer-Verlag New York Heidelberg Berlin Tokyo

Printed in Germany

Printing and binding: Beltz Offsetdruck, Hemsbach/Bergstr.
2146/3140-543210

Preface

This volume contains the Proceedings of an International School of Mathematics devoted to Optimization and Related Fields. Held from 17 to 30 September 1984, the site for the meeting was the "Ettore Majorana" Centre for Scientific Culture in Erice, Sicily. The course was conceived and directed by R. Conti, E. De Giorgi and F. Giannessi as an opportunity for promoting the exchange and for stimulating the interaction among the various branches of optimization, like calculus of variations, theory of optimal controls, mathematical programming, convex analysis, and so on. Such branches have now a fast development around the world, so that a titer interaction seems to be interesting and useful.

The school was attended by 67 people from 12 countries. In the tradition of the Ettore Majorana Centre for Scientific Culture, the participants were classified either as "invited speakers" or as "students", though in fact the latter were not students in the ordinary sense; most of them were Ph.D. holders with established publication records. Several of them contributed to the course with the following seminars:

- On point-to-set mappings in discrete programming and some applications by B. Bank;
- Stability and existence in nonlinear mixed-integer programming by B. Bank;
- Some relaxation results for integral functionals by G. Buttazzo;
- On a stability problem in Γ-convergence by R. De Arcangelis;
- Relaxation in the optimal control of abstract evolution inclusions by C. Engblom;
- Necessary conditions for constrained optimization problems: some remarks on Dubovitskii-Miljutin theory by P. Favati and S. Steffè;
- Existence and approximation of optimal solutions of infinite horizon

programs by S.D.Flam;
- Existence results and stability in Pareto problems by R.Lucchetti;
- Closure and upper semi-continuity results in mathematical programming, Nash economic equilibria by R.Lucchetti and F.Patrone;
- Stable monotone variational inequalities by L.McLinden;
- Regularity conditions for constrained extremum problems by L.Martein;
- Generalized Bellman equation for the optimal time problem in Hilbert Space by F.Mignanego;
- Efficiency and optimality of allocations in large squares economies by N.S.Papageorgiou;
- Lagrangian duality in nonlinear programming by M.Pappalardo;
- On the image of a constrained extremum problem by F.Tardella;
- Duality gap: an upper bound and a characterization of the problems with zero gap by F.Tardella;
- Non-coercive variational problems and applications to elasticity by F.Tomarelli;
- Jacobi sufficiency criterion for optimal control by V.Zeidan.

We express our sincere thanks to all those who took part in the school. Special mention should be made of the Ettore Majorana Centre in Erice which offered its facilities and stimulating environment for the meeting. We are all indebted to the Italian National Research Council (C.N.R.), and Ministry of Public Education for their financial support. Finally we want to express our special thanks to Springer-Verlag for their unfailing cooperation.

R. Conti (Florence)

E. De Giorgi (Pisa)

F. Giannessi (Pisa)

Contents

Contributing Authors

J.ABADIE	Univ. of Chicago,Graduate School of Business, 1101 East 58th Street, Chicago,Illinois 60637, U.S.A.
H.ATTOUCH	Inst. de Mathématique, Université de Paris-Sud,Centre d'Orsay, 91405 Orsay Cédex, France.
A.CAMBINI	Dip. di Matematica, Univ. di Pisa, Via Buonarroti 2, 56100 Pisa, Italia.
E.CAVAZZUTI	Ist. di Matematica, Univ. di Modena,Via Campi 213/b, 41100 Modena, Italia.
J.P.CECCONI	Ist. di Matematica, Univ. di Genova, Via L.B.Alberti 4, 16132 Genova, Italia.
H.CLARKE	Centre de Recherche de Mathématiques Appliquées , Université de Montréal, C.P. 6128, succursale A, Montréal, (Québec), Canada H3C 3J7.
G.DAL MASO	Ist. di Matematica, Univ. di Udine, Via Mantica 1, 33100 Udine, Italia.
S.DOLECKI	Dépt. de Mathématiques, Université de Limoges 123, rue Albert Thomas, 87060 Limoges Cédex, France.
I.EKELAND	Univ. de Paris 9 Dauphine, Place du Maréchal De Lattre De Tassigny, 7577 Paris 16, France.
T.FRANZONI	Ist. di Matematica Appl., Facoltà di Ingegneria, Univ. di Pisa, Via Bonanno 25bis,56100 Pisa,Italia.
R.HOPPE	Fachbereich Math., Technische Universität Berlin , Str.d.17 Juni 135, 1000 Berlin 12, West Germany.
L.LIONS	Collège de France, 11 Place Marcelin-Berthelot , 75231 Paris Cédex 05, France.
L.MANGASARIAN	Univ. of Wisconsin, Computer Sciences Dept., 1210 West Dyaton Street, Madison, WI 53706, U.S.A.
A.MIELE	Aero-Astronautics Group, 230 Ryon Building, Rice Univ., POB 1892, Houston, Texas 77251, U.S.A.
L.MODICA	Dip. di Matematica , Univ. di Pisa, Via Buonarroti 2, 56100 Pisa, Italia.
R.WETS	Dept. of Mathematics, Univ. of Kentucky Lexington, KY 40506, U.S.A.
T.ZOLEZZI	Ist. di Matematica , Univ. di Genova, Via L.B.Alberti 4, 16132 Genova, Italia.

Chapter 1

Generalized Reduced Gradient and Global Newton Methods

J. Abadie

1. INTRODUCTION

The object of this paper is to show how to solve a system of n nonlinear equations

(1) $$f(x) = 0, \qquad f : R^n \to R^n$$

by the Global Newton (GN) method, using the General Reduced Gradient (GRG) method as a numerical tool. The method thus obtained is applied to the general nonlinear programming problem with equality or inequality constraints. More than one local optimum may be obtained by the method.

We first briefly review our notations and some algebraic prerequisite (Section 2). Section 3 reviews some features of the GN method. We show in Section 4 how the GRG method is applicable to GN, then we briefly explain in Section 5 how the method of Section 4 may be used for nonlinear programming problems. We present some numerical experiments in Section 6.

2. NOTATIONS AND ALGEBRAIC PRELIMINARIES

x is any point in R^n, identified with its column-matrix of components x_i, i = 1,...,n. f is a mapping $R^n \to R^n$, $f \in C^2[R^n]$. f(x) is identified with its column-matrix of components $f_i(x)$, i = 1,..,n. f'(x) is the derivative of f(x), identified with the (n,n) matrix whose elements are $\partial f_i/\partial x_j$, the row-indices i and the column-indices j running from 1 to n. The matrix f'(x) may be written row-wise as

$$f'(x) = \begin{pmatrix} f'_1(x) \\ \vdots \\ f'_n(x) \end{pmatrix}$$

where

$$f'_i(x) = \left(\frac{\partial f}{\partial x_1}, \ldots, \frac{\partial f}{\partial x_n} \right),$$

or column-wise

$$f'(x) = \left(\frac{\partial f}{\partial x_1}, \ldots, \frac{\partial f}{\partial x_n} \right),$$

where

$$\frac{\partial f}{\partial x_j} = \begin{pmatrix} \dfrac{\partial f_1}{\partial x_j} \\ \vdots \\ \dfrac{\partial f_n}{\partial x_j} \end{pmatrix}$$

We shall need the adjoint matrix $f'(x)^a$ of f'(x), defined by its elements

$$(f'(x)^a)_{j,i} = (-1)^{i+j} \det \left| (f'(x) \backslash f'_i(x)) \backslash \frac{\partial f}{\partial x_j} \right| ,$$

where the symbol (\) means "remove", so that the right hand side is the co-factor of $\partial f_i/\partial x_j$. We then have the relation

$$f'(x)f'(x)^a = f'(x)^a f'(x) = \det\{f'(x)\} I_{n,n}$$

where $I_{n,n}$ is the (n,n) identity matrix.

x° is a particular point in R^n, which may have different meanings in different Sections.

We now recall, for completeness, some prerequisite from linear algebra.

Let

$$A = (A^{o},A^{1},\dots,A^{n})$$

be some (n,n+1) matrix. Here then, the n+1 columns $A^{o},A^{1},\dots,A^{n}$ are elements of R^{n}.

Throughout this whole paper, *we assume the rank of* A *is* n.

Removing column A^{j} gives a (n,n) matrix $B^{(j)}=A\backslash A^{j}$ (we may simply call it B if no confusion arises). We set

$$d_{j} = (-1)^{j} \det(A\backslash A^{j}) \; .$$

Not all d_j's are zero (otherwise rank (A) < n). If some $d_j = 0$, then rank $(A\backslash A^{j}) = n-1$ (this rank is less than n for a singular matrix, and cannot be less than n-1, otherwise the rank of A would be less than n).

Note the identity

$$\sum_{j=0}^{n} A^{j}d_{j} = 0 \; .$$

Suppose $y_{o},y_{1},\dots,y_{n}$ are numbers that satisfy

$$\sum_{j=0}^{n} A^{j}y_{j} = 0 \; ,$$

then there exists some number α such that

$$y_{j} = \alpha d_{j}, \quad j=0,\dots,n,$$

where α is independent of j. There are two distinct cases:

Case 1: $\alpha = 0$, i.e.: $y_{j} = 0, \quad j = 0,\dots,n;$

Case 2: $\alpha \neq 0$: then $y_{j} = 0$ if and only if $d_{j} = 0$.

In Case 2 then, all the ratios y_j/d_j have the same sign (sign of α), *and*, if either one of y_j, d_j is zero, then the other one is also zero.

3. THE GLOBAL NEWTON METHOD

In order to find solutions of

(1) $$f(x) = 0$$

in some open bounded set Ω with smooth connected boundary $\partial\Omega$, the Global Newton (GN) method introduces

(2) $$H(x,\lambda) = f(x) - \lambda f(x^\circ) = 0$$

A solution to (2) is $\lambda = 1$, $x = x^\circ$. Assuming $f(x^\circ) \neq 0$, GN tries forcing λ to 0, and consequently x to some solution to (1).

More precisely, we consider a differentiable mapping $(\lambda(t),x(t))$: $R^+ \to R \times R^n$, where $t \geq 0$ is a parameter, which we call the *time*. At time $t = 0$, we have $\lambda(0) = 1$, $x(0) = x^\circ$. GN seeks some time $t^* > 0$ such that $\lambda(t^*) = 0$, $x(t^*) = x^*$, where x^* is a solution to (1). We get from (2):

(3) $$f'(x)\,\dot{x}(t) - f(x^\circ)\,\dot{\lambda}(t) = 0,$$

where the dot $(\dot{\ })$ designates the derivative d/dt with respect to t.

Setting

$$A(x) = (A^\circ, A^1, \ldots, A^n) = \left(-f(x^\circ), \frac{\partial f}{\partial x_1}, \ldots, \frac{\partial f}{\partial x_n}\right)$$

$$x_o(t) = \lambda(t)\ ,$$

we get from Section 1 (since rank(A(x)) = n):

(4) $$\dot{x}_j(t) = \alpha(x(t))\ d_j(x(t)), \quad j = 0,\ldots,n,$$

$$\text{i.e.: } \dot{x}_j = \alpha(x)\ d_j(x) = \alpha(x)d_j \quad \text{for short},$$

where $\alpha(x)$ is independent of j.

H.B. Keller adds to (3) the equation

(5) $$\|\dot{x}(t)\|^2 + \dot{\lambda}(t)^2 = 1,$$

which avoids the case $\dot{x}_j(t) = 0$, $j = 0,1,\ldots,n$. We are then mandatorily in Case 2 of the preliminaries (Section 2). Thus

(6.i) $\alpha(x(t))$ never vanishes;

(6.ii) $\dot{x}_j(t) = 0$ for some t and some j
if and only if $d_j(x(t)) = 0$ for the same t and the same j.

(7) $$\dot{x}_j(t) = \alpha(x(t))\, d_j(x(t)), \quad j = 0,\dots,n .$$

The jacobian matrix of (5), (4) with respect to $\dot{\lambda}$, $\dot{x}$

$$J = \begin{pmatrix} \dot{x}_o & \dot{x}_1 & \dots & \dot{x}_n \\ A^o & A^1 & \dots & A^n \end{pmatrix}$$

has determinant

$$\det(J) = \sum_{j=0}^{n} \dot{x}_j d_j = \alpha(x) \sum_{j=0}^{n} d_j^2 .$$

From (4) and (5) we have

$$\alpha(x)^2 \sum_{j=0}^{n} d_j^2 = 1,$$

which gives

(8) $$\alpha(x) = \pm 1 / \sqrt{\sum_{j=0}^{n} d_j^2}$$

(9) $$\det(J) = 1/\ \alpha(x).$$

Hence det(J) never vanishes, and keeps a constant sign (the sign of α).

Initially ($t = 0$) some $\pm$ sign is chosen for $\alpha(x)$ in (8). This uniquely determines the subsequent trajectory $(\lambda(t), x(t))$ for $t > 0$, from (3) and (5). The constant sign in (8) corresponds to one of the two possible opposite initial directions for the trajectory. $\lambda(t)$, $x(t)$ have continuous derivatives.

The ratios $\dot{x}_j(t) / d_j(x(t))$ all have the same sign, independant of j and t, except when $\dot{x}_i(t) = 0$ *and* $d_i(x(t)) = 0$, for some, but not all, i. Consequently, as long as, for some given j, $d_j(x(t))$ does not change sign, then $x_j(t)$ is either strictly increasing or strictly decreasing. Moreover, if $d_j(x(t))$ changes sign at time t, then the movement of

$x_j(t)$ on the x_j- axis changes direction, and reciprocally. Equivalently, as long as $d_j(x(t))$ does not change sign, then $x_j(t)$ moves on the same direction on the x_j- axis.

The tangent direction to the projection to the trajectory in the x-space is

$$\dot{x}(t) = \dot{\lambda} f'(x)^{-1} f(x^o) = \frac{\dot{\lambda}}{\lambda} f'(x)^{-1} f(x) \tag{10}$$

provided $f'(x)$ is nonsingular. This shows that this tangent direction in this case is either the Newton direction

$$h(x) = -f'(x)^{-1} f(x) \tag{11}$$

or its opposite. We also have

$$\dot{x}(t) = -\alpha(x) \frac{d_o(x)}{\lambda} h(x). \tag{12}$$

Setting

$$v(x) = d_o(x) h(x), \tag{13}$$

we have

$$v(x) = -f'(x)^a f(x), \tag{14}$$

where $f'(x)^a$ is the adjoint matrix of $f'(x)$. $v(x)$ exists and is continuous on $\bar{\Omega}$, the closure of Ω. It is easily seen that $v(x) = 0$ if only if, either $f(x) = 0$, or $f'(x)$ is singular *and* $f(x)$ is an eigenvalue of $f'(x)^a$. It is always possible then to assume that neither situation occurs on $\partial\Omega$ (move a little $\partial\Omega$ if necessary). Consequently, there is no loss is generality on assuming that $v(x)$ *is continuous and non-zero on* $\partial\Omega$.

We are now ready to state our *boundary condition:*

(15) either $v(x)$ *points into* Ω, $\forall x \in \partial\Omega$

or $v(x)$ *points out of* Ω, $\forall x \in \partial\Omega$.

This is a more concise form of the Gould-Schmidt boundary condition; it contains the Smale boundary condition, used by Keller.

When the boundary condition (15) is not satisfied, it must exist some $x \in \partial\Omega$ such that $v(x)$ is tangent to $\partial\Omega$ (recall that $v(x)$ is continuous and never vanishes). The reciprocal is understood to be true.

Using (13), we may re-write (12) as

$$\dot{x}(t) = -\alpha(x(t)) \frac{v(x(t))}{\lambda(t)} . \tag{16}$$

Remark that $\dot{x}(t)$ is never zero (otherwise, from (5), $\dot{\lambda}(t) = 1$, a contradiction to (3)), and is continuous with respect to t, as well as $\alpha(x(t))$. Hence $v(x(t))/\lambda(t)$ is a continuous vector which is never zero.

Consider now the trajectory corresponding to some $x^{o} \in \partial\Omega$, for $t=0$. Assume the $x(t)$ - trajectory leaves Ω at $x^{F} = x(t_{F})$. Since $\dot{x}(0)$ points into Ω and $\dot{x}(t_{F})$ points out of Ω, and since $\alpha(x^{o})$, $\alpha(x^{F})$ have same sign, our boundary condition implies, from (16), that $\lambda(0)$, $\lambda(t_{F})$ are of opposite signs. Hence, from the continuity of λ, we conclude that there exists some t^{*}, $0 < t^{*} < t_{F}$, such that $\lambda(t^{*}) = 0$, and, consequently that x^{*} defined by $x^{*} = x(t^{*})$ is a solution to (1).

Sufficient conditions for the exit point $x(t_{F})$ to exist have been given by Keller, and by Gouldt and Schmidt. The existence is assumed for the rest of this paper.

Suppose one seeks a solution to (1) in Ω, where Ω is an open bounded set in R^{n}, with $\partial\Omega$ smooth and connected. Start with $t = 0$, $x(0) = x^{o} \in \partial\Omega$, and, from (3) and (5), generate the trajectory, with $\dot{x}(0)$ pointing into Ω. Continue until $x(t)$ reaches again $\partial\Omega$ or $\lambda(t^{*}) = 0$. The latter case gives a solution $x^{*} = x(t^{*})$ to (1). If other solutions to (1) are sought, continue along the trajectory as long as it lies in Ω.

4. THE GENERALIZED REDUCED GRADIENT APPROACH TO GLOBAL NEWTON

Let us now describe the GRG approach to the problem of solving (1) via the GN method.

We consider the problem

(P1) $\min \phi(z) = \lambda,$ subject to

(17) $F(z) = H(x,\lambda) = f(x) - \lambda f(x^\circ) = 0$

(18) $\lambda \geq 0$

where z is $\begin{pmatrix} \lambda \\ x \end{pmatrix} = \begin{pmatrix} x_0 \\ x \end{pmatrix} \in R^{n+1}$. We start from $z^\circ = \begin{pmatrix} 1 \\ x^\circ \end{pmatrix}$.

At any particular instant t where $\lambda(t) > 0$, there exists some $d_s(x(t)) \neq 0$ (because otherwise rank[A(x(t)) < n]). Let us set

(19) $$B = B^{(s)} = A\backslash A^s,\ N = A^s,\quad x_N = x_s,\ x_B = z\backslash x_s\ .$$

Following the GRG terminology, inspired by the simplex method for linear programming, x_B is called the basic variable, x_N the non basic one. The reader is referred to the literature (end of this paper) for an account of the GRG method for the general nonlinear programming problem. The general method is here independently applied to our particular case of solving problem (P1). This will help understanding why the GRG method is ideally suited to GN.

The implicit function theorem is applicable to (1) at point z(t). It permits considering x_B as a differentiable function of x_N in the neighbourhood of z(t). The objective function, considered as a function $\Phi(x_N)$ of x_N, has derivative

(20) $$\Phi'(x_N) = c^N - c^B B^{-1} N,$$

where c is the derivative of the objective function ϕ in (P1), decomposed into its basic and nonbasic components

$$c^N = \frac{\partial \phi}{\partial x_N} = \frac{\partial \phi}{\partial x_s}, \qquad c^B = \frac{\partial \phi}{\partial x_B} = c\backslash c^N.$$

It is convenient to set

(21) $$u = -c^B B^{-1}$$

(the row-matrix $u = (u^1,\ldots,u^n)$ is the Lagrange multiplier). Thus we have

$$\Phi'(x_N) = c^N + uN, \qquad 0 = c^B + uB. \tag{22}$$

If $\Phi'(x_N) = 0$, then λ has reached a (local) minimum (neglect any inflexion point). If not, then $\dot{x}_N = -\Phi'(x_N)$ gives the direction of move for x_N. The tangent direction of move for z is given by

$$f'(x)\dot{x} - \dot{\lambda}f(x^\circ) = 0, \tag{23}$$

from which we get

$$\dot{x}_B = -B^{-1}N\dot{x}_N. \tag{24}$$

The scaling condition $\|\dot{x}\|^2 + |\dot{\lambda}|^2 = 1$ is no more taken into account, though this may easily be done. On the other hand, we may as well set $\dot{x}_N = \pm 1$ (same sign as $-\Phi'(x_N)$).

The *tangent phase* of GRG consists on making a small step along the tangent, on the $\dot{z}(t)$ direction, from $z(t)$:

$$z^\theta = z(t) + \theta\dot{z}(t), \qquad \theta > 0. \tag{25}$$

This step should satisfy

$$\lambda^\theta \geq 0. \tag{26}$$

In order for this to be possible, it is necessary that λ be nonbasic if $\lambda = 0$.

The *restoration phase* of GRG derives, from z^θ, a point on the trajectory, by keeping constant $x_N = x_N^\theta = x_s^\theta$, and by solving, with respect to x_B, equation (17). The method explicitly used in GRG is a pseudo-Newton method

$$x_B^{(k+1)} = x_B^{(k)} - B^{-1}F(x_B^{(k)}, x_N^\theta), \quad x_B^{(0)} = x_B^0 \tag{27}$$

(Newton's method here would imply recomputing $B(x^{(k)})$, then its inverse, at each iteration k).

Upon applying the pseudo-Newton method, if λ is part of the x_B-variable and becomes negative for some value k+1, then a linear interpolation is made between $x_B^{(k)}$ and $x_B^{(k+1)}$ in order to obtain a point $\tilde{x}_B^{(k+1)}$ with $\tilde{\lambda}^{(k+1)} = 0$, another index r is chosen to replace s (one should have $d_r(x(t)) \neq 0$), and the pseudo-Newton procedure is continued again, with the new basic variable, until feasibility is achieved or some difficulty (such as non-convergence) is noted; in the latter case, the stepsize θ is reduced (divided by 10, for instance). Since convergence of the pseudo-Newton method is ascertained when θ is small enough, the restoration phase eventually gives a new feasible point.

If λ is decreased at this new feasible point, then one step of the GRG method is achieved (we omit here some refinements); otherwise the stepsize θ is reduced, and the restoration phase repeated again. If θ is small enough, the restoration phase is guaranteed to succeed with a new feasible point having a smaller λ.

Problem (P1) may terminate with a positive λ. It is then necessary to start problem (P2):

(P2) $\max \lambda$, s.t. (17),(18)

(the condition (18) does not play any active role now), starting from the solution obtained for (P1). The process explained above is used, until some maximum value is reached, from which point we return to problem (P1) again, starting from the solution obtained for (P2). By a succession of alternate problems (P1),(P2), (P1),(P2),..., we eventually reach either a solution of (1), if any exists on the trajectory, or the boundary of Ω.

At start, with $t = 0$, $\lambda = 1$, $x = x^c$, a first nonsingular matrix B is selected, and its inverse is computed. If the corresponding direction $\dot{x}(0)$ points out of Ω, then we begin by (P2) instead of (P1).

In the course of solving any of the problems (P1), (P2), if the basis matrix $B^{(s)}$ approaches being singular, then the corresponding non-basic variable $x_s(t)$ is exchanged with some basic variable $x_r(t)$, chosen in such a way that $\det(B^{(r)}) \neq 0$ (this is always possible, since the $B^{(j)}(t)$'s are continuous and not all simultaneously 0). $(B^{(r)})^{-1}$ is readily computed from $(B^{(s)})^{-1}$ by *pivoting*, as in the simplex method for linear programming (pivoting requires $O(n^2)$ operations, against $O(n^3)$ for inverting $B^{(r)}$). Pivoting is also used when λ takes value 0.

A difficulty appears when the solution to (P1) has a positive λ. In this case, this solution is a stationary point to (P2), and so it is not possible to start (P2) with it. At the solution of (P1), at time t_1, the nonbasic variable is some $x_s(t_1)$, with $d_s = \det B^{(s)} \neq 0$. This index s has remained the same since some iterations, and so $x_s(t)$ has moved in a constant direction on the x_s- axis. This direction is easy to know (record the value of $\dot{z}(t)$ at each choice of a new basis index, which includes to record it at each pivoting). Solving (P2) is then started, not from $z(t_1)$, but from a point obtained by adding some small ε to $x_s(t_1)$ ($\varepsilon > 0$ if $x_s(t)$ is increasing, $\varepsilon < 0$ otherwise). In the present case, (P2) is started with the restoration phase.

An alternative to this procedure is to replace (P2) altogether by

either $(P_{1,s})$: min x_s (if $x_s(t)$ is decreasing)

or $(P_{2,s})$: max x_s (if $x_s(t)$ is increasing)

subject to (in anyone case) (17), (18). The idea is the same as before: a succession of Problems $(P_{1,s})$ or $(P_{2,s})$, with possibly varying indices s, are solved until either $\lambda = 0$ or $x(t)$ is going out of Ω.

Once a solution x^* to (1) is achieved, we may want to compute another one. For this doing, we start from $t = t^*$, $z = z(t^*)$, the following problem:

(P'1) $\min \lambda$, s.t. (1) and (2'): $\lambda \leq 0$

(notice that (2'), which replaces (2) in (P1), does not play any active role here). Once a minimum is achieved, the optimal solution is the starting point to solving

(P'2) $\max \lambda$, s.t. (1) and (2').

(here (2') may be activated). A succession of alternate (P'1), (P'2), (P'1), (P'2),... are solved, until either $\lambda = 0$ again, or $x(t)$ is going out of Ω. In the former case, if we still need another solution, we begin with problem (P2) a succession of (P2),(P1), (P2),(P1),..., and so on. Of course the alternative procedure with $(P'_{1,s})$, $(P'_{2,s})$,....., and afterwards $(P_{2,s})$, $(P_{1,s})$,... is still applicable.

A possible complete calculation, with two solutions, is illustrated on figure 1.

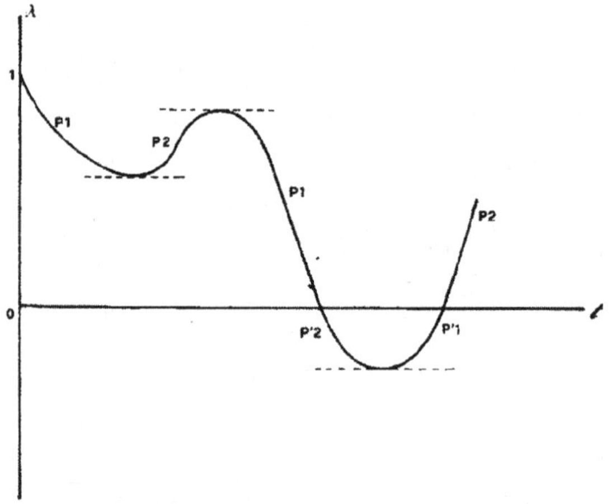

Figure 1: succession of optimization problems for Global Newton.

Instead of starting with x^o on the boundary of Ω, we may as well start with x^o *inside* of Ω, and generate *two* trajectories,starting both times with the same x^o on one direction first, and then the opposite direction.

We see that the GRG method is exactly suited to the Global Newton method, since its essence is to work some $n \times n$ nonsingular submatrix of the Jacobian matrix A(x) of (2). The GRG method solves the general nonlinear mathematical programming problem, with equalities or inequalities. In fact, we used a slightly modified GRG program to make numerical experiments (Section 6 below). Among the modifications is the following: GRG is a large step method; the stepsize must be bounded to a small value in order to closely follow the GN-trajectory. It is our pleasure here to thank Dr. G. Guerrero for the invaluable help he provided us.

5. CONSTRAINED OPTIMIZATION

The method thus far explained can be applied to constrained optimization. Suppose the problem is

$$\text{(OP)} \qquad \min f_o(x) \quad \text{s.t. } f_i(x) \le 0, \quad i = 1,\dots,m.$$

The motivation for applying GN through GRG in order to solve (OP) (instead of using GRG directly on (OP) itself) is to find more than one local solution to (OP), and should then be considered as a step towards global optimization. A good procedure might be to begin applying GRG to Problem (OP), thus generating a first (local) optimum x°. From then on, one may apply the strategy defined at the end of section 4.

The Kuhn-Tucker conditions are

$$\text{(KTC)} \quad \begin{cases} f'_o(x) + \sum_{i=1}^{m} u_i f'_i(x) = 0 \\ u_i \ge 0,\ f_i(x) \le 0,\ u_i f_i(x) = 0, \quad i = 1,\dots,m. \end{cases}$$

Setting, for some $k \ge 2$,

$$\text{(28)} \qquad \psi^+(a) = (\max\{0,a\})^k, \quad \psi^-(a) = |\min\{0,a\}|^k,$$

the system (KTC) is replaced by

$$\text{(KTC')} \quad \begin{cases} f'_o(x) + \sum_{i=1}^{m} \psi^+(y_i) f'_i(x) = 0 \\ f_i(x) + \psi^-(y_i) = 0, \quad i = 1,\dots,m, \end{cases}$$

a system of m + n equations in m + n variables x,y. Once (KTC')is solved, the multipliers u_i in (KTC) are recovered by

$$u_i = \psi^+(y_i), \quad i = 1,\dots,m. \tag{29}$$

It is worthwhile noting

$$\psi^+(a)\,\psi^-(a) = 0, \quad \forall a \in R. \tag{30}$$

There are many possible other replacements to (KTC) in the form of a system of equations. Here are two others:

$$\text{(KTC1)} \quad \begin{cases} f_o'(x) + \sum_{i=1}^{m} \frac{1}{2} z_i^2 f_i'(x) = 0 \\ f_i(x) + \frac{1}{2} y_i^2 = 0, \quad y_i z_i = 0, \quad i = 1,\dots,m; \end{cases}$$

$$\text{(KTC2)} \quad \begin{cases} f_o'(x) + \sum_{i=1}^{m} u_i f_i'(x) = 0 \\ u_i y_i = 0, \quad f_i(x) + \frac{1}{2} y_i^2 = 0, \quad i = 1,\dots,m. \end{cases}$$

System (KTC2) gives min *and* max in (OP), but this is not a serious objection when following the GN trajectory. However both (KTC1) and (KTC2) have more equations and more variables than (KTC').

6. NUMERICAL EXPERIMENTS

The three examples given here have been experimented using a modified GRG program. The first two are taken from Vignes (1980): they were used to illustrate Newton's method.

Example 1

Solve

(31) $f_1(x)=7x_2^2-6x_1x_2+4x_1+9x_2+12=0$; $f_2(x)=-7x_2^2-6x_1x_2+10x_1+x_2+30=0$

The system has three solutions:

$$A^* = \begin{pmatrix} 7.414\,656\,007 \\ 1.846\,199\,157 \end{pmatrix}, \quad B^* = \begin{pmatrix} -3 \\ 0 \end{pmatrix}, \quad C^* = \begin{pmatrix} -1.016\,505\,598 \\ -1.250\,961\,240 \end{pmatrix}$$

The general conditions are not satisfied, since $A(x) = H'(x,\lambda)$ is of rank 1 at the following three points.

$$B' = \begin{pmatrix} 43/18 \\ 0 \end{pmatrix}, \quad A' = \begin{pmatrix} -5.927\,289\,772 \\ 1.846\,199\,335 \end{pmatrix}, \quad C' = \begin{pmatrix} -1.250\,961\,240 \\ -10.05683721 \end{pmatrix}$$

(unless Ω excludes theses points). They are of the saddle point type, in the Poincaré terminology (figure 2).

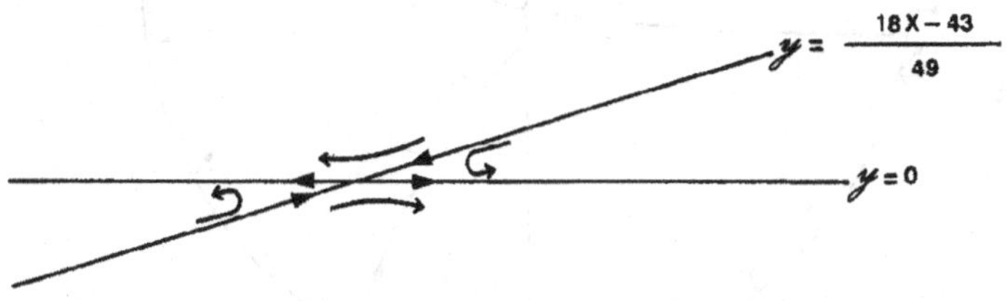

Figure 2: Some parts of trajectories near the saddle point (43/18,0).

Any trajectory is a branch of an hyperbola. Figure 3 shows two such trajectories, starting respectively at a and a': in fact, they are the two branches of the same hyperbola. The first trajectory passes through solution A^*, the second through solutions C^*, B^*. None of these trajectories contains all three solutions. Shown also on Figure 3 is v(x), which is tangent to $\partial\Omega$ at $x = w_1$ and $x = w_2$, points into Ω on the larger arc $w_1 w_2$, and points out of Ω on the smaller arc. The parabola S is the set of all x's for which det $\{f'(x)\} = 0$. There are 3 solutions A^*, B^*, C^*, and 3 singular points A', B', C' for which $f'(x)^a f(x) = 0$, $f(x) \neq 0$. The curve $f_1(x) = 0$ is an hyperbola, one branch of which contains A^* and the other B^*, C^*; the same is true for $f_2(x) = 0$. Each of these four branches of

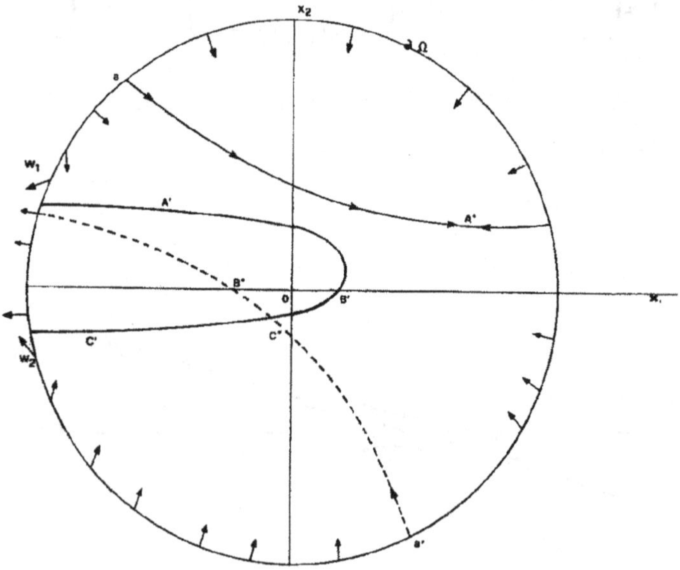

Figure 3: example 1

hyperbolas is a possible trajectory, since equation (2) shows that, if $f_i(x^o) = 0$ for some i, then $f_i(x(t)) = 0$ for the same i and all values of t. Figure 3 does not show these four branches of hyperbolas.

Figure 4 shows the behaviour of λ for various starting points.

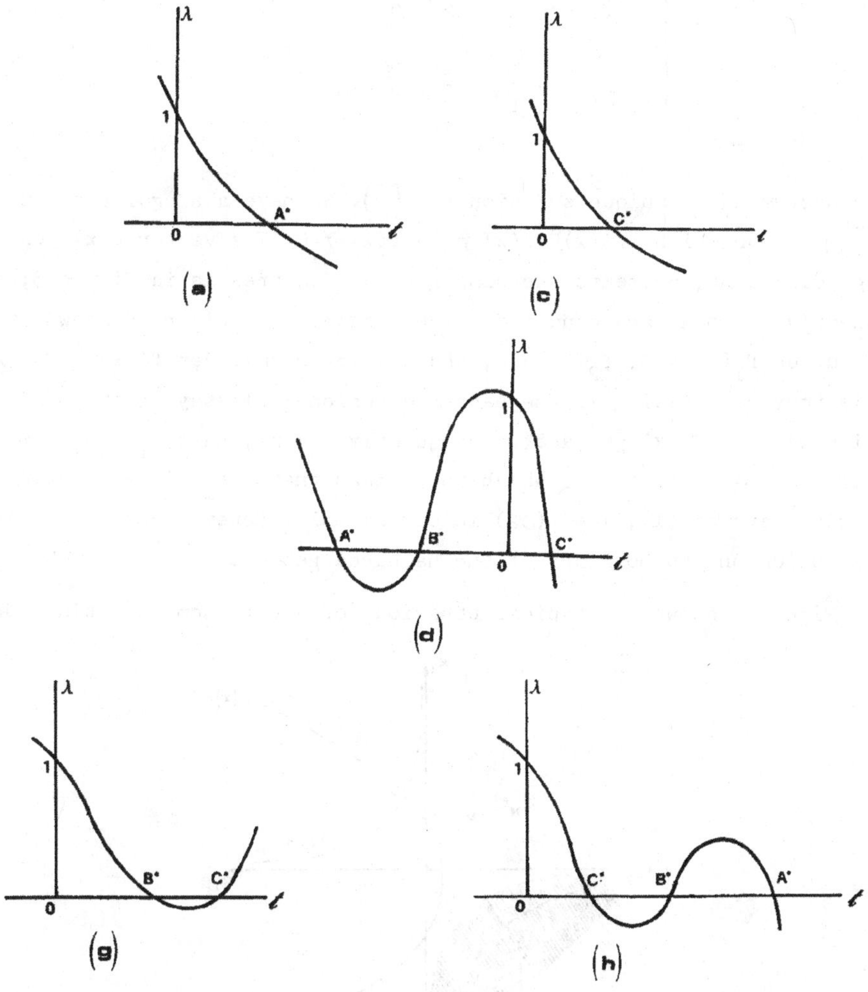

Figure 4: starting points are a=(-9,8), c=(0,-1), d=(-6,-1), g=(-6,1), h=(1,-2) in Example 1.

This example does not satisfy the boundary conditions (see $v(x)$, Figure 3).

Example 2

Solve

$$(32)\qquad \begin{cases} f_1(x) = x_1 x_2 - 2 = 0 \\ f_2(x) = 2x_1 - x_2^2 + 2 = 0. \end{cases}$$

This system has a unique solution $A^* = \begin{pmatrix} 1 \\ 2 \end{pmatrix}$. We have a singular point $A' = \begin{pmatrix} -4 \\ 2 \end{pmatrix}$, for which $f'(x)^a f(x) = 0$, $f(x) \neq 0$. The vector $v(x)$, on a large circle $\partial\Omega$, presents the same general features as in Figure 3; consequently the boundary condition is not satisfied. Figure 5 shows the two curves $f_1(x) = 0$, $f_2(x) = 0$, and the set $S = \{x: \det f'(x) = 0\}$. Any trajectory originating in the hachured region will stay in it: in this region we have $f_1(x) < 0$, and, consequently, we may have $f_1(x(t_1)) = 0$ only if $\lambda(t_1) = 0$, from (2), which would mean that $x(t_1)$ is a solution to the given system (1), i.e. (32) in our example. However, System (32) has no solution on the boundary of the hachured region.

Figure 6 shows the typical behaviour of λ with some starting points.

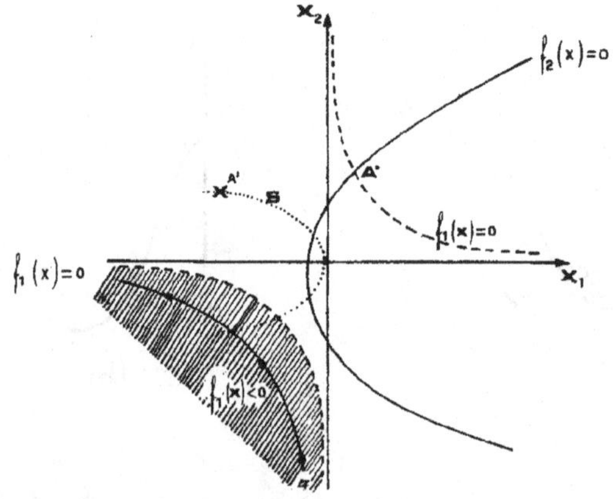

Figure 5: trajectory without solution beginning on a (Example 2)

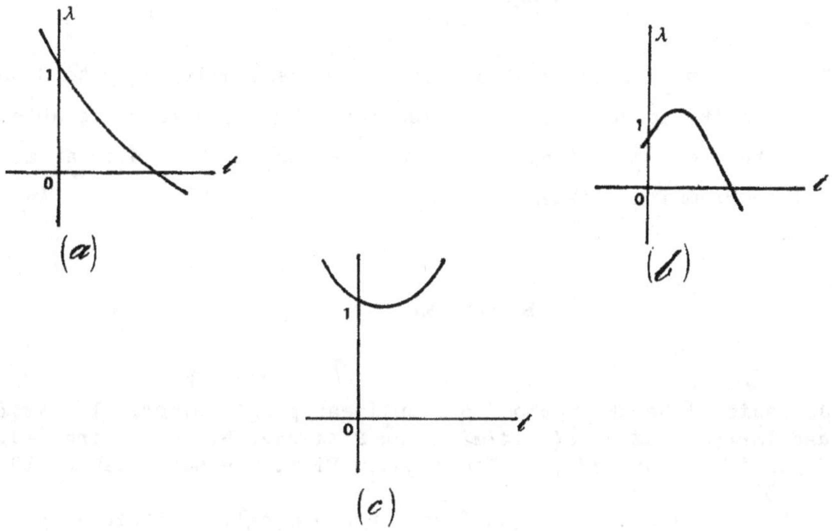

Figure 6: a=(2,0) or (4,5); b=(-1.3,0); e=(-2,-2) or (-2,-1/2),starting points for Example 2.

Example 3

Solve

$$\min\ (x_1-2)^2+(x_2-1)^2, \quad \text{s.t.}$$

$$f_1(x) = \frac{1}{4}x_1^2 + x_2^2 - 1 \le 0.$$

The solution and multiplier are

$$x^* = \begin{pmatrix} 1.664\,968\,547 \\ 0.554\,048\,675 \end{pmatrix}, \quad u^* = 0.804\,895\,572\ .$$

Those values are always obtained, whatever be the starting point. The λ parameter always have the nice behaviour of Figure 6(a) when using (KTC'). On using (KTC2) however, we have the behaviour of either Figure 6(a) or Figure 6(b), depending on the starting point.

7. CONCLUSIONS

The GRG method appears well suited to numerically apply to Global Newton method to solve systems of equations. This procedure is able to find more than one solution,and can be extended to the nonlinear mathematical programming problem.

REFERENCES

[1] J.Abadie. 'The GRG method for nonlinear programming'. In *Design and Implementation of Optimization Software*. H.J.Greenberg (ed.) Sijthoff & Noordhoff, Alphen aan den Rijn, The Netherlands (1978), 335-362.

[2] J.Abadie. 'Advances in nonlinear programming'. In *Proceedings of the Eighth I.F.O.R.S. International Conference on Operational Research*. K.B.Haley (ed.) North-Holland Publishing Company, Amsterdam-London - New York (1979), 900-930.

[3] C.B.Garcia and F.J.Gould. 'Relations between several path following algorithms and local and global Newton methods'. *SIAM Review*, 22(1980), 263-274.

[4] C.B.Garcia and W.I.Zangwill. 'Determining all solutions to certain systems of nonlinear equations'. *Math. Oper. Res*, 4(1979), 1-14.

[5] F.J.Gould and C.P.Schmidt. 'An existence result for the global Newton method'. In *Variational Inequalities and Complementarity Problems:Theory and Applications*. J.Wiley, New York (1980), 187-194.

[6] G.Guerrero, 3rd Cycle Thesis, Paris (1984).

[7] H.B.Keller. 'Global homotopies and Newton methods'. In *Numerical Analysis*. G.Golub (ed.), Academic Press, New York (1978), 72-96.

[8] J.M.Ortega and W.C.Rheinbold. 'Iterative solutions of nonlinear equations in several variables'. Academic Press, New York (1970).

[9] S.Smale. 'A convergent process of price adjustment and global Newton methods'. *J. Math. Econ.*, 3(1976), 107-120.

[10] P.Tolla. 'Linear and nonlinear programming software validity'. *Math. and Comp. in Simul.*, 25(1983), 39-42.

[11] J.Vignes. 'Algorithmes numérique, Analyse et mise en oeuvre, tome II, Equations et systèmes non linéaires'. *Editions Technip*. Paris (1980).

[12] J.Vignes. 'Implémentation des méthodes d'optimisation:test d'arrêt optimal contrôle et précision de la solution, Première partie: Aspect méthodologique'. *R.A.I.R.O.- Recherche Opérationnelle*, 18, no.1(1984), 1-18. 'Deuxième partie: Logiciel et exemple d'utilisation'. *R.A.I.R.O.- Recherche Opérationnelle*, 18, no.2(1984),103-129.

Chapter 2

Epi-Convergence and Duality. Convergence of Sequences of Marginal and Lagrangians Functions. Applications to Homogenization Problems in Mechanics.

H. Attouch

INTRODUCTION

The purpose of this lecture is to give both some theorical facts about the continuity of the Legendre-Fenchel transformation and applications to the convergence of primal and dual variables for sequences of convex minimization problems.

1. EPI-CONVERGENCE AND DUALITY. HISTORICAL INTRODUCTION

The leading idea of this introduction is that, historically, epi-convergence and the study of the continuity of the Legendre-Fenchel transformation are intimately connected.

Indeed, it is the study of the continuity of the application

$$C \to s(C,\cdot)$$

which to a closed convex set C associates its suppart function

$$s(C,x^*) = \sup_{x\in C} \langle x^*,x\rangle$$

which was at the origin of the introduction of epi-convergence.

a) A first answer to this problem was to consider sequences of closed convex *bounded* sets $\{C^n, C\,;\, n=1,2,\dots\}$. In this setting *Hausdorff metric* provides a natural measure for perturbation of sets. The classical formula

$$\text{haus}(C^n,C) = \sup_{\|x^*\|\leq 1} |s(C^n,x^*)-s(C,x^*)|$$

links the convergence of the sequence $\{C^n ; n=1,2,\ldots\}$ to C for the Hausdorff metric to the pointwise, and in fact uniform on bounded sets, convergence of their support functions.

b) Further developments of optimization theory made necessary the introduction of *unbounded* sets like cones, epigraphs... In this setting, Hausdorff metric is no more a good concept, one has to use the weaker notion of *set-convergence* (also called Kuratowski convergence)

(1) $$C = \text{Lim}\, C^n \iff \operatorname*{Limsup}_n C^n \subset C \subset \operatorname*{Liminf}_n C^n$$

where

$$\operatorname*{Liminf}_n C^n = \{x \,/\, \exists x_n \to x \text{ with } x_n \in C^n \text{ for every } n \in \mathbb{N}\}$$

$$\operatorname*{Limsup}_n C^n = \{x \,/\, \exists (n_k)_{k\in\mathbb{N}} \text{ such that } x = \lim x_k \text{ with } x_k \in C^{n_k} \text{ for every } k \in \mathbb{N}\}.$$

The question that naturally arises is:

If a sequence of closed convex sets $\{C^n ; n=1,2,\ldots\}$ is set-convergent to C, what is the corresponding notion of convergence for their support functions $\{s(C^n,\cdot) ; n=1,2,\ldots\}$?

The answer is *epi-convergence* and that is the way the notion was introduced by Wijsman [37] in 1966. He called it "infimal convergence" and was mainly motivated by considerations in statistical decision theory. Let's give the definition

A sequence of real extended valued functions F^n is said to be epi-convergent to F at x if the two following sequences hold

$$(2)\begin{cases} \text{there exists a sequence } x_n \text{ converging to } x \text{ such that } F(x)\geq\limsup_n F^n(x_n) \\ \text{for every sequence } (x_n) \text{ converging to } x, \quad \liminf_n F^n(x_n)\geq F(x). \end{cases}$$

and one writes $F(x) = \lim_e F^n(x)$.

When there is epi-convergence at every point x, the sequence F^n is said to be epi-convergent to F and

$$F = \lim_e F^n .$$

Let us introduce for every integer n=1,2,..., I_{C^n} the indicator function of C^n:

$$I_{C^n}(x) = \begin{cases} 0 & \text{if } x\in C^n \\ +\infty & \text{otherwise} \end{cases}$$

and let us notice that

$C^n \to C$ in the set-convergence sense $\Leftrightarrow I_C = \lim_e I_{C^n}$.

Since $s(C^n,\cdot)$ is nothing but the Legendre-Fenchel conjugate of I_{C^n}, the above continuity property of the mapping $C \to s(C,\cdot)$ turns to be a particular case (and in fact is equivalent) to the following basic property of epi-convergence:

For any sequences of closed convex proper functions $\{F^n,F;\ n=1,2...\}$

$$(3) \qquad F = \lim_e F^n \iff F^* = \lim_e F^{n*} .$$

At this stage the theory was developped in a satisfactoring way but only in a finite dimensional framework which blurs some important topological features.

c) Next step in the development of the theory came from a quite different orientation and can be traced to the work of G.Stampacchia, J.L.Lions ... on variational inequalities. In order to study the convergence of approximation schemes (like Galerkin approximation) and of various

perturbations Mosco [27] (1969) and Joly [21] (1973) extended the earlier notions to *infinite-dimensional spaces*.
Given X a reflexive Banach space a sequence $\{F^n : X \to]-\infty,+\infty]$; $n = 1,2,\ldots\}$ of closed convex proper functions is said to be *Mosco-convergent* to F if the two following sequences hold:

for every $x \in X$

$$\begin{cases} \text{there exists } x_n \to x \text{ strongly in } X \text{ such that } F(x) \geq \limsup_n F^n(x_n) \\ \text{for every weakly converging sequence } x_n \to x, \ \liminf_n F^n(x_n) \geq F(x) \end{cases}$$

The continuity of the Legendre-Fenchel transformation can be expressed in this setting as:

(4) $F^n \to F$ in Mosco sense $\Longleftrightarrow$ $F^{n*} \to F^*$ in Mosco sense

that is, the Legendre-Fenchel transformation is bicontinuous with respect to Mosco-convergence.
At this stage the epi-convergence theory was developped only for convex functions and intimately connected with the study of the Legendre-Fenchel transformation.

d) The concept in a *general topological setting* has been finally delineated by De Giorgi (cf [16] ..) and the Italian mathematical school (De Giorgi & Spagnolo [19] (1973), De Giorgi & Franzoni [17] (1975)...). They were mostly concerned with problems coming from calculus of variations, cf. next paragraph for definition and variational properties of epi-convergence.

e) In recent years epi-convergence has been developping in quite all branches of optimization theory. Its relations with duality theory have been intensively studied in order to cover various situations. Let us mention some of them:

Attouch & Wets [5] [6], Cavazzuti [14], Attouch, Aze & Wets [11]: Convergence notions for bivariate functions (Lagrangians...) are introduced; for example the epi-hypo convergence. They are specially designed to study the *convergence* of *saddle value problems*. They allow to attack in a unified way the convergence of primal and dual variables in mechanics (for example homogenization of composite materials, cf. section 4, reinforcement and shells...) as well as in game theory and convex programming.

Attouch & Wets [7] : the metric aspects of the Mosco-convergence are studied. It is shown that on the space of lower semicontinuous convex functions defined on $\mathbb{R}^n$, the *Legendre-Fenchel transform is an isometry* with respect to some metrics consistent with the epi-topology. In the infinite dimensional case, isometries are also exhibited but this time they correspond to topologies finer than the Mosco topology.

Back [12] : in a recent reprint gives extensions of Mosco convergence to *non reflexive Banach spaces* which preserve continuity properties of the Legendre-Fenchel transformation. These are motivated by the study of duality schemes pairing L^∞ perturbations with L^1 dual variables with in view applications to optimal economic growth program in continuous time stochastic model.

Dolecki [15], Volle [33]...have extended the duality theory to a *non convex* setting (for example duality for quasi-convex functions) and obtained in this setting continuity properties of the conjugation map with respect to epi-convergence.

Zolezzi [38], Papageorgiou [29], Luccheti & Patrone [23], Mac Linden [24], Attouch & Wets [8] ...have considered stability, perturbation problems in non linear (convex) programming and economics by means of epi-convergence tools.

f) Undoubtedly, only a small part of possible applications of the continuity properties of duality with respect to epi-convergence have been explored. In this paper we present an abstract general approach from D. Aze [9] which relies on the study of *epi-convergence* of *marginal functions*. It allows to obtain in a unified way convergence of primal and dual variables and of corresponding values. This approach bears natural relations with the epi-convergence of perturbation functions and the epi-hypo convergence of the associated Lagrangians (section 5, theorem 5.1 and section 6, theorem 6.3).

2. EPI-CONVERGENCE. DEFINITION AND VARIATIONAL PROPERTIES

(for a detailed survey refer to Attouch [2], De Giorgi [16]).

Epi-convergence is a topological notion. The only framework one needs to define it is a topological setting. So, let us denote by $(X,\mathcal{T})$ a space X equipped with a topology $\mathcal{T}$ and for every $x\in X$ by $N_{\mathcal{T}}(x)$ the neighbourhoods of x.

Definition 2.1 Given $\{F^n : X \to \bar{\bar{R}}\ ;\ n=1,2,\dots\}$ a sequence of real extended valued functions, for every $x\in X$

(5) $$\mathcal{T}\text{-}li_e\, F^n(x) = \sup_{V\in N_{\mathcal{T}}(x)}\ \liminf_n\ \inf_{y\in V} F^n(y)$$

(6) $$\mathcal{T}\text{-}ls_e\, F^n(x) = \sup_{V\in N_{\mathcal{T}}(x)}\ \limsup_n\ \inf_{y\in V} F^n(y)$$

are respectively called the

$\mathcal{T}$-epi limit inferior of the sequence F^n at x

$\mathcal{T}$-epi limit superior of the sequence F^n at x.

When these two quantities are equal the sequence F^n is said to be $\mathcal{T}$-epi convergent at x and this common value is denoted

(7) $$\mathcal{T}\text{-}lim_e\, F^n(x) = \mathcal{T}\text{-}li_e\, F^n(x) = \mathcal{T}\text{-}ls_e\, F^n(x).$$

When this equality holds for every $x\in X$, the sequence F^n is said to be $\mathcal{T}$-epi convergent and we write

$$F = \mathcal{T}\text{-}lim_e\, F^n .$$

Proposition 2.2

When the space (X,T) is metrizable the following characterization of epi-convergence holds:

(i) $F(x) = T\text{-}\lim_e F^n(x)$

$\Updownarrow$

(ii) $\begin{cases} \text{there exists a sequence } x_n \overset{T}{\to} x \text{ such that } & F(x) \geq \limsup_n F^n(x_n) \\ \text{for every converging sequence } x_n \overset{T}{\to} x & F(x) \leq \liminf_n F^n(x_n) . \end{cases}$

Remark 2.3 It is the above sequential formulation (ii) of epi-convergence which is convenient for applications. Given a sequence of functions $\{F^n ; n=1,2,...\}$ one has to verify, in order to prove its T-epi convergence to a function F, *two* sentences:

a) for every $x \in X$, one has to construct an approximating sequence $x_n \overset{T}{\to} x$ such that $F(x) \geq \limsup_n F^n(x_n)$.

b) for every $x_n \overset{T}{\to} x$ one has to verify that $F(x) \leq \liminf_n F^n(x_n)$.

Such formulation is equivalent to T-epi convergence when the topology T is metrizable. When (X,T) is not metrizable (which is the case when working with weak topologies on Banach spaces) the two notions do not coincide. In that case it will be useful to consider the following sequential epi-convergence notions:

$$(8) \begin{cases} \text{seq-}T\text{-li}_e\, F^n(x) = \inf\{\liminf F^n(x_n) \,/\, x_n \overset{T}{\to} x\} \\ \text{seq-}T\text{-ls}_e\, F^n(x) = \inf\{\limsup F^n(x_n) \,/\, x_n \overset{T}{\to} x\} \quad \text{and} \\ F(x) = \text{seq-}T\text{-lim}_e\, F^n(x) \quad \text{when the two above quantities are equal.} \end{cases}$$

The following fundamental *variational properties of epi-convergence* are direct consequences of formulation (ii) Prop. 2.2 of epi-convergence.

Theorem 2.4

Let $\{F^n : X \to \bar{\mathbb{R}}\}$; $n \in \mathbb{N}$, be a sequence of real extended valued functions and for every $n \in \mathbb{N}$, let x_n be an ε_n-minimizer of F^n that is

$$F^n(x_n) \leq \inf_{y \in X} F^n(y) + \varepsilon_n \tag{9}$$

with $\varepsilon_n \to 0$ as $n \to +\infty$.

Let us assume that the sequence $\{x_n ; n \in \mathbb{N}\}$ is $\mathcal{T}$-relatively compact with respect to a topology $\mathcal{T}$ on X.

Then the following implication holds: (i) $\Rightarrow$ (ii)

$$\text{(i)} \quad F = \mathcal{T}\text{-}\lim_e F^n$$

$$\Downarrow$$

$$\text{(ii)} \quad \inf_{y \in X} F^n(y) \xrightarrow[n \to +\infty]{} \min_{y \in X} F(y)$$

and every $\mathcal{T}$-cluster point of an ε_n-minimizing sequence (that is verifying (9)) with $\varepsilon_n \to 0$ does minimize F over X.

Theorem 2.5

Let $(X,\mathcal{T})$ be a topological space and $F = \mathcal{T}\text{-}\lim_e F^n$. For every $\mathcal{T}$-continuous function $G : X \to \mathbb{R}$

$$F + G = \mathcal{T}\text{-}\lim_e (F^n + G).$$

Comments: a) When studying the limit behaviour of a sequence of minimizing problems

$$\inf_{y \in X} F^n(y) \qquad n = 1,2,\ldots$$

there is no a priori given topological structure. It is the analysis of the relative compactness properties of corresponding minimizing sequences which yields the topology $\mathcal{T}$. Then $\mathcal{T}$-epi-convergence is the right concept in order to go to the limit (as $n \to +\infty$) on the above sequence of minimization problems.

b) The terminology is justified by the following geometric characterization of epi-convergence:

$$F = \mathcal{T}\text{-}\lim_e F^n \iff \operatorname{epi} F = \mathcal{T}\text{-}\operatorname{Lim}(\operatorname{epi} F^n)$$

that is, $\mathcal{T}$-epi-convergence of F^n to F is equivalent to the set-convergence of the sequence of epigraphs $\{\operatorname{epi} F^n ; n=1,2,\dots\}$ to the epigraph of F in the product space $X \times \mathbb{R}$ (equipped with the product topology $\mathcal{T}\times d$, d usual topology of $\mathbb{R}$).

c) It is a natural question to ask: What is the relation between epi-convergence and pointwise convergence. Indeed none of these notions implies nor is implied by the other. This appears clearly on formulation (ii)(Prop. 2.2) of $\mathcal{T}$-epi-convergence. There is indeed an important case where the two notions coincide: that is monotone convergence. This explain the success of all monotone approximation schemes in optimization theory.

d) When $F^n \equiv F$ for every $n \in \mathbb{N}$, then

$$\mathcal{T}\text{-}\lim_e F^n(x) = \sup_{V \in N_{\mathcal{T}}(x)} \inf_{y \in V} F(y)$$

is equal to the $\mathcal{T}$-lower semicontinuous regularization of F.
More generally when $F = \mathcal{T}\text{-}\lim_e F^n$, then F is $\mathcal{T}$-lower semicontinuous.
When $(X,\mathcal{T})$ is metrizable

$$F(x) = \inf\{ \lim_{n\to+\infty} F^n(x_n) \;;\; x_n \xrightarrow{\mathcal{T}} x\}.$$

So $\mathcal{T}$-epi-convergence can be viewed as an extension of the Γ-closure operation (that is the $\mathcal{T}$-lower semicontinuous regularization for convex functions) which justifies the equivalent terminology of Γ-convergence introduced by De Giorgi [16].

e) Epi-convergence can indeed be characterized by its variational properties:

Taking $X = \mathbb{R}^n$ with its usual topology and $\{F^n : X \to]-\infty,+\infty]; n=1,2,\dots\}$ the following equivalence holds:

$$F = \lim_e F^n \iff \forall\lambda>0 \quad \forall x\in X \quad \inf_{u\in X} \{F^n(u)+\frac{1}{2\lambda}|x-u|^2\} \xrightarrow[n\to+\infty]{} \inf_{u\in X} F(u)+\frac{1}{2\lambda}|x-u|^2\}.$$

So epi-convergence can be characterized by means of *convergence of the infimum for a whole family of quadratic perturbations*. Let us notice that, in this case, the compactness property which allows to pass from epi-convergence to convergence of the infima is satisfied because of coerciveness of the functions

$$u \to F^n(u) + \frac{1}{2\lambda}|x-u|^2$$

and the resulting compactness (X is finite dimensional) of their level sets.

For every $\lambda>0$, the function F_λ defined by

$$(10) \qquad F_\lambda(x) = \inf_{u\in X} \{F(u) + \frac{1}{2\lambda}|x-u|^2\}$$

is called the *Moreau-Yosida* approximate of index λ of F. More generally one can prove that epi limit functions (cf.(5) and (6)) can be expressed in terms of pointwise convergence of the Moreau-Yosida approximates cf. Attouch [1] [2].

This combined with the regularity property of the Moreau-Yosida approximates (locally lipschitz) explains the important role this approximation (F_λ increases to F as λ decreases to zero) plays in the epi-convergence theory.

The same type of idea explains the continuity properties of the Fenchel transformation

$$F = \lim_e F^n \iff F^* = \lim_e F^{n*} :$$

When working with *convex* functions, epi-convergence can be characterized by means of convergence of the infimum

$$\inf_{u \in X} \{F^n(u) + \langle x^*, u\rangle\}$$

for the whole family of *linear continuous perturbations* $(x^* \in X^*)$. This will be made clear in the next section.

3. CONTINUITY OF THE LEGENDRE-FENCHEL TRANSFORMATION WITH RESPECT TO EPI-CONVERGENCE

Let X be a reflexive[2] Banach space and $F^n, F : X \to]-\infty,+\infty]$ a sequence of closed convex proper functions ($\not\equiv +\infty$).
The topologies which play a basic role are the weak and the strong topologies of X (resp. X^*).
We denote $\text{seq}\, X_w\text{-li}_e F^n$, $\text{seq}\, X_w\text{-ls}_e F^n$, $\text{seq}\, X_w\text{-lim}_e F^n$ the sequential epi-limits for the weak topology of X of the sequence F^n and $X_s\text{-li}_e F^n$, $X_s\text{-ls}_e F^n$, $X_s\text{-lim}_e F^n$ the corresponding notions for the strong topology of X. Similar notions are introduced for the conjugate functions F^{n*} with X^* replacing X. The following theorems 3.1, 3.2, 3.3 and 3.4 come from H. Attouch [2]. They improve earlier results from J.L. Joly [21] and U. Mosco [28].

Theorem 3.1

Let X be a reflexive Banach space and $F^n : X \to]-\infty,+\infty]$ a sequence of closed convex functions which is uniformly proper, that is

(11) There exists a bounded sequence $\{u_{on} ; n\in\mathbb{N}\}$ in X such that $\sup_{n\in\mathbb{N}} F^n(u_{on}) < +\infty$.

Then,

$$(\text{seq}\, X_w\text{-li}_e F^n)^* = X^*_s\text{-ls}_e (F^n)^*.$$

(2) Most results are formulated for reflexive Banach spaces. Theorem 3.3 holds for general Banach spaces.

Proof of theorem 3.1 Let us write for simplicity $w\text{-}li_e F^n$ instead of $\text{seq}\, X_w\text{-}li_e F^n$ and call $F = w\text{-}li_e F^n$ this function.
The inequality $F^* \le s\text{-}ls_e (F^n)^*$ is immediate.
The difficult point is to prove the opposite inequality, that is:

$$(12) \qquad \forall x^* \in X^* \ \exists x_n^* \to x^* \quad \text{such that} \quad F^*(x^*) \ge \limsup_n F^{n^*}(x_n^*) \ .$$

When the functions $\{F^{n^*} ; n \in \mathbb{N}\}$ are equicoercive one can take $x_n^* = x^*$, in which case (12) becomes

$$\inf_{u \in X} \{F(u) - \langle x^*, u\rangle\} \le \liminf_n \inf_{u \in X} \{F^n(u) - \langle x^*, u\rangle\}.$$

Introducing u_n minimizing $F^n(u) - \langle x^*, u\rangle$ over X, from equi-coercivness of the F^n, the sequence $\{u_n ; n \in \mathbb{N}\}$ remains bounded in X and hence weakly relatively compact in X. Conclusion follows from inequality

$$F \le w\text{-}li_e F^n \ .$$

In the general case the idea is to reduce to the above situation by *coercifying* functions F^n: let us introduce for every $\lambda > 0$

$$F^{n,\lambda} = F^n + \frac{\lambda}{2} |\cdot|^2$$

$$F^\lambda = F + \frac{\lambda}{2} |\cdot|^2$$

By lower semicontinuity of the norm for the weak topology of X we still have

$$F^\lambda \le w\text{-}li_e F^{n,\lambda} \ .$$

For every $\lambda > 0$ the sequence $\{F^{n,\lambda} ; n \in \mathbb{N}\}$ is uniformly coercive: this follows from coercivness of the quadratic perturbation $\frac{\lambda}{2}|\cdot|^2$ and from the fact that, without restriction, we can assume the F^n to satisfy a uniform minorization

$$F^n(u) \geq - r(|u| + 1) \quad \text{for every} \quad n \in \mathbb{N} \text{ and } u \in X.$$

Applying the preceding argument to the sequence $\{F^{n,\lambda} ; n \in \mathbb{N}\}$, $\lambda > 0$ fixed, we have

$$(13) \qquad (F^{\lambda})^* (x^*) \geq \limsup_n (F^{n,\lambda})^* (x^*).$$

Computation of $(F^{n,\lambda})^*$ yields

$$(F^{n,\lambda})^* (x^*) = (F^{n^*})_{\lambda} (x^*) = \inf_{u^* \in X^*} \left\{ F^{n^*}(u^*) + \frac{1}{2\lambda} |x^* - u^*|^2 \right\}$$

that's precisely the Moreau-Yosida approximation of index λ of F^{n^*}. Noticing that

$$F^* \geq (F^{\lambda})^*$$

and that inequality (13) holds for all $\lambda > 0$, we obtain

$$F^*(x^*) \geq \limsup_{\lambda \to 0} \limsup_{n \to +\infty} (F^{n^*})_{\lambda} (x^*) .$$

A classical diagonalization lemma (cf. Attouch [1]) yields the existence of a mapping $n \to \lambda(n)$ with $\lambda(n) \to 0$ as $n \to +\infty$ such that

$$F^*(x^*) \geq \limsup_n (F^{n^*})_{\lambda(n)} (x^*).$$

Introducing x_n^* realizing the inf-convolution in $(F^{n^*})_{\lambda(n)} (x^*)$ we obtain

$$(14) \qquad F^*(x^*) \geq \limsup_n \left\{ F^{n^*}(x_n^*) + \frac{1}{2\lambda(n)} |x^* - x_n^*|^2 \right\}.$$

It follows

$$F^*(x^*) \geq \limsup_n F^{n^*}(x_n^*) .$$

We just have to prove that $x_n^* \to x^*$ strongly in X^* in order to complete the proof. From uniform proper assumption (11)

$$F^{n^*}(x_n^*) \geq \langle x_n^*, u_{on} \rangle - F^n(u_{on}) \geq -c(1 + |x_n^*|).$$

Returning to (14), for n sufficiently large

$$F^*(x^*) + 1 + c(1 + |x_n^*|) \geq \frac{1}{2\lambda(n)} |x^* - x_n^*|^2 .$$

If $F^*(x^*) = +\infty$ there is nothing to prove. Otherwise one easily derives from the fact that $\lambda(n) \to 0$ that $x_n^* \to x^*$ strongly in X^*.

The following result follows in a straight way from theorem 3.1.

Theorem 3.2

Let F^n, $F : X \to]-\infty, +\infty]$ be a sequence of closed convex proper functions, where X is a reflexive Banach space. The following implication holds (i) $\Rightarrow$ (ii)

$$\text{(i)} \quad F = \text{seq}\, X_w\text{-lim}_e\, F^n$$
$$\Downarrow$$
$$\text{(ii)} \quad F^* = X_s^*\text{-lim}_e\, F^{n^*}$$

Proof of theorem 3.2 By assumption

$$\text{seq}\, X_w\text{-ls}_e\, F^n \leq F \leq \text{seq}\, X_w\text{-li}_e\, F^n .$$

Thus

$$(15) \qquad (\text{seq}\, X_w\text{-li}_e\, F^n)^* \leq F^* \leq (\text{seq}\, X_w\text{-ls}_e\, F^n)^* .$$

We can apply theorem 3.1 since $F = \text{seq}\, X_w\text{-lim}_e\, F^n$ has been assumed to be proper. The uniform proper assumption (11) is consequently satisfied and thus

$$(\text{seq}\, X_w\text{-li}_e\, F^n)^* = X_s^*\text{-ls}_e\, F^{n^*}$$

Let us verify that

$$(16) \qquad (\text{seq}\, X_w\text{-ls}_e\, F^n)^* \leq X_s^*\text{-li}_e\, F^{n^*} .$$

It will follow from (15)

$$X_s^*\text{-ls}_e F^{n^*} \le F^* \le X_s^* - \text{li}_e F^{n^*}$$

that is

$$F^* = X_s^*\text{-lim}_e F^{n^*} .$$

Let us verify (16). Equivalently we have to prove that for every $x^* \epsilon X^*$, for every strongly convergent sequence $x_n^* \to x^*$ in X^*, for every $x \epsilon X$

$$(17) \qquad \langle x^*, x\rangle - (\text{seq}\, X_w\text{-ls}_e F^n)(x) \le \liminf_n F^{n^*}(x_n^*).$$

By definition of $\text{seq}\, X_w\text{-ls}_e F^n$, for every $\varepsilon>0$, there exists a weakly converging sequence $x_n^\varepsilon \to x$ such that

$$(18) \qquad \varepsilon + (\text{seq}\, X_w\text{-ls}_e F^n)(x) \ge \limsup_n F^n(x_n^\varepsilon).$$

By definition of F^{n^*}

$$F^{n^*}(x_n^*) + F^n(x_n^\varepsilon) \ge \langle x_n^*, x_n^\varepsilon\rangle .$$

Passing to the limit as $n \to +\infty$

$$\liminf_n F^{n^*}(x_n^*) + \limsup_n F^n(x_n^\varepsilon) \ge \langle x^*, x\rangle .$$

From (18)

$$\liminf_n F^{n^*}(x_n^*) + (\text{seq}\, X_w\text{-ls}_e F^n)(x) + \varepsilon \ge \langle x^*, x\rangle .$$

This being true for every $\varepsilon>0$ and every $x \epsilon X$, conclusion (17) follows.

Theorem 3.2 allows to pass from epi-convergence of sequences of closed convex proper functions for the weak topology to epi-convergence of their conjugates for the strong topology.
The following theorem gives an answer to the converse problem which consists to pass by duality from epi-convergence for the strong topology of X to epi-convergence for the $\sigma(X^*,X)$ weak topology of X^*.

Its extension to non reflexive Banach spaces has been obtained by D. Aze [9].

Theorem 3.3

Let X be a separable Banach space and $F^n : X \to]-\infty,+\infty]$ a sequence of closed convex functions which satisfies the following "uniform coercivness" property:

(19) for every sequence $\{x_n^* ; n \in \mathbb{N}\}$ in X^* such that $\sup_{n\in\mathbb{N}} F^{n^*}(x_n^*) < +\infty$,

then $\sup_{n\in\mathbb{N}} |x_n^*| < +\infty$.

Then the following equalities hold

(20) $$(X_s\text{-}li_e\, F^n)^* = \text{seq}\, X_w^*\text{-}ls_e\, F^{n^*} = X_w^*\text{-}ls_e\, F^{n^*}$$

Comments

If one drops assumption (19) equality

$$(X_s\text{-}li_e\, F^n)^* = \text{seq}\, X_w^*\text{-}ls_e\, F^{n^*}$$

may fail to be true. Take $F^{n^*} = I_{K^n}$ the indicator functional of a closed convex set K^n such that the sequential weak limit of $\{K^n ; n\in\mathbb{N}\}$ is not closed.

Since epi-convergence can be expressed by means of set-convergence of level sets, cf. R. Wets [36], the uniform coercivness assumption (19) allows to work only on bounded subsets of X^* on which, thanks to the separability assumption of X, the weak topology $\sigma(X^*,X)$ is metrizable.

This is an heuristic explanation to (20).

Proof of theorem 3.3 follows the lines of proof of theorem 3.1. But now we have to coercify functions F^n in order to make them inf-compact for the strong topology of X. To that end let us introduce $\{x_1,x_2,\ldots,x_k,\ldots\}$ a dense denumerable subset of X, $E_k = \text{span}\{x_1,\ldots,x_k\}$ the finite dimensional subspace generated by $x_1 \ldots x_k$, and B_k the closed ball of radius k in E_k. Then take

$$F^{n,k} = F^n + I_{B_k}$$

$$F^k = F + I_{B_k}$$

We can now state the following basic result

Theorem 3.4

Let X be a reflexive separable Banach and $F^n : X \to]-\infty,+\infty]$ a sequence of closed convex proper functions which is equicoercive, that is

$$\sup_n F^n(x_n) < +\infty \quad \Rightarrow \quad \sup_n |x_n| < +\infty \ .$$

Then, the following equivalence holds:

$$F = \text{seq}\, X_w\text{-lim}_e\, F^n \quad \Leftrightarrow \quad F^* = X_s\text{-lim}_e\, F^{n^*}$$

The following corollary is useful for applications

Corollary 3.5 Let X be a reflexive separable Banach space and $F^n : X \to]-\infty,+\infty]$ a sequence of closed convex functions which is "strongly" equicoercive, that is

$$F^n(x) \geq c(|x|) \quad \text{with} \quad \lim_{r \to +\infty} \frac{c(r)}{r} = +\infty$$

Then the following statements are equivalent: (i) $\Leftrightarrow$ (ii) $\Leftrightarrow$ (iii) $\Leftrightarrow$ (iv)

(i) $F = \text{seq}\, X_w\text{-lim}_e\, F^n$

(ii) $F^* = X^*_s\text{-lim}_e\, F^{n^*}$

(iii) $\forall x^* \in X^* \quad F^*(x^*) = \lim_{n \to +\infty} F^{n^*}(x^*)$

(iv) there exists a dense subset $D \subset X^*$ such that for every $x^* \in D$

$$F^*(x^*) = \lim_{n \to +\infty} F^{n^*}(x^*) \ .$$

Let us end this section and say a few words about *Mosco convergence.* Preceding theorems (thm 3.1 - thm 3.4) tell us that by conjugation $F \to F^*$, epi-convergence is preserved but topologies (weak and strong) are exchanged. It is natural to introduce Mosco convergence which is epi-convergence for both weak and strong topologies (introduced by Mosco [1]):

Definition 3.6 Let $F^n, F : X \to]-\infty, +\infty]$ be a sequence of closed convex proper functions and X a reflexive Banach space. The sequence F^n is said to be Mosco-convergent to F if the two following sequences hold:

(i) for every $x \in X$ there exists $x_n \to x$ strongly in X such that $F^n(x_n) \to F(x)$

(ii) for every $x \in X$, for every weakly converging sequence $x_n \to x$, $\liminf_n F^n(x_n) \geq F(x)$.

Theorem 3.7
Let X be a reflexive Banach space. The Fenchel transformation $F \to F^*$ is a one to one mapping from $\Gamma(X)$ onto $\Gamma(X^*)$ which is bicontinuous for Mosco convergence. In other words

$$\begin{array}{c} F^n \to F \quad \text{in Mosco sense} \\ \Updownarrow \\ {F^n}^* \to F^* \quad \text{in Mosco sense.} \end{array}$$

Proof of theorem 3.7 It is a direct consequence of theorem 3.1: Let us assume that $F^n \to F$ in Mosco sense; then $X_s\text{-}ls_e\, F^n \leq F \leq \text{seq}\, X_w\text{-}li_e\, F^n$.

By duality, inequalities are reversed

$$(\text{seq}\, X_w\text{-}li_e\, F^n)^* \leq F^* \leq (X_s\text{-}ls_e\, F^n)^*.$$

From theorem 3.1

$$(\text{seq}\, X_w\text{-}li_e\, F^n)^* = X_s^*\text{-}ls_e\, {F^n}^*$$

and

$$(X_s\text{-}ls_e\, F^n)^* = (\text{seq}\, X_w^*\text{-}li_e\, {F^n}^{*})^{**} \leq \text{seq}\, X_w^*\text{-}li_e\, {F^n}^*.$$

Finally

$$X_s^*\text{-ls}_e F^{n^*} \le F^* \le \text{seq}\, X_w^*\text{-li}_e F^{n^*}$$

which means that

$$F^{n^*} \to F^* \text{ in Mosco sense.}$$

Remark One can indeed prove that Mosco convergence is equivalent to the pointwise convergence of all the Moreau-Yosida approximations. By the way we obtain that Mosco convergence is attached to a topology cf. H. Attouch [1], [2]. Mosco convergence plays a decisive role in the study of stability, approximation... of optimization problems in convex analysis, cf. Attouch [1], Mac Linden [24], Sonntag [31], Zolezzi [38]...

4. CONVERGENCE OF DUAL VARIABLES. AN EXAMPLE OF THE EPI-CONVERGENCE APPROACH ON AN HOMOGENIZATION PROBLEM

The following example describes the homogenization asymptotic analysis of composite materials. Let us first describe its "primal" version. Composite materials play an important role in modern mechanical engineering. The parameters which describe the physical properties (like conductivity, elasticity...) of such materials are discontinuous and oscillate between the different values characterizing each component. When these components are intimately mixed these parameters oscillate very rapidly and it becomes very difficult to describe the microscopic behaviour of such materials. On the opposite, from a macroscopic point of view the material tends to behave like an homogenous, ideal one: its determination is the purpose of the homogenization theory.

Let us illustrate this limit analysis procedure for the electrostatic potential equation. Let us assume that the material has a periodic structure of period ε in each directions. Introducing $Y=(0,1)^N$ a basic cell in $\mathbb{R}^N$ and $a(\cdot) : \mathbb{R}^N \to \mathbb{R}^+$ a Y-periodic function, the function $a(\frac{\cdot}{\varepsilon})$ is εY-periodic and describes at a microscopic scale the conductivity of the material.

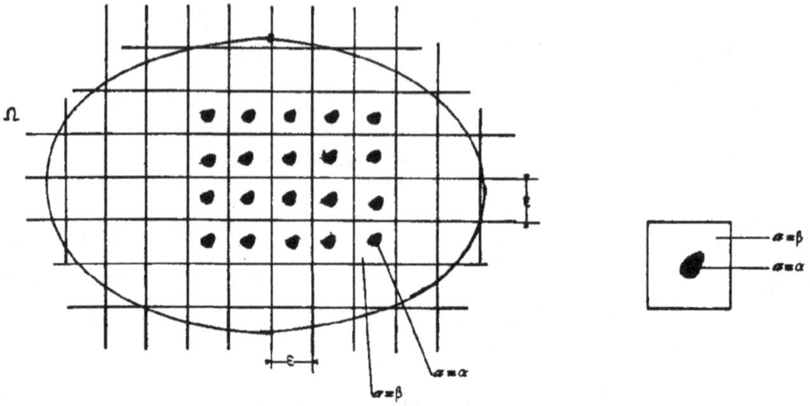

In the above picture is considered the case of two components of respective conductivity α and β.

The material occupies a volume Ω in $\mathbb{R}^N$, where Ω is a bounded open domain. For every distribution potential u on Ω the corresponding stored energy is given by

$$F^{\varepsilon}(u) = \int_{\Omega} a(\frac{x}{\varepsilon})\,|\mathrm{grad}u|^2\,dx$$

and at equilibrium for a given density of charge f on $\Omega, f \in L^2(\Omega)$ (we suppose the boundary of Ω maintained at potential u=0) the potential u_ε is solution of the variational problem

$$(21)\qquad \min_{u\in H_0^1(\Omega)} \left\{ \frac{1}{2} \int_{\Omega} a(\frac{x}{\varepsilon})\,|\mathrm{grad}u|^2\,dx - \int_{\Omega} f(x)\,u(x)\,dx \right\}.$$

Its Euler equation is

$$\left|\begin{array}{l} -\mathrm{div}\,(a(\frac{x}{\varepsilon})\,\mathrm{grad}u_\varepsilon) = f \text{ on } \Omega \\ u_\varepsilon = 0 \quad \text{on } \partial\Omega \end{array}\right.$$

The real physical problem is at $\varepsilon=\varepsilon_0$ fixed but small (with respect to $|\Omega|$). As already explained the homogenization procedure consists taking ε as a parameter, with corresponding solution u_ε, and let ε tend to zero

in the corresponding equations.
Relying on the variational formulation (21) of these equations, this problem can be attacked by means of epi-convergence theory:
Let us notice that the sequence $\{u_\varepsilon\ ;\ \varepsilon \to 0\}$ remains bounded in $H_0^1(\Omega)$ (we assume $a(\cdot) \geq \lambda_0 > 0$) and hence weakly relatively compact in this space, and that the function $u \to \int fu$ is a continuous perturbation for this topology.
Relying on theorems 2.4 and 2.5, the convergence of the sequence of solutions $\{u_\varepsilon\ ;\ \varepsilon \to 0\}$ follows from the following epi-convergence theorem (we state it in a slightly more general form) cf. Marcellini [26], Attouch [2], [3], [4].

Theorem 4.1
Let $\{F^\varepsilon : H^1(\Omega) \to \mathbb{R}^+\ ;\ \varepsilon \to 0\}$ be the sequence of convex continuous functions from $H^1(\Omega)$ into $\mathbb{R}^+$ defined by

$$F^\varepsilon(u) = \int_\Omega j(\tfrac{x}{\varepsilon}, \operatorname{grad} u(x))dx$$

where $j : \mathbb{R}^N \times \mathbb{R}^N \to \mathbb{R}^+$, $(y,z) \to j(y,z)$ is Y-periodic with respect to y, convex continuous with respect to z and satisfies $(0 < \lambda_0 \leq \Lambda_0 < +\infty)$

$$\lambda_0 |z|^2 \leq j(y,z) \leq \Lambda_0(1+|z|^2) \text{ for all } y \in \mathbb{R}^N, \text{ for all } z \in \mathbb{R}^N .$$

Taking T equal to the weak topology of $H^1(\Omega)$, for every $u \in H^1(\Omega)$

$$T\text{-}\lim_e F^\varepsilon(u) = F^{hom}(u) \quad \text{exists}$$

with

$$F^{hom}(u) = \int_\Omega j^{hom}(\operatorname{grad} u)\, dx$$

and

$$j^{hom}(z) = \min_{\{w\ Y\text{-periodic}\}} \int_Y j(y, \operatorname{grad} w(y)+z)\, dy . \tag{22}$$

The situation described above corresponds to $j(y,z) = \frac{1}{2} a(y) |z|^2$. As a corollary we obtain the convergence of the solutions $\{u_\varepsilon\ ;\ \varepsilon \to 0\}$ of the corresponding minimization problems

$$\min_{u\in H_0^1(\Omega)} \left\{\int_\Omega j\left(\frac{x}{\varepsilon},\text{grad}u\right) dx - \int_\Omega f(x)\,u(x)\,dx\right\}$$

or under the equivalent form

$$\begin{cases} -\text{div}\ \partial j\left(\frac{x}{\varepsilon},\text{grad}u_\varepsilon\right) = f \text{ on } \Omega \\ u_\varepsilon = 0 \quad \text{on } \partial\Omega \end{cases}$$

that is:

(23) $$u_\varepsilon \to u \quad \text{weakly in } H_0^1(\Omega)$$

$$\int_\Omega j\left(\frac{x}{\varepsilon},\text{grad}u_\varepsilon\right) dx \to \int_\Omega j^{hom}(\text{grad}u)\,dx \quad \text{(convergence of the energies)}$$

where u is the solution of the limit homogenized problem

(24) $$\begin{cases} -\text{div}\ \partial j^{hom}(\text{grad}u) = f \quad \Omega \\ u = 0 \quad \partial\Omega\ . \end{cases}$$

This can be viewed as the primal form of the homogenization problem since variables u_ε, u can be considered as primal variables.
The dual variables $\sigma_\varepsilon = \partial j\left(\frac{x}{\varepsilon},\text{grad}u_\varepsilon\right)$ play an important role too! For example in the electrostatic problem the primal variable u_ε is the potential, the dual variable σ_ε is the vector density of current. In the corresponding elasticity model, u_ε is the deplacement vector while $\sigma_\varepsilon = \partial j\left(\frac{x}{\varepsilon}\ e(u_\varepsilon)\right)$ is the strain tensor ($e(u)$ is the linear deformation tensor). The puzzling question is that the convergence of the dual variables cannot be derived in an elementary way from the convergence of the primal ones. For example in the electrostatic above model, taking $j(y,z)=\frac{1}{2}a(y)|z|^2$,

$$\sigma_\varepsilon = a\left(\frac{x}{\varepsilon}\right).\text{grad}u_\varepsilon\ .$$

It is not even obvious that the sequence σ_ε, as a product of two weakly converging sequences in $L^2(\Omega)$, is convergent in distribution sense!
We are going to raise this question in a quite general way, taking advantage of the dual role of the variables σ_ε and of the continuity properties of the Fenchel duality transformation with respect to epi-convergence. (section 3).

Let us take $X=H_o^1(\Omega)$, for every $\varepsilon>0$, u_ε is the solution of the primal problem

(25) $$(P_\varepsilon) \quad \min_{u\in X} F^\varepsilon(u)$$

where $F^\varepsilon(u) = \int_\Omega j\left(\frac{x}{\varepsilon}, \text{grad}u(x)\right) dx - \int_\Omega fu\, dx$.

The *dual problem* (following Ekeland & Temam [20]) is attached to the *following perturbation functional:*

(26) for every $u\in X$, for every $\tau\in Y$ (where $Y=L^2(\Omega)$)

$$F^\varepsilon(u,\tau) = \int_\Omega j\left(\frac{x}{\varepsilon}, \text{grad}u(x)+\tau\right) dx - \int_\Omega fu\, dx .$$

The *marginal function* (also called value function) h^ε is given by

(27) $$h^\varepsilon(\tau) = \min_{u\in X} F^\varepsilon(u,\tau)$$

So,

$$h^\varepsilon(0) = \min_{u\in X} F^\varepsilon(u) .$$

The dual formulation of (P_ε) is attached to the regularity of the marginal function h^ε at $\tau=0$. When the convex function h^ε is continuous at $\tau=0$ (or at least lower semicontinuous) which requires qualification assumptions, clearly satisfied here, we obtain

(28) $$\begin{aligned} h^\varepsilon(0) &= h^{\varepsilon^{**}}(0) \\ &= \sup_\sigma \{-h^{\varepsilon^*}(\sigma)\} \\ &= -\inf_\sigma \{F^{\varepsilon^*}(0,\sigma)\} . \end{aligned}$$

In our situation

$$h^{\varepsilon^*}(\sigma) = \int_\Omega j^*\left(\frac{x}{\varepsilon},\sigma\right) dx + I_{\{-\text{div}(\cdot)=f\}}(\sigma) .$$

The dual problem is

(29) $$(P_\varepsilon^*) \quad -\inf_\sigma \left\{ \int_\Omega j^*\left(\frac{x}{\varepsilon},\sigma\right) dx + I_{\{-\text{div}(\cdot)=f\}}(\sigma) \right\}$$

whose Euler equation is

$$(30)\qquad \begin{cases} \partial j^*\left(\frac{x}{\varepsilon}, \sigma_\varepsilon\right) = \operatorname{grad} u_\varepsilon \\ -\operatorname{div} \sigma_\varepsilon = f \\ u_\varepsilon = 0 \quad \text{on } \partial\Omega \end{cases}$$

u_ε appears as the Lagrange multiplier of the constraint $-\operatorname{div}\sigma = f$. So

$$\sigma_\varepsilon = \partial j\left(\frac{x}{\varepsilon}, \operatorname{grad} u_\varepsilon\right) \text{ and}$$

$$-\operatorname{div}\, \partial j\left(\frac{x}{\varepsilon}, \operatorname{grad} u_\varepsilon\right) = f .$$

Finally the two quantities u_ε and σ_ε appear as respective solutions of the primal and dual problems (P_ε) and (P_ε^*).

Since σ_ε is a solution of a minimization problem (P_ε^*), let us follow the epi-convergence approach in order to study the convergence of the sequence $\{\sigma_\varepsilon ; \varepsilon \to 0\}$. Because of the equi-coercivness of the integrands $j_\varepsilon^* = j^*\left(\frac{x}{\varepsilon}, \cdot\right)$ (which follows from the uniform bound from above on the j_ε) the sequence $\{\sigma_\varepsilon ; \varepsilon \to 0\}$ remains bounded in $L^2(\Omega)^N$. Thus we are led to consider the sequence $\{G^\varepsilon ; \varepsilon \to 0\}$ given by

$$(31)\qquad G^\varepsilon(\sigma) = \int_\Omega j^*\left(\frac{x}{\varepsilon}, \sigma(x)\right) dx + I_{\{-\operatorname{div}(\cdot) = f\}}(\sigma)$$

and study its epi-convergence for the weak topology of $Y = L^2(\Omega)^N$.

The point is that G^ε is equal to $h^{\varepsilon *}$ the Legendre-Fenchel transform of the marginal function h^ε! Thanks to the equi-coercivness of the sequence $\{h^{\varepsilon *} ; \varepsilon \to 0\}$ on $L^2(\Omega)^N$, by relying on theorem 3.4 what we have to prove is

(32) *the epi-convergence for the strong topology of* Y *of the sequence of marginal functions* $\{h^\varepsilon ; \varepsilon \to 0\}$.

Thanks to corollary (3.5) and the strong uniform coercivness of the sequence (h_ε^*) this is equivalent to prove the pointwise convergence of the sequence of marginal functions:

$$(33)\qquad \forall \tau \in L^2(\Omega)^N \quad h^\varepsilon(\tau) \to h^{hom}(\tau) .$$

Noticing that

$$h^{\varepsilon}(\tau) = \inf_{u \in X} F^{\varepsilon}(u,\tau)$$

and thanks to the uniform coercivity on $X=H^1_o(\Omega)$ of functionals $\{F^{\varepsilon}(\cdot,\tau)\,;\ \varepsilon \to 0\}$ this is equivalent to prove

(34) $\forall \tau \in L^2(\Omega)^N$ $F^{\varepsilon}(\cdot,\tau)$ epi-converges to $F^{hom}(\cdot,\tau)$ for the weak topology of $X=H^1_o(\Omega)$.

Let us summarize the above considerations and say that:
Since one has been able to solve the epi-convergence problem for each perturbed sequence $\{F^{\varepsilon}(\cdot,\tau)\,;\ \varepsilon \to 0\}$ one obtains automatically, by using continuity properties of the Legendre-Fenchel transformation with respect to epi-convergence in an abstract setting, *both* convergences of primal and dual variables (attached to this perturbation).
So the right problems one has to solve in order to obtain the whole information is the *epi-convergence of the perturbed functionals*.
Fortunately it is not a real difficulty to extend the unperturbed epi-convergence theorem 4.1 to the perturbed case. Indeed, the perturbation function $\tau(x)$ can be "frozen" (it does not play any role with respect to high oscillations of variable y): By a direct proof relying just on verification of definition of epi-convergence D. Aze [9] has obtained the following result (cf. also Bensoussan, J.L. Lions, Papanicolaou [13]):

Theorem 4.2

With the same assumptions as in theorem 4.1 let consider for every $\tau \in Y = L^2(\Omega)$ the sequence

$$F^{\varepsilon}_{\tau}(u) = \int_{\Omega} j\left(\frac{x}{\varepsilon}, \mathrm{grad}\,u(x)+\tau(x)\right) dx$$

Then for every $u \in X=H^1_o(\Omega)$, taking T=weak topology of X

$$T\text{-}\lim_e F^{\varepsilon}_{\tau}(u) = F^{hom}_{\tau}(u)$$

where

$$F^{hom}_{\tau}(u) = \int_{\Omega} j^{hom}(\mathrm{grad}\,u(x)+\tau(x))\, dx$$

and

$$j^{hom}(z) = \min_{\{w \text{ Y-periodic}\}} \int_Y j(y, \operatorname{grad} w(y)+z)\, dy .$$

Thanks to theorem 4.2 and the above considerations we can now formulate the following dual homogenization theorem (D. Aze [9], P. Suquet [32]).

Theorem 4.3

Let $\varphi : \mathbb{R}^N \times \mathbb{R}^N \to \mathbb{R}$, $(y,z) \to \varphi(y,z)$ which satisfies: φ is Y-periodic with respect to y, convex with respect to z and $c_o\left(|z|^2-1\right) \le \varphi(y,z) \le C_o |z|^2$ (where $0<c_o \le C_o < +\infty$).

Let us consider the sequence of closed convex functions $\{G^\varepsilon : L^2(\Omega) \to]-\infty,+\infty];\ \varepsilon \to 0\}$ defined by

$$G^\varepsilon(\sigma) = \int_\Omega \varphi\left(\frac{x}{\varepsilon}, \sigma(x)\right) dx + I_{\{\operatorname{div}(\cdot)=-f\}}(\sigma)$$

Then for every $\sigma \in L^2(\Omega)$, taking τ equal to the weak topology of $L^2(\Omega)^N$

$$\tau\text{-}\lim_e G^\varepsilon(\sigma) = G^{hom}(\sigma) \quad \text{exists}$$

and G^{hom} is given by the following formulas:

$$G^{hom}(\sigma) = \int_\Omega \varphi^{hom}(\sigma(x))dx + I_{\{-\operatorname{div}(\cdot)=f\}}(\sigma)$$

(35) $$\varphi^{hom}(z) = \min_{\left\{\begin{array}{l} .\ \operatorname{div}\sigma=0 \\ .\ \fint_Y \sigma = 0 \\ .\ \sigma.n \text{ takes opposite values} \\ \quad \text{on opposite faces of } \partial Y. \end{array}\right.} \int_Y \varphi(y, \sigma(y)+z)\, dy$$

Remark 4.4 In order to obtain the above formulation of φ^{hom} just notice that $\varphi^{hom} = \left((\varphi^*)^{hom}\right)^*$ where $(\varphi^*)^{hom}$ is given by primal homogenization formula (22) with $j=\varphi^*$.

Corollary 4.5 Denoting u_ε the solution of

$$\begin{cases} -\operatorname{div} \partial j\left(\frac{x}{\varepsilon}, \operatorname{grad} u_\varepsilon\right) = f \text{ on } \Omega \\ u_\varepsilon = 0 \quad \text{on } \partial\Omega \end{cases}$$

<u>and</u> $\sigma_\varepsilon = \partial j\left(\frac{x}{\varepsilon},\text{grad}u_\varepsilon\right)$ <u>the "dual variable", the following convergences hold</u>

(36)
$$\begin{cases} u_\varepsilon \to u \quad \text{weakly in } H^1_0(\Omega) \\ \sigma_\varepsilon \to \sigma = \partial j^{hom}(\text{grad}u) \quad \text{weakly in } L^2(\Omega)^N \end{cases}$$

<u>where</u> u <u>is the solution of the homogenized problem</u>:

$$\begin{cases} -\text{div}\ \partial j^{hom}(\text{grad}u) = f \quad \text{on } \Omega \\ u = 0 \quad \text{on } \partial\Omega \end{cases}$$

<u>Moreover there is convergence of the energies, that is</u>,

(37)
$$\begin{cases} \int_\Omega j\left(\frac{x}{\varepsilon},\text{grad}u_\varepsilon\right)dx \xrightarrow[(\varepsilon\to 0)]{} \int_\Omega j^{hom}(\text{grad}u)\,dx \\ \int_\Omega j^*\left(\frac{x}{\varepsilon},\dot{\sigma}_\varepsilon(x)\right)dx \xrightarrow[(\varepsilon\to 0)]{} \int_\Omega \left(j^{hom}\right)^*(\sigma(x))dx \end{cases}$$

5. CONVERGENCE OF PRIMAL AND DUAL VARIABLES. A GENERAL APPROACH VIA EPI-CONVERGENCE OF MARGINAL FUNCTIONS

The approach followed in the preceding paragraph can be generalized and written in the abstract setting of perturbation and duality theory in order to cover various situations in convex optimization.
We follow the lines of D. Aze recent paper [9] and first recall some classical results about duality (refer to Ekeland & Temam [20],...).

X and Y are topological vector spaces, X^* and Y^* their duals (or more generally consider a duality pairing).

$F: X\times Y \to]-\infty,+\infty]$ is a closed convex proper function (usually called the perturbation function)

The primal problem is

(P)
$$\inf_{x\in X} F(x,0)$$

For every $y\in Y$,

(38) $$h(y) = \inf_{x \in X} F(x,y)$$

is called the marginal function (or value function). It is a convex function which is not necessarily closed.
Let us notice that h^* the Legendre-Fenchel transform of h is equal to

(39) $$h^*(y^*) = F^*(0,y^*)$$

The problem

(P^*) $$\inf_{y^* \in Y^*} F^*(0,y^*)$$

is called the dual problem of (P).
The inequality $\inf P^* + \inf P \geq 0$ is always true.
Under some qualification assumption, h turns to be continuous at the origin and the above inequality turns to be an equality: equality between primal and dual problems.
More precisely let us assume that

(i) X is a reflexive Banach space
(ii) $F(x,\cdot)$ is coercive
(iii) there exists $x_o \in X$ such that $F(x_o,\cdot)$ is finite and continuous at the origin

then

- $\inf P^* = -\inf P$
- P and P^* have solutions
- $\bar{x}$ solution of P, $\bar{x}^*$ solution of P^* iff $\bar{x}$ and $\bar{x}^*$ are tied by the extremality relation: $F(\bar{x},0) + F^*(0,\bar{x}^*) = 0$.

Let us consider the parametrized version of this duality scheme and consider

$$\{F^n : X \times Y \to]-\infty,+\infty] \; ; \; n = 1,2,\ldots\}$$

a sequence of perturbation functions, $F^n \in \Gamma_o(X \times Y)$

$$\inf_{x \in X} F^n(x,0)$$

the corresponding primal problems,

$$\inf_{y^* \in Y^*} F^{n^*}(0,y^*)$$

the corresponding dual problems, h^n the associated marginal functions... We look for "*minimal*" *assumptions* on the $\{F^n ; n \in \mathbb{N}\}$ which guaranty the existence of solutions $\overline{x}_n$ and $\overline{y}^*_n$ to the primal and dual problems and and their convergence as n goes to $+\infty$.

Theorem 5.1

Let X be a reflexive Banach space and Y a separable Banach space (not necessarily reflexive). Let us consider

$$\{F^n : X \times Y \to]-\infty,+\infty] \; ; \; n=1,2,\ldots\}$$

a sequence of closed convex proper functions which is assumed to satisfy:

(i) $(X_w \times Y_s)\text{-}\lim_e F^n = F$ exists

(ii) "a uniform qualification assumption":
There exists $r_o > 0$ such that, for every sequence $\{y_n ; n=1,2,\ldots\}$ in Y with y_n belonging to $B_{r_o} = \{y \in Y ; \|y\| \le r_o\}$ for every $n \in \mathbb{N}$, there exists a bounded sequence $\{x_n ; n=1,2,\ldots\}$ in X such that

$$\limsup_n F^n(x_n,y_n) < +\infty.$$

(iii) a "uniform coercivness assumption":

$$F^n(x,0) \ge c(\|x\|)$$

with c independent of n, satisfying $\lim_{r \to +\infty} c(r) = +\infty$,

then,

. $Y_s\text{-}\lim_e h^n = h$

. $Y^*_w\text{-}\lim_e h^{n^*} = h^*$

where h is the marginal function associated to F.

Moreover

$$F(\cdot,0) = X_w\text{-lim}_e\, F^n(\cdot,0)\ .$$

Let us denote by x_n an ε_n-minimizer for P_n, by y_n^* an ε_n-minimizer for P_n^*, with $\varepsilon_n \to 0$ as $n \to +\infty$, then x_n and y_n^* are bounded. Considering x and y^* any weak limit point of such sequences

$$\begin{cases} x \text{ minimizes } P \\ y^* \text{ minimizes } P^* \\ \inf P + \inf P^* = 0 \\ \inf P^n \to \inf P \quad \text{as} \quad n \to +\infty \\ \inf P^{n^*} \to \inf P^* \quad \text{as} \quad n \to +\infty\ . \end{cases}$$

Comments

The above theorem allows to derive from the epi-convergence of the perturbation functions $\{F^n\ ; n=1,2,\dots\}$ the convergence of all the mathematical objects attached to the duality scheme.

In some particular situations like the one described in the preceding section thanks to the equi-continuity property of functions F^n with respect to the perturbation variable (for the strong topology of Y) one can freeze this variable and reduce proving the epi-convergence for every $y \in Y$

$$F(\cdot,y) = X_w\text{-lim}_e\, F^n(\cdot,y)\ .$$

The generality of the "qualification assumption" allows to cover various optimization problems with constraints (D. Aze [10]) in mechanics, convex programming...

6. A UNIFYING POINT OF VIEW: EPI/HYPO-CONVERGENCE OF LAGRANGIANS

Let us complete the duality scheme exposed in the preceding section and introduce the *Lagrangian function* $K : X \times Y \to \overline{\mathbb{R}}$ classicaly defined by

$$K(x,y^*) = -(F(x,\cdot))^*(y^*).$$

In other words $-K$ is the partial Legendre-Fenchel transform of the perturbation function F with respect to the perturbation variable.
The Lagrangian function K is convex-concave with respect to (x,y^*) and upper semicontinuous with respect to y^*.
A couple of solutions of primal and dual problems can be characterized as a saddle point of the Lagrangian K; more precisely

$$(x,y^*) \text{ is a saddle point of } K$$
$$\Updownarrow$$
$$\begin{cases} x \text{ is a solution of } \mathcal{P} \\ y^* \text{ is a solution of } \mathcal{P}^* \\ \inf \mathcal{P} + \inf \mathcal{P}^* = 0 . \end{cases}$$

Let us now examine the parametrized version of this duality scheme and formulate directly on the sequence of Lagrangians functions $\{K^n : X\times Y^* \to \overline{\mathbb{R}} / n=1,2,\dots\}$ the minimal assumptions which guaranty the convergence of the corresponding saddle points and saddle values.
We know from Attouch & Wets [5][6] that the good concept of convergence for a sequence of bivariate functions which allows to derive the convergence of corresponding saddle points is *epi-hypo convergence*. Let us first recall its definition in an abstract topological setting:

Given (X,τ) and (Y,σ) two topological spaces and

$$\{K^n : X\times Y \to \overline{\mathbb{R}} \quad ; \quad n=1,2,\dots\}$$

a sequence of bivariate functions

Let us introduce for every (x,y) belonging to $X\times Y$

$$(40) \qquad \left(e_\tau/h_\sigma\text{-ls } K^n\right)(x,y) = \sup_{\{y_n \xrightarrow{\sigma} y\}} \inf_{\{x_n \xrightarrow{\tau} x\}} \limsup_n K^n(x_n,y_n)$$

$$(41) \qquad \left(h_\sigma/e_\tau\text{-li } K^n\right)(x,y) = \inf_{\{x_n \xrightarrow{\tau} x\}} \sup_{\{y_n \xrightarrow{\sigma} y\}} \liminf_n K^n(x_n,y_n)$$

These are the sequential notions which coincide, when (X,τ) and (Y,σ) are metrizable, with the topological ones. Indeed, as in the case of epi-convergence these are the ones we are going to use here.

Definition 6.1 The sequence $\{K^n ; n\in\mathbb{N}\}$ is said to be τ-epi/σ-hypo convergent and K is an epi/hypo limit if the two following inequalities hold:

$$e_\tau/h_\sigma\text{-ls}\, K^n \le K \le h_\sigma/e_\tau\text{-li}\, K^n \tag{42}$$

Let us notice that in general the two above functions $e_\tau/h_\sigma\text{-ls}\, K^n$ and $h_\sigma/e_\tau\text{-li}\, K^n$ are not comparable. Consequently when inequalities (42) hold it may exist a whole interval $\left[e_\tau/h_\sigma\text{-ls}\, K^n,\ h_\sigma/e_\tau\text{-li}\, K^n\right]$ of epi-hypo limit functions: there is not in general a unique limit function. The variational properties of epi/hypo convergence, that motivated the definition are formalized by the next theorem.

Theorem 6.2

Let $K^n : X\times Y \to \overline{\mathbb{R}}$ be a sequence of bivariate functions and for each $n\in\mathbb{N}$, let (x_n,y_n) be a saddle point of K^n.

Let us assume that the sequence (x_n,y_n) is relatively compact in $(X,\tau)\times(Y,\sigma)$.

Then, the following implication hold

(i) K is a τ-epi/σ-hypo limit of the sequence K^n

$\Downarrow$

(ii) every τ/σ limit point $(\bar{x},\bar{y})$ of the sequence (x_n,y_n) is a saddle point of K and $K(\bar{x},\bar{y}) = \lim_n K^n(x_n,y_n)$.

Let us now return to the duality scheme and the pairing via the partial Legendre-Fenchel transform between the convex functions with respect to the two variables and the convex-concave functions.

In order to make this correspondance a one to one correspondance one has to introduce classes of convex-concave functions:

Following Rockafellar [30], with any convex-concave function $K : X\times Y^* \to \overline{\mathbb{R}}$ we associate its convex and concave parents defined by

$$F(x,y) = \sup_{y^*\in Y^*} [K(x,y^*)+\langle y^*,y\rangle]$$

$$G(x^*,y^*) = \inf_{x\in X} [K(x,y^*)-\langle x^*,x\rangle]$$

We have the following relations between these functions

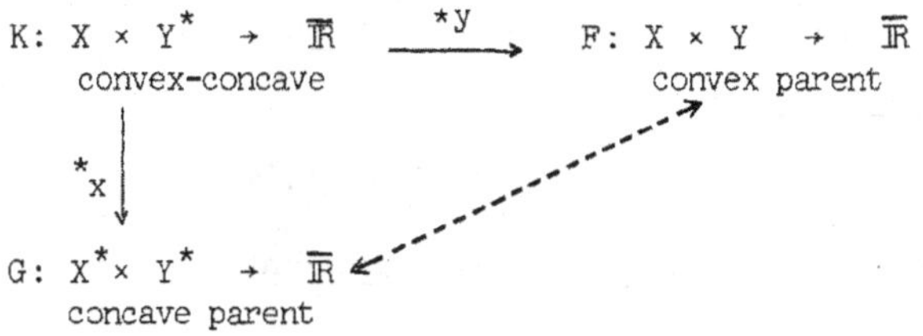

Two convex-concave functions are said to be equivalent if they have the same parents. A bivariate function K is said to be closed if its parents are conjugate of each other, i.e. if the above diagram can be closed through the classical Legendre-Fenchel transform.

For closed convex-concave functions K, the associated equivalence class is an interval denoted by $\left[\underline{K},\overline{K}\right]$ with

$$\underline{K}(x,y^*) = (K(\cdot,y^*))^{**}(x) = (-G(\cdot,y^*))^*(x) = \underline{cl}_x K$$

$$\overline{K}(x,y^*) = -(-K(x,\cdot))^{**}(y^*) = -(F(x,\cdot))^*(y^*) = \overline{cl}_{y^*}K$$

where $\underline{cl}_x$ denotes the extended lower closure with respect to x and $\overline{cl}_{y^*}$ the extended upper closure with respect to y^*.

From Rockafellar [30]

(43) The map $K \xrightarrow{*y} F$ establishes a one to one correspondance between closed convex-concave (equivalence) class and closed convex functions.

We can now state the equivalent version of theorem 5.1 but now expressed in terms of Lagrangian functions (Attouch, Aze & Wets [11], Attouch & Wets [6]).

Theorem 6.3

Let X be a reflexive Banach space and Y a separable Banach space (not necessarily reflexive). Let us consider $\{F^n : X\times Y \to]-\infty,+\infty] \; ; \; n=1,2,\dots\}$ a sequence of closed convex proper functions which satisfies a) and b)

a) "*uniform qualification assumption*": there exists r_o such that for any sequence $\{y_n \; ; n\in\mathbb{N}\}$ in Y with $\|y_n\| \le r_o$ there exists a bounded sequence $\{x_n \; ; n\in\mathbb{N}\}$ in X which satisfies

$$\limsup_n F^n(x_n,y_n) < +\infty .$$

b) "*uniform coerciveness assumption*": there exists $c:\mathbb{R}^+ \to \mathbb{R}^+$ a coercive function such that

$$F^n(x,0) \geq c(\|x\|).$$

Then, the two following sentences are equivalent

(i) $$F = (X_w \times Y_s)\text{-lim}_e F^n$$

$$\Updownarrow$$

(ii) $$\underline{cl}_x\left(e_w/h_{w^*}\text{-ls } K^n\right) \leq K \leq \overline{cl}^{y^*}\left(h_{w^*}/e_w\text{-li } K^n\right)$$

holds for any functions $K^n \in \left[\underline{K}^n, \overline{K}^n\right]$ and $K \in \left[\underline{K}, \overline{K}\right]$.

Remark For the equivalence class of closed convex-concave Lagrangians functions K^n, the right notion of convergence as expressed above is the extended notion of epi/hypo convergence for the weak topologies of X and Y^*

$$\underline{cl}_x\left(e_w/h_{w^*}\text{-ls } K^n\right) \leq K \leq \overline{cl}^{y^*}\left(h_{w^*}/e_w\text{-li } K^n\right),$$

which involves, in addition to epi/hypo-convergence, the above extended closure operations.

This notion as explained in Attouch, Aze & Wets [11] does not satisfy any more the local properties of epi/hypo-convergence. On the other hand it is well fitted to the Legendre-Fenchel transformation which is a non local one! Indeed the basic variational properties (theorem 6.2), as one can easily verify, are still satisfied by the extended notion of epi/hypo-convergence.

REFERENCES

[1] H. Attouch. 'Familles d'operateurs maximaux monotones et mesurabilité'. *Ann. Mat. Pura Appl.* 120 (1979), 35-111.

[2] H. Attouch. 'Variational convergence for functions and operators'. *Applicable Mathematics Series*. Pitman Advanced Publishing Program (1984).

[3] H. Attouch. 'Theorie de la Γ-convergence. Applications à des inéquations variationnelles de la mécanique'. *Seminaire Goulaouic - Meyer-Schwartz* (1982-83). Publications Ecole Polytechnique (Palaiseau).

[4] H. Attouch. 'Variational properties of epi-convergence'. Proceedings of the international congress on multifunctions and normal integrands, stochastic analysis, approximation and optimization'. Catane (Sicilia) (1984), G. Salinetti (ed.), *Lecture Notes in Math*, 1091 Springer Verlag, Berlin.

[5] H. Attouch and R. Wets. 'A convergence theory for saddle functions'. *Trans. Amer. Math. Soc.* Vol 280, n.1, Nov.(1983).

[6] H. Attouch and R. Wets. 'A convergence for bivariate functions aimed at the convergence of saddle value'. Proceedings S.Margherita Ligure on 'Mathematical theories of optimization'. Edited by J.P.Cecconi and T. Zolezzi. *Lecture Notes in Math.* 979, Springer Verlag, (1981).

[7] H. Attouch and R. Wets. 'Isometries for the Legendre-Fenchel transform'. Publications Ceremade Paris-Dauphine (1984) (to appear).

[8] H. Attouch and R. Wets. 'Approximation and convergence in non linear optimization', in Nonlinear Programming 4, (Eds. O. Mangasarian, R. Meyer, S. Robinson) *Academic Press*, New York, 367-394, (1981).

[9] D. Aze. 'Epi-convergence et dualité. Application à la convergence des variables primales et duales pour des suites de problèmes en optimisation convexe'. Publication AVAMAC (Univ. Perpignan) 1984-85 (to appear).

[10] D. Aze. 'Deux exemples de convergence d'infima de problèmes d'optimisation sous leur forme duale par des méthodes d'epi-convergence'. Publication AVAMAC (Univ. Perpignan) 1984-85 (to appear).

[11] H. Attouch, D. Aze and R. Wets. 'Convergence of convex-concave saddle functions. Publication AVAMAC (Univ. Perpignan) 1984-85 (to appear).

[12] T. Back. 'Continuity of the Fenchel transform of convex functions'. *Tech. Report*, Northwestern University, Nov. 1983.

[13] A.Bensoussan, J.L. Lions and G. Papanicolaou. 'Asymptotic analysis for periodic structures'. North Holland (1978).

[14] E. Cavazzuti. 'Alcune caratterizzazioni della Γ-convergenza multipla'. *Annali di Matematica pura ed applicata* (1982) (IV), Vol. XXXII, pp. 69-112.

[15] S. Dolecki. 'Duality in optimization and continuity of polarities; *International School of Math.* " G. Stampacchia ", Erice (1984).

[16] E. De Giorgi. 'Convergence problems for functionals and operators'. Proceedings of the international meeting on recent methods in nonlinear analysis. Rome, May(1978). Edited by E. De Giorgi, E. Magenés, U. Mosco. Pitagora. Editrice Bologna.

[17] E. De Giorgi and T. Franzoni. 'Su un tipo di convergenza variazionale'. *Rend. Acc. Naz. Lincei*, 58 (1975), 842-850.

[18] E. De Giorgi and G. Dal Maso. 'Γ-convergence and calculus of variations'. Proceedings S.Margherita Ligure (1981) 'Mathematical Theories of Optimization'. Edited by J.P. Cecconi and T. Zolezzi. *Lecture Notes in Math.* 979, Springer Verlag .

[19] E. De Giorgi and S. Spagnolo. 'Sulla convergenza degli integrali dell'energia per operatori ellittici del II ordine'. *Boll. Un. Mat. Ital.* (4) 8, 391-411 (1973).
[20] I. Ekeland and R. Temam. 'Convex analysis and variational problems'. North Holland (1978).
[21] J.L. Joly. 'Une famille de topologies sur l'ensemble des fonctions convexes pour lesquelles la polarité est bicontinue'. *J. Math. Pures Appl.*, 52, 421-441 (1973).
[22] J.L. Lions. 'Some methods in the Mathematical analysis of systems and their control'. *Science Press*, Pekin, China. Gordon and Breach, Science Publishers, Inc. New York.
[23] R. Lucchetti and F. Patrone. 'Closure and upper semicontinuity results in mathematical programming'. *Nash and economic equilibria* (to appear).
[24] L. Mac Linden. 'Successive approximation and linear stability involving convergent sequences of optimization problems'. *J. of Approximation theory* 35, 311-354 (1982).
[25] L. Mac Linden and R.C. Bergstrom. 'Preservation of convergence of convex sets and functions in finite dimensions'. *Transactions of the American Math. Soc.* Vol. 268, n.1, (1981).
[26] P. Marcellini. 'Periodic solutions and homogenization of non linear variational problems'. *Ann. Mat. Pura. Appl.* (4), 117, 139-152 (1978).
[27] U. Mosco. 'Convergence of convex sets and of solutions of variational inequalities'. *Advances in Math.*, 3, 510-585 (1969).
[28] U. Mosco. 'On the continuity of the Young-Fenchel transformation'. *J. Math. Anal. Appl.* 35, 518-535 (1971).
[29] N. Papageorgiou. 'Stochastic nonsmooth analysis and optimization'. Thesis University of Illinois.
[30] R.T. Rockafellar. 'A general correspondance between dual minimax problems and convex programs'. *Pacific J. Math.*, 25, 597-611 (1968).
[31] Y. Sonntag. 'Convergence au sens de Mosco'...Thèse d'état. Université de Provence (Marseille) (1982).
[32] P. Suquet. 'Plasticité et homogénéisation'. Thèse d'état. Paris (1982).
[33] M. Volle. 'Conjugaison par tranches'. *Ann. Mat. Pura Appl.* (4), 139, 279-311, (1985).
[34] D. Walkup and R. Wets. 'Continuity of some convex-cone valued mappings'. *Proceed. Amer. Math. Soc.* 18 (1967), 229-235.
[35] R. Wets. 'Convergence of convex functions, variational inequalities and convex optimization problems in Variational Inequalities and Complementarity problems'. Eds. P. Cottle, F. Giannessi, J.L. Lions, Wiley, Chichester (UK) 375-403 (1980).
[36] R. Wets. 'A formula for the level sets of epi-limits and some applications'. *Working paper*, I.I.A.S.A. (Laxenburg, Austria) Sept. 1982.
[37] R. Wijsman. 'Convergence of sequences of convex sets, cones and functions II. *Transactions Amer. Math. Soc.* 123, 32-45 (1966).
[38] T. Zolezzi. 'On stability analysis in mathematical programming'. *Mathematical Programming studies*. Fiacco editor, (to appear).

Chapter 3

Non-Linear Separation Theorems, Duality and Optimality Conditions

A. Cambini

1. INTRODUCTION

Recently [15, 16] a theorem of the alternative has been stated for generalized systems and it has been shown how to deduce, from such a theorem, known optimality conditions like saddle-point one, regularity conditions, known theorems of the alternative and new ones; furthermore connections among optimality conditions, duality, penalty functions have been shown. Some of these new ideas have been deepened by other people and new interesting results have been obtained [6, 22, 26, 27, 28, 29]. The aim of this paper is to deepen this unifying approach in such a way to give a survey of these recent studies and, at the same time, to obtain some new results.

2. THEOREMS OF THE ALTERNATIVE AND SEPARATION FUNCTIONS

Assume that we are given the positive integers n and ν, the non-empty sets $H \subset \mathbb{R}^{\nu}$, $X \subseteq \mathbb{R}^{n}$, and the real-valued function $F: X \to \mathbb{R}^{\nu}$. We want to study conditions for the generalized system

$$(1) \qquad F(x) \in H \;, \quad x \in X$$

to have (or not to have) solutions.

To this aim we introduce the following

Definition 1.1: $w : \mathbb{R}^{\nu} \to \mathbb{R}$ is called *weak separation function* iff[1]

(2a) $$\mathrm{lev}_{>0}\, w \supseteq \mathcal{H}$$

$s: \mathbb{R}^{\nu} \to \mathbb{R}$ is called *strong separation function* iff

(2b) $$\mathrm{lev}_{>0}\, s \subseteq \mathcal{H}$$

The following theorem holds:

Theorem 1.1

Let the sets $\mathcal{H}$, X and the function F be given.

i) The systems (1) and (3a)

(3a) $$w(F(x)) \leq 0 \quad , \quad \forall x \in X$$

are not simultaneously possible, whatever the weak separation function w might be.

ii) The systems (1) and (3b)

(3b) $$s(F(x)) \leq 0 \quad , \quad \forall x \in X$$

are not simultaneously impossible, whatever the strong separation function s might be.

(1) If ρ is a real-valued function $\rho: \mathbb{R}^n \to \mathbb{R}$, we set $\mathrm{lev}_{>0}\, \rho = \{y \in \mathbb{R}^n : \rho(y) > 0\}$. Sets $\mathrm{lev}_{\geq 0}\rho$, $\mathrm{lev}_{<0}\rho$, $\mathrm{lev}_{\leq 0}\rho$ are defined in a similar way.

Proof: i) If (1) is possible, there exists $\bar{x} \in X$ such that $F(\bar{x}) \in H$; from (2a) we have $w(F(\bar{x})) > 0$ so that (3a) is false. ii) If (1) is impossible, i.e., if $F(x) \subset H$, $\forall x \subset X$ then (2b) implies $s(F(x)) \leq 0 \quad \forall x \in X$ so that (3b) is true. This completes the proof. #

Let us note that the impossibility of (1) is not equivalent to the possibility of (3a) (or(3b)). For this reason we say that *weak* (strong) *alternative* holds between (1) and (3a)(or(3b)).

When the possibility of (3b) implies the impossibility of (1) we say that *alternative* holds between (1) and (3b). Set $K = F(X)$; it is obvious that system (1) is impossible iff $K \cap H = \emptyset$. Thus, if we refer to $\mathbb{R}^{\nu}$ as the *image space*, the impossibility of (1) is equivalent to the disjunction between K and H in the image space.

When $K \subseteq \text{lev}_{\leq 0} w$, that is (3a) is possible so that (1) is impossible, we say that *alternative* holds between (1) and (3a) or, equivalently, that w is a weak separation function which *guarantees alternative.*

In this order of ideas two important questions consist in choosing, for any given triplet (F, X, H), an appropriate class of weak separation functions and in finding, in this class, an element, if one exists, which guarantees alternative. These questions will be analyzed for a wide class of systems.

From now on we suppose that H is a convex cone, satisfying the following property[2]

$$(4) \qquad H + \text{cl}\, H = H$$

For instance, (4) is verified in the following cases: a) H is open

[2] cl A denotes the closure of set A.

or H is closed; b) cl H is pointed, i.e. $(\mathrm{cl}\ H) \cap (-\mathrm{cl}\ H) = \{0\}$ and[3] $H = (\mathrm{cl}\ H) \setminus \{0\}$; c) H is the Cartesian product of two convex cones H_1, H_2 with H_1 and/or H_2 satisfying a) or b).

Consider now the *conic extension* of K with respect to cl H, i.e., the set $E \triangleq K - \mathrm{cl}\ H$.

The following lemma holds:

Lemma 1.1 System (1) is impossible iff $E \cap H \neq \emptyset$.

Proof: Sufficiency: $H \cap E = \emptyset \Rightarrow K \cap H = \emptyset \Rightarrow$ (1) impossible. *Necessity:* assume that (1) is impossible and that $E \cap H \neq \emptyset$. Then there exist $\bar{x} \in X$, $h \in \mathrm{cl}\ H$ such that $F(\bar{x}) - h \in H$ that is $F(\bar{x}) \in (\mathrm{cl} H) + H = H$ and this is absurd. #

The reason of introducing E is based on the fact that E can have some properties which are not valid for K; for instance E turns out to be convex even if K is not. E is a key set in further analysis and we will see that its properties are very important in order to study optimality conditions, duality, regularity and so on.

As regards to the convexity of E, the following lemma holds:

Lemma 1.2 Let X be convex. The following conditions are equivalent:

i) F is (cl H)-convex-like[4]

ii) $E = F(X) - \mathrm{cl}\ H$ is convex.

[3] $A \setminus B = \{a \in A: a \notin B\}$.

[4] Let $C \subset \mathbb{R}^s$ be a convex cone, and $X \subseteq \mathbb{R}^n$ a convex set. A function $F: X \to \mathbb{R}^s$ is C-convexlike iff $\forall x, y \in X$, there exist $z \in X$ such that $F(z) - (1-\alpha)F(x) - \alpha F(y) \in C$, $\forall \alpha \in [0,1]$.

Proof: We have

i) $\Leftrightarrow \forall x,y \in X,\ \forall \alpha \in [0,1],\ (1-\alpha)F(x)+\alpha F(y) \in F(X)-\mathrm{cl}\ H$

$\Leftrightarrow \forall x,y \in X,\ \forall p_1,p_2 \in \mathrm{cl}\ H,\ \forall \alpha \in [0,1]$

$(1-\alpha)(F(x)-p_1)+\alpha(F(y)-p_2) \in (F(X)-\mathrm{cl}\ H \Leftrightarrow$ ii). #

Lemma 1.1 shows that the impossibility of (1) is equivalent to the disjunction between E and H in the image space; when E is convex it is natural to study this disjunction by means of the class of linear functionals $W = \{w: \mathbb{R}^{\nu} \to \mathbb{R},\ w \in H^*\}$ [5].

The following theorem of the alternative holds:

Theorem 1.2

Consider system (1)

(1) $$F(x) \in H\ ,\quad x \in X$$

and suppose that F is (cl H)-convexlike. Then i) and ii) hold:

i) if (1) is impossible then

(5) $$\exists \bar{w} \subset H^* : \bar{w}(F(x)) \leq 0\quad ,\quad \forall x \in X$$

ii) if (5) holds and moreover

$$\{x \in X: \bar{w}(F(x)) = 0\} = \emptyset \quad \text{when} \quad \bar{w} \notin \mathrm{int}\ H^*,$$

then system (1) is impossible.

(5) A^* denotes the *polar* of $A \subset \mathbb{R}^s$, i.e. the set $A^* = \{\lambda \in \mathbb{R}^s : \langle \lambda, y\rangle \geq 0\ \forall y \in A\}$ where $\langle\ ,\ \rangle$, denotes the scalar product.

Proof: i) According to Lemma 1.2 and Lemma 1.1, E is convex and $E \cap H = \emptyset$. Since ri $E \cap$ ri(cl H) $= \emptyset$[6] there exists a hyperplane which separates E and cl H properly, that is there exists a linear functional $\bar{w}$ such that $\mathrm{lev}_{\geq 0}\bar{w} \supseteq \mathrm{cl}\, H$, $\mathrm{lev}_{\leq 0}\bar{w} \supseteq E$. The first inclusion implies $\bar{w} \in H^*$ and the second the inequalities $\bar{w}(F(x) \leq 0 \quad \forall x \in X$. ii) If $\bar{w} \in \mathrm{int}\, H^*$[6] then $\bar{w}(h) > 0 \quad \forall h \in H$ so that $\bar{w}$ is a weak separation function and the thesis follows by Theorem 1.1. Suppose now $\bar{w} \notin \mathrm{int}\, H^*$ and that there exists $\bar{x} \in X$ with $F(\bar{x}) \in H$ so that $\bar{w}(F(\bar{x}) \geq 0$; then by (5), $\bar{w}(F(\bar{x}))=0$, and this is absurd. #

Consider now the following important particular case of (1):

(6)
$$v = \ell + m \quad , \quad H = (\mathrm{int}\ U) \times V$$
$$f: X \to \mathbb{R}^{\ell} \quad , \quad g: X \to \mathbb{R}^{m} \quad , \quad F(x)=(f(x),g(x))$$

where the positive integers ℓ and m, the closed convex cones $U \subset \mathbb{R}^{\ell}, V \subset \mathbb{R}^{m}$, with int $U \neq \emptyset$ (otherwise $H = \emptyset$), and the functions f, g are given.

The generalized system (1) becomes

(7)
$$f(x) \in \mathrm{int}\ U \quad , \quad g(x) \in V \quad , \quad x \in X$$

Let W_1 the class of functions

$$W_1=\{w:\mathbb{R}^{\ell} \times \mathbb{R}^{m} \to \mathbb{R} : w(u,v,\theta,\lambda)=\langle \theta,u\rangle+\langle \lambda,v\rangle, \theta \in U^*, \lambda \in V^*\}.$$

It is easy to show that $w \in W_1$ is a weak separation function when

(6) ri A, int A denote, respectively, the relative interior and the interior of A.

$\theta \in U^* \setminus \{0\}$. We may show that it is possible to choose $w \in W_1$ in such a way that it guarantees alternative for a wide class of systems (7).

As a consequence of Theorem 1.2 we have the following Corollaries.

Corollary 1.1 Let $F(x) = (f(x), g(x))$ be cl H-convexlike. Then i) and ii) hold:

i) if (7) is impossible then:

(8a) $$\exists \bar\theta \in U^*, \ \exists \bar\lambda \in V^* \text{ with } (\bar\theta, \bar\lambda) \neq 0 \text{ such that}$$

(8b) $$\langle \bar\theta, f(x) \rangle + \langle \bar\lambda, g(x) \rangle \le 0 \quad , \quad \forall x \in X$$

ii) if (8) holds and moreover

$$\{x \in X: \ f(x) \in \operatorname{int} U, \ g(x) \in V \ , \ \langle \bar\lambda, g(x) \rangle = 0\} = \emptyset$$

when $\bar\theta = 0$, then system (7) is impossible.

Proof: Similar to the one given in Theorem 1.2. #

Corollary 1.2 Let X be convex, f be a U-function[7] and g be a V-function. Then i) and ii) hold:

[7] Let C be a convex cone. F is said to be a C-function on a convex set X, iff $\forall x, y \in X$

$$F[(1-\alpha)x+\alpha y]-(1-\alpha)F(x)-\alpha F(y) \in C \quad , \quad \forall \alpha \in [0,1].$$

Note that a $(\mathbb{R}^n_+)$-function is a concave function and a $(\mathbb{R}^n_-)$-function is a convex function.

i) <u>if</u> (7) <u>is impossible then</u>:

(9a) $$\exists\bar{\theta} \in U^*, \quad \exists\bar{\lambda} \in V^*, \text{ with } (\bar{\theta},\bar{\lambda}) \neq 0 \quad \text{such that}$$

(9b) $$<\bar{\theta}, f(x)> + <\bar{\lambda}, g(x)> \leq 0 \quad , \quad \forall x \in X$$

ii) <u>suppose that</u> (9) <u>holds and moreover</u>

$$\{x \in X : f(x) \in \text{int } U \ , \ g(x) \in V \ , \ <\bar{\lambda}, g(x)> = 0\} = \emptyset$$

<u>when</u> $\bar{\theta} = 0$. <u>Then system</u> (7) <u>is impossible</u>.

Proof: It is sufficient to note that $F(x) = (f(x),g(x))$ is (cl H)-convexlike. #

Some sufficient conditions for the convexlikeness of pairs of two functions can be found in Ref.12.

As outlined in Ref. 16, when $U = \mathbb{R}^{\ell}_+$, Corollary 1.2 becomes Theorem 1 of Ref. 3; if, in addition f and g are concave in the ordinary sense and $V = \mathbb{R}^m_+$, then Corollary 1.2 becomes Theorem 3 of Ref. 15.

Further instances of how theorems of the alternative can be derived from Corollary 1.2 are found in Ref. 15, 16.

3. WEAK ALTERNATIVE AND OPTIMALITY CONDITIONS

In this section we will see how weak alternative can be used to study optimality conditions.

Consider the following extremum problem

(10) $$P: \min \phi(x) \quad , \quad x \in R \stackrel{\Delta}{=} \{x \in X : g(x) \geq 0\}$$

where $X \subseteq \mathbb{R}^n$, $\phi: X \to \mathbb{R}$, $g: X \to \mathbb{R}^m$.

A feasible solution $\bar{x} \in R$ is optimal for problem (10) iff the system

(11) $$f(x) \stackrel{\Delta}{=} \phi(\bar{x}) - \phi(x) > 0 \quad , \; g(x) \geq 0 \quad , \; x \in X$$

has not solutions.

Note that (11) is a particular case of (7) and (1).

Taking into account that for i) of Theorem 1.1, systems (1) and (3a) cannot be both possible, any assumption which ensures that an element of a given class of weak separation functions guarantees alternative, becomes a sufficient condition for $\bar{x}$ to be optimal.

In this way we can obtain some optimality conditions such as a generalized saddle-point condition.

With this aim consider the set of functions

(12) $$W_2 = \{w: \mathbb{R} \times \mathbb{R}^m \to \mathbb{R} \, , \; w(u,v,\theta,\omega) = \theta u + \gamma(v,\omega) \, , \; \theta \geq 0 \; , \quad \omega \in \Omega\}$$

where $\gamma : \mathbb{R}^m \to \mathbb{R}$ and Ω is the domain of parameter ω such that:

(13a) $$\forall \theta \geq 0 \; , \quad \forall \omega \in \Omega \; , \; \text{lev}_{\geq 0} w \supseteq \text{cl}\, H$$

(13b) $$\bigcap_{\theta \geq 0, \omega \in \Omega} \text{lev}_{\geq 0} \, w = \text{cl}\, H$$

(13c) $$w \in W_2 \; , \; k > 0 \quad \underline{\text{implies}} \; kw \in W_2$$

where $\text{cl}\, H = \mathbb{R}_+ \times \mathbb{R}^m_+$.

It is easy to verify that the class of linear functionals

(14) $$w(u,v,\theta,\omega) = \theta u + \langle \omega , v \rangle \; , \quad \theta \in \mathbb{R}_+ \, , \; \omega \in \mathbb{R}^m_+$$

satisfies properties (13a, b , c).

The following Lemma shows some connections between the properties of w and γ.

Lemma 2.1 Consider the set of functions W_2. Then (13) is equivalent to (15):

(15a) $$\forall\ \omega \in \Omega\ ,\ \mathrm{lev}_{\geq 0}\ \gamma \supseteq \mathbb{R}^m_+$$

(15b) $$\bigcap_{\omega \in \Omega} \mathrm{lev}_{\geq 0}\ \gamma = \mathbb{R}^m_+$$

(15c) $$\forall \bar{\omega} \in \Omega,\ \forall k > 0\ ,\ \exists \hat{\omega} \in \Omega : \gamma(\cdot, \hat{\omega}) = k\gamma(\cdot, \bar{\omega})$$

Proof: (13a) $\Leftrightarrow$ (15a) .

(13a) $\Rightarrow w(u,v,\theta,\omega) \geq 0 \quad \forall u,\ \theta \geq 0\ , \quad \forall v,\omega \in \mathbb{R}^m_+$

$\Rightarrow w(0,v,\theta,\omega) = \gamma(v,\omega) \geq 0 \quad \forall v,\ \omega \in \mathbb{R}^m_+ \ \Rightarrow$ (15a) .

(15a) $\Rightarrow \gamma(v,\omega) \geq 0 \quad \forall v,\omega \in \mathbb{R}^m_+ \ \Rightarrow\ \theta u + \gamma(v,\omega) \geq 0 \quad \forall \theta, u \geq 0$

$\forall v,\ \omega \in \mathbb{R}^m_+ \ \Rightarrow$ (13a).

(13b) $\Leftrightarrow$ (15b). Suppose that (13b) holds and (15b) is false. Then $\exists \bar{v} \notin \mathbb{R}^m_+$ such that $\gamma(\bar{v},\omega) \geq 0 \quad \forall \omega \in \Omega$ and, consequently, $w(0,\bar{v},\theta,\omega) = \gamma(\bar{v},\bar{\omega}) \geq 0$ $\forall \theta \geq 0,\ \forall \omega \in \Omega$ so that $(0,\bar{v}) \in \mathrm{cl}\ H$. Thus $\bar{v} \in \mathbb{R}^m_+$ and this is absurd. Suppose now that (15b) holds and (13b) is false. Then $\exists (\bar{u},\bar{v}) \notin \mathrm{cl}\ H$ such that $\theta\bar{u} + \gamma(\bar{v},\omega) \geq 0 \quad \forall \theta \geq 0,\ \forall \omega \in \Omega$ and this relation implies $\bar{u} > 0$, otherwise $\theta\bar{u} \to -\infty$ when $\theta \to +\infty$. On the other hand $w(\bar{u},\bar{v},0,\omega) = \gamma(\bar{v},\omega) \geq 0\ \forall \omega \in \Omega$ so that $\bar{v} \in \mathbb{R}^m_+$ for (15b), and this is absurd.

(13c) $\Leftrightarrow$ (15c). $\forall \bar{\omega} \in \Omega\ ,\ \forall k > 0$ consider the function $w(u,v,\theta,\bar{\omega}) = (\theta/k)u + \gamma(v,\bar{\omega})$. Then $kw = \theta u + k\gamma(v,\bar{\omega}) \in W_2$ iff (15c) holds. #

The following Lemma gives conditions under which $w \in W_2$ guarantees weak alternative between (1) and (3a) where, now, system (1) is of the form (11).

Lemma 2.2 i) When $\theta > 0$ or $\theta = 0$ and $\text{lev}_{\geq 0}\gamma \supseteq \mathbb{R}^m_+$, the function $w \in W_2$ guarantees weak alternative between (1) and (3a), with $Z =]-\infty, 0]$.
ii) When $\theta = 0$ and $\text{lev}_{>0}\gamma \not\supseteq \mathbb{R}^m_+$, w guarantees weak alternative between (1) and (3a), with $Z =]-\infty, 0[$.

Proof: (i) In the present case, namely (6) with $\ell = 1$, (2a)becomes $\text{lev}_{>0}w \supseteq]0,+\infty[\times \mathbb{R}^m_+$, or

$$(u,v) \in]0,+\infty[\times \mathbb{R}^m_+ \Rightarrow \theta u + \gamma(v;\omega) > 0 .$$

This relationship holds since now we have either $\theta > 0$ and $\text{lev}_{\geq 0}\gamma \supseteq \mathbb{R}^m_+$, or $\theta = 0$ and $\text{lev}_{>0}\gamma \supseteq \mathbb{R}^m_+$. Thus, the thesis follows from (i)of theorem 1.1.

(ii) In the present case, namely (4.2)-(5.3), (2a) becomes:

$$\text{lev}_{\geq 0}w \supseteq]0,+\infty[\times \mathbb{R}^m_+ ,$$

or:

$$(u,v) \in]0,+\infty[\times \mathbb{R}^m_+ \Rightarrow \theta u + \gamma(v;\omega) \geq 0 .$$

Since $\theta = 0$, this relationship is an obvious consequence of (15a).Again weak alternative follows from (i) of theorem 1.1. This completes the proof.

Taking into account Lemma 2.2, it is immediate to interprete (i) of Theorem 1.1 as a sufficient optimality condition for problem (10); this is contained in the following:

Corollary 2.1 If $\bar{x} \in \mathbb{R}^n$ fulfils conditions: (i) $\bar{x} \in R$; (ii) there exist $\bar{\theta} \in \mathbb{R}_+$ and $\bar{\omega} \in \Omega$, such that

$$\bar{\theta}\,[\phi(\bar{x}) - \phi(x)] + \gamma(g(x);\bar{\omega}) \leq 0 , \quad \forall x \in X , \tag{16}$$

and moreover

$$\{x \in X:\ \phi(x) < \phi(\bar{x}), g(x) \geq 0\ ,\ \gamma(g(x),\bar{\omega})=0\}=\emptyset$$

when $\theta=0$ and $\mathrm{lev}_{>0}\gamma \not\supseteq \mathbb{R}^m_+$; then $\bar{x}$ is a global minimum point of (10).

Now, introduce the function

$$L(x;\theta,\omega) \triangleq \theta\phi(x) - \gamma(g(x);\omega)\ ,$$

and let us prove the following:

Theorem 2.1

Condition (i)-(ii) of Corollary 2.1 is equivalent to the other one : there exist $\bar{x} \in X$, $\bar{\theta} \in \mathbb{R}_+$ and $\bar{\omega} \in \Omega$, such that

(17) $$L(\bar{x};\bar{\theta},\omega) \leq L(\bar{x};\bar{\theta},\bar{\omega}) \leq L(x;\bar{\theta},\bar{\omega})\ ,\quad \forall x \in X\ ,\ \forall \omega \in \Omega,$$

and moreover

$$\{x \in X:\ \phi(x) < \phi(\bar{x}), g(x) \geq 0\ , \gamma(g(x),\bar{\omega}) = 0\}=\emptyset$$

if $\theta = 0$ and $\mathrm{lev}_{>0}\gamma \not\supseteq \mathbb{R}^m_+$.

Proof: Let us prove that (i)-(ii) of Corollary 2.1 $\Rightarrow$(17). $\bar{\omega}\in\Omega$ and $\bar{x} \in R$ imply that $\gamma(g(\bar{x});\bar{\omega})\geq 0$ (since $\mathrm{lev}_{\geq 0}\gamma \supseteq \mathbb{R}^m_+$); at $x = \bar{x}$ (16) implies $\gamma(g(\bar{x});\bar{\omega}) \leq 0$; it follows $\gamma(g(\bar{x});\bar{\omega}) = 0$. Hence, (16) is equivalent to the 2-nd of (17).

Let us prove, now, that (17) implies i)-ii) of Corollary 2.1. The 1-st of (17) implies

(18) $$\gamma(g(\bar{x}),\bar{\omega}) \leq \gamma(g(\bar{x}),\omega)\ ,\quad \forall \omega \in \Omega\ .$$

Suppose that $g(\bar{x}) \not\geq 0$. Then, by (15b) there exists $\hat{\omega}\in\Omega$ such that $\gamma(g(x),\bar{\omega})<0$;

hence by (15c)

$$\gamma(g(\bar{x}),\bar{\omega}) \le K(g(\bar{x}),\hat{\omega}) \quad \forall K > 0 \quad \text{and this is absurd.}$$

Thus $g(\bar{x}) \ge 0$ and i) of Corollary 2.1 is proven. Because of (15b) we have $\gamma(g(\bar{x}),\bar{\omega}) \ge 0$. Suppose that $\gamma(g(\bar{x}),\bar{\omega}) > 0$; by (15c) and (18) we have $\gamma(g(\bar{x}),\bar{\omega}) \le \frac{1}{2}\, \gamma(g(\bar{x}),\bar{\omega})$, so that $\gamma(g(\bar{x}),\bar{\omega}) = 0$. Account taken of this equality, it is easy to show that the second part of (17) implies (16). This completes the proof. #

Now, note that (17) can be regarded as a generalized saddle-point condition and L as a generalized Lagrangean function. When $\gamma(v,\omega) = \langle \omega, v \rangle$, (17) becomes the well-known John saddle-point condition and L the classic Lagrangean function.

Note that, when (16) or (17) holds, from the proof of Theorem 2.1 we have that $(\bar{x},\bar{\omega})$ fulfils the generalized complementarity condition $\gamma(g(x),\omega) = 0$ which collapses to the well-known ordinary one, when γ is linear and $\Omega = \mathbb{R}^m_+$.

Now, suppose that $\bar{x}$ is an optimal solution for problem (10), so that system (1) in the form of (11) has not solutions. The impossibility of (1) does not imply the possibility of (3a). In this sense, any condition which guarantees, in a given, suitable class of weak separation functions, the existence of an element w satisfying (3a) becomes a necessary optimality condition (besides the weak alternative).

When E is convex, choosing the class of linear functionals, for (i) of Corollary 1.1, (16) and (17) hold.

Consequently, when X is convex, $-\phi$, g are concave, (16) and (17) become necessary otimality conditions too.

When E is not convex the above mentioned class is not useful since a linear functional does not guarantee, in general, alternative.

A general approach is to consider a transformation of the constraints $T = (T_1,\dots,T_m)$ such that

$$T_i(v_i,\mu_i) \gtreqless 0 \quad \text{according to} \quad v_i \gtreqless 0 \;,\; \forall \mu_i \geq 0$$

and to choose, properly, w in the class of weak separation functions W_3, where

$$W_3 = \{w: \mathbb{R} \times \mathbb{R}^m \to \mathbb{R} \;,\; w(u,v,\lambda,\mu) = u + \sum_{i=1}^{m} \lambda_i T_i(v_i,\mu_i)\} \;.$$

Note that $T(g(x),\mu)$ may have a certain property, for instance concavity, differentiability, which does not hold for g.

An appropriate choice of W_3 may be the exponential one,that is the case where $T_i(v_i,\mu_i) = v_i \exp(-\mu_i v_i)$ $i = 1,\dots,m$.By means of this transformation it has been possible (Ref.27,28) to characterize a wide class of problems for which there exists $w \in W_3$ which guarantees alternative, i.e. such that (19) holds:

(19a) $$\exists \bar{\lambda},\bar{\mu} \geq 0, \; (\bar{\lambda},\bar{\mu}) \neq 0 \quad \text{such that}$$

(19b) $$\phi(\bar{x}) - \phi(x) + \sum_{i=1}^{m} \bar{\lambda}_i g_i(x) \exp(-\bar{\mu}_i g_i(x)) \leq 0 \;,\; \forall x \in X$$

Let us note that (16), (17), (19) are global optimality conditions since X is not specified. Some other results can be obtained by means of a further analysis of the weak alternative, based on local arguments (Ref. 16,27); in particular it is possible to deduce, in a very simple way, the classic necessary conditions of the Lagrange,Karush,Kuhn-Tucker type. With this end, consider a class of weak separation functions $W = \{w(u,v,\omega), \omega \in \Omega\}$, satisfying the following properties:

(20a) $$\bigcap_{\omega \in \Omega} \text{lev}_{\geq 0} w(u,v,\omega) = \text{cl}\, H$$

(20b) $$\bigcap_{\omega \in \Omega} \text{lev}_{> 0} w(u,v,\omega) = \text{int}\, H$$

It is easy to prove that the class of the linear functionals

$$w(u,v,\lambda,\mu) = \lambda u + \langle\mu,v\rangle \ , \quad \lambda \in \mathbb{R}^{+} \ , \quad \mu \in \mathbb{R}^{m}_{+}$$

verifies (20).

The following Lemma plays a fundamental role in the derivation of F. John conditions, and for necessary conditions of isoperimetric kind in the calculus of variations (see [17]). To this end denote by E_G the conic extension (in the sense of sect.1) of the cone generated by K at the origin 0 (i.e. the union of rays having 0 as common origin and non-empty intersection with K), and by E_G^* the (non-negative polar) of E_G.

Lemma 2.3 Let W be a class of weak separation functions satisfying property (20), and assume that the condition:

$$E_G^* \neq \{0\}$$

is fulfilled. If $\bar{x}$ is a local minimum point for problem (10), then there exists $\bar{\omega} \in \Omega$ such that

$$\limsup_{x \to \bar{x}} \frac{w(f(x),g(x),\bar{\omega}) - w(f(\bar{x}),g(\bar{x}),\bar{\omega})}{\|x - \bar{x}\|} \leq 0 \tag{21}$$

Remark If $w(u,v,\omega) = 0$ when $\omega = 0$, Lemma 2.3 holds with $\bar{\omega} \neq 0$.

Now we are able to prove the following classic theorem:

Theorem 2.2

Let $\bar{x}$ be a local optimal solution for problem (10) and set $I=\{i:g_i(\bar{x})=0\}$. Suppose that X is an open neighbourhood of $\bar{x}$, ϕ_i, g_j $i=1,\ldots,\ell$ $j \in I$ are differentiable at $\bar{x}$ and g_j, if $j \notin I$ is continuous at $\bar{x}$. Then there exist

(23a) $\bar{\lambda} \geq 0$, $\bar{\mu} \geq 0$, $(\bar{\lambda},\bar{\mu}) \neq 0$ such that

(23b) $$\bar{\theta}\nabla\phi(\bar{x}) - \sum_{i=1}^{m} \bar{\mu}_i \nabla g_i(\bar{x}) = 0$$

(23c) $$\bar{\mu}_i \cdot g_i(\bar{x}) = 0 \quad i=1,\dots,m$$

Proof: The continuity of $g_j(x)$ at $\bar{x}$, $j \notin I$ implies that $\bar{x}$ is a local optimal solution for problem P'

$$P': \min \phi(x) \;,\; g(x) \geq 0 \;,\; x \in U(\bar{x})$$

where $U(\bar{x})$ is a suitable neighbourhood of $\bar{x}$ such that $g_j(x)>0$, $\forall x \in U(\bar{x})$. Consider the class of functions

$$w(u,v,\lambda,\mu) = \lambda u + \langle \mu, v\rangle \;,\; \lambda \geq 0,\; \mu \in \mathbb{R}^m_+$$

and the Lagrangean function associated with P'

(24) $$L(x,\lambda,\mu) = \lambda\phi(x) - \langle \mu, g(x)\rangle.$$

Taking into account that $\inf_{x \to \bar{x}} \sup \frac{\psi(x)-\psi(\bar{x})}{\|x - \bar{x}\|} = |\nabla\psi(\bar{x})|$ when ψ is differentiable at $\bar{x}$, for Lemma 2.3 and (24), there exist $\bar{\lambda} \geq 0$, $\bar{\mu} \geq 0$ such that $|\nabla L(\bar{x},\bar{\lambda},\bar{\mu})| \leq 0$, i.e., $\nabla L(\bar{x},\bar{\lambda},\bar{\mu}) = 0$. Setting $\bar{\lambda}_i = 0$, $i \notin I$, the proof of the theorem is complete. #

4. STRONG ALTERNATIVE AND OPTIMALITY CONDITIONS

Let $s: \mathbb{R} \times \mathbb{R}^m \to \mathbb{R}$, $s(u,v) = u - \delta(v)$ be a strong separation function with $\delta: \mathbb{R}^m \to \mathbb{R}$. It is easy to show that the condition $\mathrm{lev}_{>0} s \subseteq \mathcal{H}$ implies $\delta(v) \geq 0 \;\; \forall v \geq 0$ and $\delta(v) = +\infty \;\; \forall v \ngeq 0$.

In order to avoid this kind of restrictions we will give a more general definition of a strong separation function than the one given in section 1.

Consider system (1) and let $\bar{\mathcal{K}} \subseteq \mathbb{R}^\nu$ be such that $\mathcal{K} \subseteq \bar{\mathcal{K}}$, namely

(25) $$F(x) \in \bar{K} \quad , \quad \forall x \in X \quad .$$

$\bar{K}$ trivially exists since (25) is satisfied by at least $\bar{K} = \mathbb{R}^{\nu}$.

We say that $s: \mathbb{R}^{\nu} \to \mathbb{R}$ is a *strong separation function* iff we have:

(26) $$\mathrm{lev}_{>0} s \cap \bar{K} \subseteq H$$

Let us note that (26) reduces to (2b) when $\bar{K} = \mathbb{R}^{\nu}$.

The following Lemma generalizes ii) of Theorem 1.1.

Lemma 3.1 Let s be a strong separation function. The systems (1) and the following one:

(27) $$s(F(x)) \leq 0 \quad , \quad \forall x \in X$$

cannot be both impossible.

Proof: Suppose that (27) is impossible i.e. there exists $\bar{x} \in X$ such that $s(F(\bar{x})) > 0$; from (27) it results $F(\bar{x}) \in \mathrm{lev}_{>0} s \cap \bar{K} \subseteq H$ and thus (1) is possible. #

Consider now problem (10) and let $\bar{K} \subseteq \mathbb{R} \times \mathbb{R}^m$ be such that

(28) $$(\phi(\bar{x})-\phi(x), g(x)) \in \bar{K} \quad , \quad \forall x \in X$$

For instance we can set $\bar{K} = [-\rho,\rho]^{1+m}$ if there exists a positive real number ρ such that $\|\phi(x)\| < \rho/2$, $\|g(x)\| \leq \rho \quad \forall x \in X$ or we can set $\bar{K} = \{(u,v) \in \mathbb{R} \times \mathbb{R}^m : u \leq M, \ v \leq M\}$ if $\phi(x) \leq M$, $g_i(x) \leq M \quad i=1,..,m$, $\forall x \in X$.

Let

(29) $$S=\{s:\mathbb{R}\times\mathbb{R}^m \to \mathbb{R}, s(u,v,\omega)=u-\delta(v,\omega), \ \omega \in \Omega\}$$

be a class of strong separation functions, where $\delta(\cdot,\omega): \mathbb{R}^m \to \mathbb{R}$ and Ω is a set of parameters. For Lemma 3.1, the systems (11) and (30)

$$s(f(x),g(x),\omega)=\phi(\bar{x})-\phi(x)-\delta(g(x),\omega) \le 0 \quad , \forall x \in X \ , \ \forall \omega \in \Omega \tag{30}$$

cannot be both impossible, so that the optimality of $\bar{x}$ implies the validity of (30). Consequently (30) becomes a necessary optimality condition too, when the class S satisfies (31)

$$\bigcup_{\omega \in \Omega} (\bar{K} \cap \text{lev}_{>0} s(u,v,\omega)) = \bar{K} \cap H \ . \tag{31}$$

The following theorem holds :

Theorem 3.1
Consider the class of strong separation functions (29) satisfying (31). Then (30) is a sufficient condition for $\bar{x}$ to be optimal.

Proof: Suppose that there exists $\hat{x} \in X$ such that $g(\hat{x}) \ge 0$, $\phi(\hat{x}) < \phi(\bar{x})$. Then $(\phi(\bar{x})-\phi(\hat{x}),g(\hat{x})) \in \bar{K} \cap H$ and, for (31), there exists $\bar{\omega} \in \Omega$ such that $(\phi(\bar{x})-\phi(\hat{x}),g(\hat{x})) \in \text{lev}_{>0} s(u,v,\bar{\omega})$ and this contradicts (30) for $x = \hat{x}$. #

Consider now the case $\bar{K} = \mathbb{R} \times \mathbb{R}^m$ and suppose that the class (29) satisfies the following properties:

$$\delta(v,\omega) = +\infty \ , \quad \forall v \not\geq 0 \ , \ \forall \omega \in \Omega \tag{32}$$

$$\bigcup_{\omega \in \Omega} \text{lev}_{>0} s(u,v,\omega) = \text{int } H \tag{33}$$

The following Lemma holds :

Lemma 3.2 Conditions (33) and (34) are equivalent

$$\forall v > 0 \ , \ \inf_{\omega \in \Omega} \delta(v,\omega) = 0 \tag{34}$$

Proof: (33) $\Rightarrow$ (34). Ab absurdo, suppose that there exists $\bar{v}>0$ such that $\inf_{\omega} \delta(\bar{v},\omega) = \ell > 0$. Let $\bar{u}$ be such that $0< \bar{u} < \ell$. We have $\bar{u}-\delta(\bar{v},\omega) <0$ $\forall\, \omega \in \Omega$ and thus $(\bar{u},\bar{v}) \notin \mathrm{lev}_{>0} s(u,v,\omega)$ $\forall\, \omega \in \Omega$ with $(\bar{u},\bar{v}) \in \mathrm{int}\, H$ and this contradicts (33).

(34) $\Rightarrow$ (33). The condition $\mathrm{lev}_{>0} s \subseteq H$ implies $\delta(v,\omega) \geq 0$ $\forall v > 0$, $\forall \omega \in \Omega$. By definition and from (32) we have $\bigcup_{\omega\in\Omega} \mathrm{lev}_{>0} s \subseteq \mathrm{int}\, H$, and thus it is sufficient to show that $\mathrm{int}\, H \subseteq \bigcup_{\omega} \mathrm{lev}_{>0} s$.

Let $(\bar{u},\bar{v}) \in \mathrm{int}\, H$. Since $\inf_{\omega} \delta(\bar{v},\omega) = 0$, there exists $\bar{\omega}$ such that $0 \leq \delta(\bar{v},\bar{\omega}) < \bar{u}$; consequently $(\bar{u},\bar{v}) \in \mathrm{lev}_{>0} s(u,v,\bar{\omega})$ and this implies $\mathrm{int}\, H \subseteq \bigcup_{\omega} \mathrm{lev}_{>0} s$. #

Consider again problem (10) and set $R^{\circ} = \{x \in X: g(x) > 0\}$, $R^{*} = \{\bar{x} \in R: \phi(\bar{x}) = \min_{x\in R} \phi(x)\}$.

The following theorem gives a necessary and sufficient condition for $\bar{x}$ to be optimal which is weaker than the one stated in Theorem 3.1.

Theorem 3.2

Consider problem (10) and assume that ϕ is continuous, $R^{*} \neq \emptyset$, $R^{\circ} \neq \emptyset$, $R = \mathrm{cl}\, R^{\circ}$. Let S be the class of strong separation functions (29) satisfying (32) and (33). Then i) and ii) hold:

i) $\bar{x}$ is an optimal solution for (10) iff

(35)
$$\sup_{\omega\in\Omega} \sup_{x\in X} s(f(x),g(x),\omega) \leq 0$$

ii) if $\hat{x}$ is a feasible solution of (10) such that

(36)
$$\phi(\hat{x}) = \inf_{\omega\in\Omega} \inf_{x\in X} [\phi(x)+\delta(g(x),\omega)]$$

then $\hat{x}$ is an optimal solution for problem (10).

Proof: i) *Necessity.* Suppose that $\bar{x}$ is an optimal solution for (10). Then (30) holds and this implies (35). *Sufficiency.* Ab absurdo , suppose that there exists $\hat{x}$ such that $g(\hat{x}) \geq 0$, $\phi(\hat{x}) < \phi(\bar{x})$. Since ϕ is continuous at $\hat{x}$ and $R = \text{cl } R^o$, there exist a suitable neighbourhood $U(\hat{x})$ of $\hat{x}$ and $x^o \in R^o \cap U(\hat{x})$ such that $\phi(x^o) < \phi(\bar{x})$. Setting $\bar{u}=\phi(\bar{x})-\phi(x^o) > 0$, $\bar{v}=g(x^o)>0$, we have $(\bar{u},\bar{v}) \in \text{int } H$ and, from (33), there exists $\bar{\omega}$ such that $(\bar{u},\bar{v}) \in \text{lev}_{>0}\ s(u,v,\bar{\omega})$. Consequently $\sup_{x \in X} s(f(x),g(x),\bar{\omega}) > 0$ and this contradicts (35).

ii) Let $\bar{x}$ be an optimal solution for (10) and set $m = \phi(\bar{x})$, $\ell = \inf_{\omega \in \Omega} \inf_{x \in X} [\phi(x)+\delta(g(x),\omega)]$. Since ϕ is continuous at $\bar{x}$ and $R = \text{cl } R^o$, for every $\varepsilon > 0$ there exist a suitable neighbourhood $U(\bar{x})$ of $\bar{x}$ and $x^o \in R^o \cap U(\bar{x})$, such that $\phi(x^o) \leq \phi(\bar{x}) + \varepsilon = m + \varepsilon$. From (34) there exists $\bar{\omega}$ such that $0 \leq \delta(g(x^o),\bar{\omega}) < \varepsilon$ and, consequently, $\ell \leq \inf_{x \in X} (\phi(x)+\delta(g(x),\bar{\omega}) \leq \phi(x^o)+\delta(g(x^o),\bar{\omega}) \leq m + 2\varepsilon$ and this implies $\ell \leq m$.

On the other hand, since $\delta(g(x),\omega) \geq 0$ $\forall x \in X$, $\forall \omega \in \Omega$ we have $m \leq \ell$. It follows $\ell = m$. #

5. LANGRANGEAN PENALTY APPROACHES

Penalty approaches are a natural extension of the original Lagrangean method and aim to get an optimal solution of a constrained extremum problem by solving a sequence of unconstrained ones.

More exactly, exterior penalty function methods usually solve problem (10) by a sequence of unconstrained minimization problems whose optimal solutions approach the solution of (10) *outside* the feasible set so that the sequence of unconstrained minima converges to a feasible point of the constrained problem that satisfies some sufficient optimality conditions. On the contrary, interior penalty function methods solve (10) through a sequence of unconstrained optimization problems whose minima are at points in the interior of feasible set; staying in the interior is ensured by formulating a barrier function by which an infinitely large penalty is imposed for crossing the boundary of the feasible set from the inside.

In this section it will be shown that these approaches can be viewed in terms of weak and strong separation functions. To this end consider problem (10), with $X = \mathbb{R}^n$ and the continuous functions $p_r: \mathbb{R}^n \to \mathbb{R}$, $r = 1,2,..$ such that

(37)
$$p_r(v) = 0 \text{ if } v \geq 0; \; p_r(v) > 0 \text{ if } v \ngeq 0$$
$$p_{r+1}(v) > p_r(v) \; ; \; \lim_{r\to+\infty} p_r(v) = +\infty \text{ if } v \ngeq 0$$

The function $w(u,v,r) = u-p_r(v)$ is, for any r, a weak separation function and, moreover, it is easily seen that

(38a) $$w(\cdot,r) \text{ is continuous for any } r$$

(38b) $$\text{lev}_{>0} w(\cdot,r) \supset \text{lev}_{>0} w(\cdot,r+1)$$

(38c) $$\bigcap_{r=1}^{+\infty} \text{lev}_{>0} w(\cdot,r) = H$$

(38d) $$\forall h \in H, \; \exists K(h) > 0 \text{ such that } w(h,r) \geq K(h) \quad \forall r$$

The following theorem holds (Ref.16).

Theorem 4.1

Let W be a class of weak separation functions satisfying (38). Then system (1) is impossible iff

(39) $$\inf_{r} \sup_{x\in X} w(F(x),r) \leq 0$$

Since $w(u,v,r)$ is, for any r, a weak separation function, i) of Theorem 1.1 can be applied and (3a) becomes

(40) $$\phi(\bar{x})-\phi(x)-p_r(g(x)) \leq 0 \quad , \quad \forall x \in \mathbb{R}^n ,$$

and is a sufficient condition for the feasible $\bar{x}$ to be optimal . Such a condition can be weakened by applying theorem 4.1; (39) becomes

(41) $$\lim_{r\to+\infty} \inf_{x\in\mathbb{R}^n} [\phi(x)+p_r(g(x))] \geq \phi(\bar{x})$$

and is a sufficient condition weaker than (41). Denote by Φ_r the infimum in (41). From (37) we deduce

(42) $$\Phi_1 \leq \Phi_2 \leq .... \leq \Phi \stackrel{\Delta}{=} \inf_{x\in R} \phi(x)$$

Assume that $\exists\bar{r}$ such that $\Phi_{\bar{r}} > -\infty$, and that there is a proper $x^r \in R^n$ such that $\phi(x^r) = \Phi_r$, $\forall r \geq \bar{r}$. If $\bar{x}$ is any limit point of sequence $\{x^r\}$, then condition (41) is fulfilled and theorem 1.1 gives the optimality of $\bar{x}$. The construction of sequence $\{x^r\}$ by solving the infimum problems in (41) is the well-known *exterior penalty method* and p_r is said a *penalty function*; if the above convergence can be ensured after a finite number of steps, i.e. if $\exists\bar{r}$ such that (40) is fulfilled at $r = \bar{r}$, then p_r is said *exact penalty function* (Ref.20). Hence, the conditions for a penalty function to be exact can be regarded as conditions which ensure (40) instead of (42).

A particular case of (38), corresponding to a well-known penalty function is

$$w(u,v;r,\alpha) = u-r \sum_{i=1}^{m} (-\min\{0,v_i\})^{\alpha} \quad , \quad \alpha \geq 1 .$$

A more general class of functions satisfying (38) is contained in Ref. 20, where the case of both equality and inequality constraints is considered. The latter requires only formal changes in the above reasoning. In fact, if the constraints of (10) are $g(x) = 0$, it is enough to replace in (7) $V = \mathbb{R}^m_+$ with $V = \{0\}$, so that now $H = \{(u,v) \in \mathbb{R}^{1+m} : u>0; v=0\}$.

In such a case a weak separation function is for instance the following one

$$w(u,v,\lambda,r) = u + \langle\lambda, v\rangle - r\langle v,v\rangle, \quad \text{with} \quad \lambda \in \mathbb{R}^m_+ \ , \ r \in \mathbb{R}_+ \ ,$$

which corresponds to the so-called augmented Lagrangean approach (Ref. 31).

It follows that *exterior penalty approach can be formulated in terms of weak separation.*

Now, consider again problem (10) and assume that ϕ, g are continuous, $R = \text{cl}\, R^\circ$, $R^* \neq \emptyset$.

Let $\{\mu_k\}$ be a sequence of real numbers tending to infinity such that, for each k, $k = 1,2,..,$ $\mu_k > 0$, $\mu_{k+1} > \mu_k$. Assume that, for each k, problem (43) has a solution

$$(43) \qquad \min_{x \in R^\circ} [\phi(x) + \frac{1}{\mu_k} \delta(g(x))]$$

where δ is a continuous function such that $\delta(g(x)) \geq 0$ if $g(x) > 0$; $\delta(g(x)) = +\infty$ if $g(x) \not> 0$. Interior penalty function methods solve, for each k, problem (43) obtaining the point x_k; any limit point of the sequence $\{x_k\}$ is a solution to problrm (10). This procedure corresponds to find a feasible solution $\hat{x}$ satisfying (37).

It follows that *interior penalty approach can be formulated in terms of strong separation.*

6. DUALITY

In this section it is shown that a dual problem naturally arises when optimality is studied through alternative. In this way some generalizations are easily achieved.

Assume that $\bar{x}$ is an optimal solution for problem (10)and,moreover, that a constraint qualification holds.

Consider the class of weak separation functions (12) satisfying (13), with $\theta = 1$, and the class of strong separation functions (29).

Since $w(f(x),g(x),\omega) = \phi(\bar{x}) - \phi(x) + \gamma(g(x),\omega)$,it results $w(f(\bar{x}), g(\bar{x}),\omega) = \gamma(g(\bar{x}),\omega) \geq 0 \quad \forall\omega \in \Omega$, so that

$$\sup_{x \in X} w(f(x),g(x),\omega) \geq 0 \qquad \forall\omega \in \Omega$$

or, equivalently,

$$\phi(\bar{x}) \geq \inf_{x \in X}[\phi(x) - \gamma(g(x),\omega)] \qquad \forall\omega \in \Omega.$$

It follows

(44)
$$\phi(\bar{x}) \geq \sup_{\omega \in \Omega} \inf_{x \in X}[\phi(x) - \gamma(g(x),\omega)] ;$$

In a similar way we can use strong alternative: since $s(f(x),g(x),\omega) = \phi(\bar{x}) - \phi(x) - \delta(g(x),\omega)$, it results $s(f(\bar{x}),g(\bar{x}),\omega) = -\delta(g(\bar{x}),\omega) \leq 0$ $\forall\omega \in \Omega$ so that $\inf_{x \in X} s(f(x),g(x),\omega) \leq 0 \quad \forall\omega \in \Omega$. It follows

(45)
$$\phi(\bar{x}) \leq \inf_{\omega \in \Omega} \inf_{x \in X}[\phi(x) + \delta(g(x),\omega)]$$

Set

$$L(x,\omega)=\phi(x)-\gamma(g(x),\omega) \quad ; \quad L_S(x,\omega)=\phi(x)+\delta(g(x),\omega)$$

and assume, for sake of simplicity, that any infimum (supremum) which appears in (44), (45), is achieved as a minimum (maximum).

Then two news problems can be associated to P:

$$D: \max_{\omega\in\Omega} \min_{x\in X} L(x,\omega) \quad ; \quad D_s: \min_{\omega\in\Omega} \min_{x\in X} L_s(x,\omega)$$

Problem D is called the generalized Lagrangian dual an it reduces to the usual Lagrangian dual when γ is linear. We refer to D and D_s, respectively, as the *weak dual* and the *strong dual* of the *primal* problem P.

Relation (44) is known as the *weak duality theorem*. The difference Δ between the left-hand side and the right-hand side of (44) is the *duality gap*.

The following theorem is a general formulation of the so-called strong duality theorem.

Theorem 5.1

Consider the pair of problems P and D. Then $\Delta = 0$ iff there exists $\bar{w}\in W_2$ which guarantees alternative.

Proof: $$\Delta = 0 \Leftrightarrow \phi(\bar{x}) \le \max_{\omega\in\Omega} \min_{x\in X} L(x,\omega)$$

$$\Leftrightarrow \exists\bar{\omega}\in\Omega: \phi(\bar{x}) \le \min_{x\in X} (\phi(x)-\gamma(g(x),\bar{\omega})$$

$$\Leftrightarrow \exists\bar{\omega}\in\Omega: \phi(\bar{x})-\phi(x)+\gamma(g(x),\bar{\omega}) \le 0 \quad , \quad \forall x \in X$$

or, equivalently, iff the function $\bar{w}(u,v,\bar{\omega}) = u-\gamma(v,\bar{\omega})$ is a weak separation function which guarantees alternative. #

Let us note the strict connection between the strong duality theorem and the necessary optimality conditions. To this end note that ny condition which guarantees the validity of a necessary optimality condition is also a condition which ensures to be zero the duality gap, when the same class of weak separation functions is adopted. As a consequence, the results obtained in section 2 can be used to characterize classes of problems having $\Delta = 0$.

For instance, when P is convex and a constraint qualification holds, if we choose the class of linear weak separation functions $w(u,v,\omega) = u + \langle\omega,v\rangle$, that is $L(x,\omega) = \phi(x) - \langle\omega,g(x)\rangle$, it results $\Delta = 0$; when P is the linear fractional problem or when the objective function ϕ is concave and g is affine, if we choose the class of exponential weak separation functions $w(u,v,\lambda,\mu) = u + \sum_{i=1}^{m} \lambda_i \exp(-\mu_i v_i)$, that is $L(x,\lambda,\mu) = \phi(x) - \sum_{i=1}^{m} \lambda_i \exp(-\mu_i g_i(x))$, it results $\Delta = 0$ (Ref.28).

Consider now problem D in the case where $L(x,\omega)$ is the usual Lagrangean function. It results, in general, $\Delta > 0$. In the image space the duality gap can be easily characterized.

For $\mu \in \mathbb{R}^m$ and $\alpha \in \mathbb{R}$ let us set

$$H(\mu,\alpha) = \{(u,v) \in \mathbb{R} \times \mathbb{R}^m : u + \langle\mu,v\rangle \leq \alpha\}$$

$$A = \{\alpha \in \mathbb{R}_+ \ / \ \exists \bar{\mu} \in \mathbb{R}^m_+ : H(\bar{\mu},\alpha) \supseteq E\}$$

$$\alpha_0 = \inf A \quad \text{if} \quad A \neq \emptyset$$

The following theorem holds (Ref. 29).

Theorem 5.2

i) <u>If</u> $A = \emptyset$, <u>then</u> $\Delta = +\infty$. ii) <u>If</u> $A \neq \emptyset$, <u>then</u> $\Delta = \alpha_0$.

Working in the image space it is possible to establish an upper bound for Δ.

Setting $v^* = \max_{\omega\in\Omega} \min_{x\in X} L(x,\omega)$, it results (Ref.29)

$$v^* \geq F(-\sigma(g)) - \rho(\phi) \tag{46}$$

where F is the perturbation function

$$F(\varepsilon) \overset{\Delta}{=} \min \phi(x) \quad , \; x \in R_{\varepsilon} = \{x \in X: \; g(x) \geq \varepsilon\}$$

and $\rho(\phi)$, $\sigma(g)$ are, respectively, the lack of convexity of function ϕ and the lack of concavity of function g.

As a particular case of (46) we have, when g is concave,

$$(47) \qquad 0 \leq \Delta \leq \rho(\phi) \quad .$$

Now, consider again the strong dual D_s and set

$$\Phi(\omega) \overset{\Delta}{=} \min_{x \in X}[\phi(x) + \delta(g(x),\omega)].$$

We refer to (45) as the *strong duality theorem* .

The difference between the right-hand side and the left-hand one of (45) is called *strong duality gap*.

As an obvious consequence of Theorem 3.1, we have the following :

Theorem 5.3

Consider the pair of problems P and D_s and assume that the class of strong separation functions (29) satisfies (31). Then, the strong duality gap is zero.

7. REGULARITY

In section 2 we have pointed out that, if $w(f(x),g(x),\bar{\theta},\bar{\lambda})=\bar{\theta}(\phi(\bar{x})-\phi(x)+\langle\bar{\lambda},g(x)\rangle$ is a weak separation function which guarantees alternative, then

$$(48) \qquad \bar\theta(\phi(\bar x)-\phi(x)) + <\bar\lambda, g(x)> \le 0 \quad , \quad \forall x \in X$$

becomes a sufficient condition for $\bar x$ to be optimal for problem (10).For this optimality condition and any other which involves Lagrange multipliers the problem of *regularity* arises, that is the problem of finding conditions which guarantees that θ is different from zero. When such conditions involve only the constraints, they are referred to as *constraint qualifications*. Now, we will see how the image space suggests some simple ideas concerning conditions under which $\theta \neq 0$.

Let us note that the validity of (48) is equivalent to state the existence of an hyperplane $\Gamma \subset \mathbb{R} \times \mathbb{R}^m$ which separates E and H,that is, $E \subset \Gamma^-$, $H \subset \Gamma^+$ where

$$\Gamma^- = \{(u,v) \in \mathbb{R} \times \mathbb{R}^m : \bar\theta u + <\bar\lambda, v> \le 0\} \ ;$$

$$\Gamma^+ = \{(u,v) \in \mathbb{R} \times \mathbb{R}^m : \bar\theta u + <\bar\lambda, v> \ge 0\}, \ \Gamma = \Gamma^- \cap \Gamma^+ \ .$$

From a geometrical point of view, a regularity condition for (48)is equivalent to the one which ensure that Γ does not contain the line $r = \{(u,0: u \in \mathbb{R}\}$.

Consider now the simplest case where E is convex, that is the function $F(x) = (f(x),g(x))$ is cl H-convexlike and let T be the tangent cone[8] of E at the origin. It is easy to show that Γ separates E and

[8] The tangent cone $T(\bar h)$ to A at $\bar h \in A$ is defined as the set of h for which there exist a sequence $\{h^r\} \subset A$ and a positive sequence $\{\alpha_r\} \subset \subset \mathbb{R}_+$, such that $\lim_{r\to+\infty} h^r = \bar h$, $\lim_{r\to+\infty} \alpha_r(h^r - \bar h) = h - \bar h$.

$\mathcal{H}$ iff Γ separates T and $\mathcal{H}$. The reason of introducing T is given by the fact that we can characterize regularity in terms of disjunction between T and int $U = \{(u,0) \in \mathbb{R} \times \mathbb{R}^m : u > 0\}$.

The following theorem holds (Ref.16).

Theorem 6.1

Consider problem (10) and assume that $F(x) = (f(x),g(x))$ is cl $\mathcal{H}$-convexlike. Then (48) is fulfilled with $\bar{\theta} \neq 0$ iff

$$T \cap \text{int } U = \emptyset \,. \tag{49}$$

Let us note that, in the convex case, constraint qualifications are sufficient conditions for (49) to be satisfied. Condition (49) is given in the image space and it is equivalent to the following one given in the original space (Ref.25).

Condition 1. For every sequence $\{x^r\} \subset X$ and for every positive sequence $\{\alpha_r\} \subset \mathbb{R}^+$, we have

$$\lim_{r \to +\infty} \alpha_r (f(x^r), g(x^r)) \neq (u°, 0), \quad u° \in \text{int } U \tag{50}$$

or such a limit does not exist.

In the case where E is not convex, (49) becomes a necessary condition to have regularity. Conditions under which (49) is sufficient too, and some other regularity conditions for the differentiable case, can be found in Ref.25.

The following theorem [35] states a necessary and sufficient regularity condition which generalizes (49).

Theorem 6.2

Consider problem (10) and assume that (48) holds. Then (48) is fulfilled with $\bar{\theta} \neq 0$ iff

(51) $$\bar{E} \cap \text{int } U = \emptyset$$

where(9) $\bar{E} = \text{cl}(\text{conv}(\text{con } E))$.

8. THE IMAGE OF A CONSTRAINED PROBLEM

In this section we will see how the image space can be used in order to find necessary and/or sufficient conditions under which an optimal solution for a constrained extremum problem exists.

With this end let m and n be positive integers; assume we are given $X \subseteq \mathbb{R}^n$, $\phi: X \to \mathbb{R}$, $g: X \to \mathbb{R}^m$; and let $V \subseteq \mathbb{R}^m$ be a closed convex cone, containing the origin 0 of $\mathbb{R}^m$.

We consider the following constrained extremum problem:

(P) $$\min \phi(x) \quad ; \quad x \in R \triangleq \{x \in X : g(x) \in V\} ,$$

and we assume that $R \neq \emptyset$.

Note that, when $V = \mathbb{R}^m_+$, (P) collapses to (10).

(9) con A denotes the cone generated by A.
conv B denotes the convex hull of B.

Given a point $\bar{x} \in R$, we set $f_{\bar{x}}(x) = \phi(\bar{x}) - \phi(x)$ and $F_{\bar{x}}(x) = (f_{\bar{x}}(x), g(x))$, so that $F_{\bar{x}}: X \to \mathbb{R} \times \mathbb{R}^m$. Moreover, we define $K_{\bar{x}} \triangleq \{(u,v) \in \mathbb{R} \times \mathbb{R}^m : u = f_{\bar{x}}(x);\ v = g(x);\ x \in X\}$ and we call *image* of problem (P), with respect to the point $\bar{x}$, the problem:

$$(P_{\bar{x}}) \qquad \max\ (u) \quad , \quad \text{s.t.} \quad (u,v) \in R_{\bar{x}} \triangleq \{(u,v) \in K_{\bar{x}} : v \in V\}$$

Set $E_{\bar{x}} = K_{\bar{x}} - \text{cl}\, H$, we shall call *extended image* of problem (P), with respect to the point $\bar{x}$, the problem:

$$(P^{(e)}_{\bar{x}}) \qquad \max\ (u) \ , \quad \text{s.t.} \quad (u,v) \in R^{(e)}_{\bar{x}} \triangleq \{(u,v) \in E_{\bar{x}} : v \geq 0\}.$$

The sets $K_{\bar{x}}$ and $E_{\bar{x}}$, and therefore the problems $(P_{\bar{x}})$ and $(P^{(e)}_{\bar{x}})$, obviously depend on the choice of $\bar{x}$ in R; even if such dependence is of a very particular kind. Indeed, if $\hat{x} \in R$, it is easily seen that:

$$K_{\hat{x}} = K_{\bar{x}} + \{(\phi(\hat{x}) - \phi(\bar{x}), 0)\} \ ; \quad E_{\hat{x}} = E_{\bar{x}} + \{(\phi(\hat{x}) - \phi(\bar{x}), 0)\}.$$

Moreover, it can be easily verified that, if $\bar{x}$, $\hat{x} \in R$, problem $(P_{\bar{x}})$ [or $(P^{(e)}_{\bar{x}})$] has an optimal solution, iff $(P_{\hat{x}})$ [respect. $(P^{(e)}_{\bar{x}})$] does have.

For this reason, when we consider properties which hold independently on the choice of $\bar{x}$ over R, we shall drop $\bar{x}$ from the corresponding notation.

We shall now give some general results, which should be of some help in analysing the relations that hold between a constrained extremum problem and its (extended) image.

Lemma 7.1 The following equalities hold:

$$\phi(\bar{x}) - \inf_{x \in R} \phi(x) = \sup_{(u,v) \in \mathcal{R}} (u) = \sup_{(u,v) \in \mathcal{R}^{(e)}} (u) = \sup_{\substack{(u,v) \in \mathcal{E} \\ v=0}} (u) ,$$

where the image and the extended image are considered with respect to the point $\bar{x} \in R$.

Proof: The first equality follows easily, observing that $x \in R \Rightarrow$ $\Rightarrow (\phi(\bar{x})-\phi(x), g(x)) \in K$ and $(u,v) \in K \Rightarrow \exists\, x \in R$ such that $u = \phi(\bar{x}) - \phi(x)$, $v = g(x)$. To prove the other equalities, note first of all that, since $(u,v) \in \mathcal{R} \Rightarrow (u,0) = (u,v) - (0,v) \in \mathcal{E}$, we have $\sup_{(u,v) \in \mathcal{R}} (u) \le \sup_{\substack{(u,v) \in \mathcal{E} \\ v=0}} (u)$.

From the inclusion $\{(u,v) \in \mathcal{E} : v = 0\} \subseteq \mathcal{R}^{(e)}$, we then get $\sup_{\substack{(u,v) \in \mathcal{E} \\ v=0}} (u) \le$

$\le \sup_{(u,v) \in \mathcal{R}^{(e)}} (u)$. Observe finally that, for every $(u,v) \in \mathcal{R}^{(e)}$, there must be, by definition of $\mathcal{E}$, a point $(\tilde{u},\tilde{v}) \in \mathcal{R}$ with $\tilde{u} \ge u$. From this we get $\sup_{(u,v) \in \mathcal{R}^{(e)}} (u) \ge \sup_{(u,v) \in \mathcal{R}} (u)$, and this is the last inquality needed to complete the proof. #

Theorem 7.1

The following conditions are equivalent:

(i) problem (P) has a global minimum point;

(ii) problem $(\mathcal{P})$ has a global minimum point;

(3i) problem $(\mathcal{P}^{(e)})$ has a global minimum point.

Moreover, if one of the above conditions is verified, and if the images are considered with respect to $\bar{x}$, we have:

(52) $$\min_{x\in R} \phi(x) = \phi(\bar{x}) - \max_{(u,v)\in \mathcal{R}} (u) = \phi(\bar{x}) - \max_{(u,v)\in \mathcal{R}^{(e)}} (u).$$

Proof: (i) $\Leftrightarrow$ (ii). Let $\hat{x} \in X$, $\hat{u} = f(\hat{x})$, $\hat{v} = g(\hat{x})$. We have $\hat{x} \in P$, iff $(\hat{u},\hat{v}) \in \mathcal{R}$. Let now $\tilde{x} \in R$, $\tilde{u} = f(\tilde{x})$, $\tilde{v} = g(\tilde{x})$, we then have $\phi(\tilde{x}) < \phi(\hat{x})$, iff $\tilde{u} > \hat{u}$. The thesis follows easily from the above remarks. (ii) $\Rightarrow$(3i). It is a consequence of lemma 7.1 noting that $\mathcal{R} \subseteq \mathcal{R}^{(e)}$.
(3i) $\Rightarrow$(2i). Let $(\hat{u},\hat{v}) \in \mathcal{R}^{(e)}$ be such that $\hat{u} = \max_{(u,v)\in\mathcal{R}^{(e)}} (u)$, then there exists $(\tilde{u},\tilde{v}) \in \mathcal{R}$, such that $\tilde{u} \geq \hat{u}$, hence the thesis follows from lemma 7.1. The second part of the theorem is a consequence of the first part and of lemma 7.1. #

Now we shall study the existence of the minimum of problem (P),by means of its relationships with the extended image problem ($\mathcal{P}^{(e)}$).More precisely, we shall give some sufficient condition, one of which generalizes the well-known Weierstrass condition (of semicontinuity of functions and compactness of domain) to the kind of problem that we consider.

First of all observe that, as an easy consequence of Theorem 7.1, we get the following:

Theorem 7.2

Problem (P) has a global minimum point, iff the set $D=\mathcal{E}\cap\{(u,v)\in\mathbb{R}\times\mathbb{R}^m : v=0\}$ is closed and one has $D \neq \emptyset$ and $\sup_{(u,v)\in D} (u) < +\infty$.

Note that the assumption $D \neq \emptyset$ is equivalent to requiring $R \neq \emptyset$, whereas, account taken of Lemma 7.1, the assumption $\sup_{(u,v)\in D} (u) < +\infty$ is equivalent to $\inf_{x\in R} \phi(x) > -\infty$.

From this proposition we get immediately the following:

Corollary 7.1 Let $D \neq \emptyset$ and $\sup_{(u,v) \in D} (u) < +\infty$. Then, if E is closed, problem (P) has a global minimum point.

This corollary allows us to claim that, once sure that $D \neq \emptyset$ and $\sup_{(u,v) \in D} (u) < +\infty$, every condition ensuring the closure of E, or of a subset containing D, ensures also the existence of the minimum of (P). Sufficient conditions for the closure of E can be found in Ref.7,8,33.

The following theorem holds (Ref.33).

Theorem 7.3
Let $R \neq \emptyset$. Then, if X is compact and $F = (f,g)$ is $(\mathrm{cl}\, H)$-upper semicontinuous, problem (P) has a global minimum point.

Note that, when $V = \mathbb{R}^m_+$, and, therefore $\mathrm{cl}\, H = \mathbb{R}^{1+m}_+$, this result collapses to the well-known Weierstrass condition.

Some other necessary and/or sufficient conditions can be found in Ref.33.

9. CONCLUDING REMARKS

Theorems of the alternative can be considered as a general framework within which optimality conditions and related topics can be studied. By means of a theorem of the alternative we have introduced weak and strong alternative and we have shown how sufficient optimality conditions of the saddle-point type, exterior penalty method and weak duality can be deduced and/or interpreted by weak alternative (see Table 1). Furthermore, we have seen that strong alternative (see Table 1)produces strong duality, penalty interior scheme and necessary optimality conditions (different from the stationary type; these, on the contrary, turn out to be a further deepending of weak analysis and are based on local arguments). Connections among the topics which appear in the first (second) column of table 1 are shown.

The general approach consists in introducing the image space, in studying a certain question on it, and then, when a result has been obtained on the image space, to obtain its counterimage, namely the corresponding result in the original space.

Other kinds of problems, which can be reduced to the above scheme are combinatorial problem, discrete optimization problems, variational inequalities (Ref.16). This general scheme can be also extended to vector extremum problems (Ref.9) and to multifunctions.

LOGICAL CORRESPONDENCE

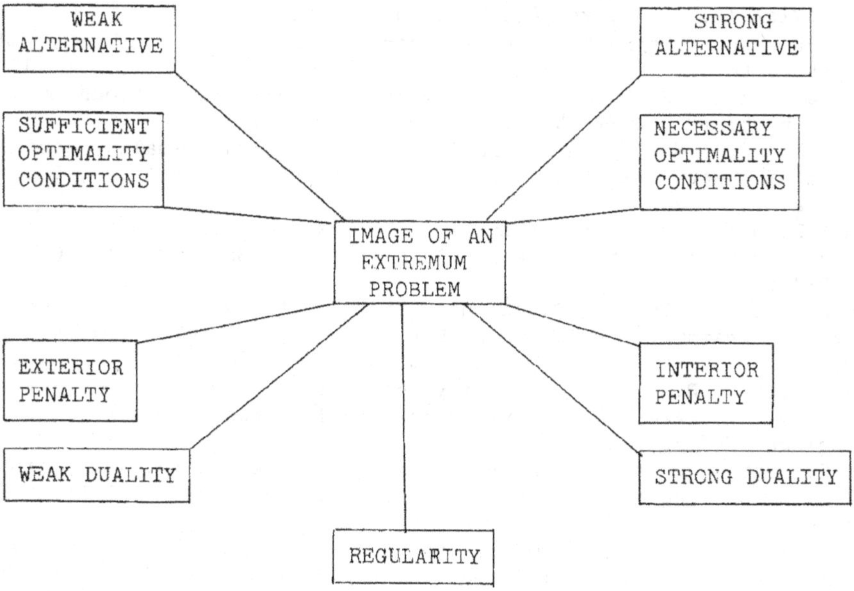

TABLE 1

REFERENCES

[1] M.Avriel. 'Non linear programming. Analysis and methods'.Prentice Hall, (1976).

[2] M.S.Bazaraa, C.M.Shetty. 'Foundations of optimizations'. *Lecture Notes in Economics and Mathematical Systems*. Springer-Verlag(1976).

[3] M.S.Bazaraa. 'A theorem of the alternative with applications to convex programming: optimality, duality and stability'. *J. of Math. Analysis and Appl.*, vol.41, (1973), 701-715.

[4] A.Ben-Israel, A.Ben-Tal, and S.Zlobec. 'Optimality in nonlinear programming. A feasible directions approach'. *J.Wiley*, New York , (1981).

[5] A.Ben-Israel, A.Charnes, and K.O.Kortanek. 'Asymptotic duality over closed convex sets'. *J. Math. Analysis Appl.*, vol.35, (1971),677-691.

[6] C.Beoni. 'A generalization of Fenchel duality theory'.*Jou Optimization Th. Appl.*, vol.49, N.3, June 1986.

[7] G.R.Bitran, T.Magnanti. 'The structure of admissible points with respect to Cone Dominance'. *J.O.T.A.*, vol.29, n.4, (1979),573-614.

[8] O.Calligaris, P.Oliva. 'Necessary and sufficient conditions for Pareto problems'. *Boll. UMI* (5), 18B (1981), 177-216.

[9] A.Cambini, L.Martein. 'Separation functions and optimality conditions in vector extremum problems'. Technical report A-120, Dept. of Math., Univ. of Pisa.

[10] A.Y.Dubovitskii and A.A.Milyutin. 'The extremum problem in the presence of constraints'. Doklady Akademiie Nauk SSSR, vol.149,(1963), 759-762.

[11] E.Eisenberg. 'On cone functions'. In "Recent advances in mathematical programming". P.Wolfe and R.L.Graves (eds.),*McGraw-Hill*,(1963), 27-33.

[12] K.H.Elster, R.Nehse. 'Optimality conditions for some non - convex problems'. *Lecture Notes in Control and Information Sciences*,n.23, Springer-Verlag, (1980).

[13] H.Everett. 'Generalized Lagrange multiplier method for solving problems of optimum allocation of resources'. *Oper. Res.*, vol.11,(1963), 399-417.

[14] A.V.Fiacco,and G.P.McCormick. 'Nonlinear programming: sequential unconstrained minimization techniques'. J. *Wiley*,New York, (1968).

[15] F.Giannessi. 'Theorems of alternative, quadratic programs and complementarity problems'. In "Variational inequalities and complementarity problems". Cottle-Giannessi-Lions (eds);*J.Wiley*, New York, (1980), 151-186.

[16] F.Giannessi. 'Theorems of the alternative and optimality conditions'. *J.O.T.A.*, vol.42, n.3, March (1984).

[17] F.Giannessi. 'On Lagrangian non-linear multipliers theory for constrained optimization and related topics'. Dept. Math., Appl.Math. Sect. (Optimization Group), Res. Report n.123, (1984).

[18] F.J.Gould. 'Extensions of Lagrange multipliers in nonlinear programming'. *SIAM J. Applied Math.*, vol.17, (1969), 1280-1297.
[19] F.J.Gould. 'Nonlinear pricing: applications to concave programming'. *Operat. Research*, vol.19, n.4, (1971), 1026-1035.
[20] S.P.Han, and O.L.Mangasarian. 'Exact penalty functions in nonlinear programming'. *Math. Programming*, vol.17, (1979), 251-269.
[21] M.Hayashi, H.Komiya. 'Perfect duality for Convexlike programs'.*J. O.T.A.*, vol.38, n.2 (1982).
[22] V.Jeyakumar. 'Convexlike alternative theorems and mathematical programming'. Dept. Math., Univ. of Melbourne, Res.Rep. n.13,(1984). To appear on "*Optimization, Mathematische Operationsforschung und Statistik*".
[23] L.McLinden. 'Duality theorems and theorem of the alternative'.*Proc. Ann. Math. Soc.*, vol.53, n.1, (1975), 172-175.
[24] O.L.Mangasarian. 'Nonlinear programming'. *McGraw-Hill*, New York , (1969).
[25] L.Martein. 'Regularity conditions for constrained extremum problems'.*Jou.Optimization Th.Appl.*,vol.47,N.2,Oct.1985.
[26] L.Martein. 'A necessary and sufficient regularity condition for convex extremum problems'. Technical Report A-90, Dept. of Math., Univ. of Pisa.
[27] L.Martein. 'Sulla separabilità locale in problemi di estremo vincolato'. Proceed. of the VI Meeting "Associazione per la Matematica Applicata alle Scienze Economiche e Sociali", (1982).
[28] L.Martein. 'Sulla dualità lagrangiana esponenziale'. Technical Report A-114, Dept. of Math., Univ. of Pisa.
[29] M.Pappalardo. 'On the duality gap in non convex optimization'. *Mathematics of Operations Research* (to appear).
[30] R.T.Rockafellar. 'Convex Analysis'. Princeton, (1970).
[31] R.T.Rockafellar. 'Penalty methods and augmented Lagrangians in nonlinear programming'. Proceedings of 5-th Conference on Optim.Tech., Springer-Verlag, Berlin, (1973), 418-425.
[32] R.T.Rockafellar. 'Augmented Lagrange multiplier functions and duality in nonconvex programmin'. *SIAM J. on Control*, vol.12, (1974), 268-283.
[33] F.Tardella. 'On the image of a constrained extremum problems'.Tech. Report A-106, Dept. of Math., Univ. of Pisa.
[34] J.Tind,and L.A.Wolsey.'An elementary survey of general duality theory'. *Mathematical Programming*, vol.21, n.3, (1981),241-261.
[35] F.Tardella. (private communication).

Chapter 4

Convergence of Equilibria in the Theory of Games

E. Cavazzuti

1. INTRODUCTION

Aim of this note is to expose some, recently introduced, definitions of convergence, that reveal to be useful to study convergence of equilibria in game theory, and illustrate the more significant properties. Only for brevity, we shall examine two person games exclusively and for these games saddle points and Nash equilibria.

The most useful tool to study the quoted convergences, is the general theory of Γ-limits of De Giorgi; for these we send back to [8],[7],[12], [2].

A prominent space will be reserved to variational aspect to the studied convergences. For other applications of remarkable interest we send back to references.

List of simbols and notations.

N positive integers

Q rationals

R real numbers

$\tilde{R}$ extended reals

X_1, X_2 sets

$(X_1,\tau_1),(X_2,\tau_2)$ topological spaces

$A_1(X_1)$, $A_2(X_2)$ families of open sets in the topologies τ_1, τ_2

$\mathcal{U}_1(x_1), \mathcal{U}_2(x_2)$ neighbourhood systems of x_1, in τ_1 and x_2 in τ_2

$\text{ext}^- = \inf$

$\text{ext}^+ = \sup$

$\tilde{R}^{X_1 \times X_2}$ = {functions from $X_1 \times X_2$ to $\tilde{R}$}

$C^\circ(X_1 \times X_2, R) = \{f: X_1 \times X_2 \to R$, f continuous$\}$

The use of indexed spaces is useful in considering non zero sum games.

2. REGULARIZATION OPERATORS, STABLE PAIRS, CLOSED FUNCTIONS

The concept of closed saddle functions has been introduced by R.T. Rockafellar in the duality theory for saddle functions and used to give general existence theorems for saddle points.

Here we shall consider, with Rockafellar's one, other closure concepts with local character: they are strictly related to Γ-limits.

For semplicity we shall consider functions of two variables, $h: X_1 \times X_2 \to \tilde{R}$, and we will be always interested in minimizing with respect to x_2 and maximizing with respect to x_1.

All regularizing operators considered here will produce the lower semicontinuity with respect to x_2 or the upper semicontinuity with respect to x_1. All lower regularizing operators will be indexed by ℓ, the upper regularizing by u.

All topological spaces will be Hausdorff spaces, for simplicity.

Definition 1.1: Let (X_i, τ_i) be topological spaces and $h_i: X_i \to \tilde{R}$. Let us define Γ_ℓ, Γ_u, cl_ℓ, cl_u by:

(1.1)

$$(\Gamma_\ell h_2)(\bar{x}_2) = \sup_{U_2 \in \mathcal{U}_2(\bar{x}_2)} \inf_{x_2 \in U_2} h_2(x_2) = \text{s.c.}^- h_2(\bar{x}_2)$$

$$(\text{cl}_\ell h_2)(\bar{x}_2) = \begin{cases} -\infty & \text{if } \exists y_2 \mid (\Gamma_\ell h_2)(y_2) = -\infty \\ (\Gamma_2 h_2)(\bar{x}_2) & \text{otherwise} \end{cases}$$

$$(\Gamma_u h_1)(\bar{x}_1) = \inf_{U_1 \in \mathcal{U}_1(\bar{x}_1)} \sup_{x_1 \in U_1} h_1(x_1) = \text{s.c.}^+ h_1(\bar{x}_1)$$

$$(\text{cl}_u h_1)(\bar{x}_1) = \begin{cases} -\infty & \text{if } \exists y_1 \mid (\Gamma_u h_1)(y_1) = +\infty \\ (\Gamma_u h_1)(\bar{x}_1) & \text{otherwise} \end{cases}$$

Remark Γ_ℓ is the usual lower semicontinuity (ℓ.s.c.) operator. cl_ℓ is considered in duality theory of optimization and called extended lower closure [13]. The operator cl_ℓ is a global operator in the sense that it depends on all values assumed by the function; Γ_ℓ is a local operator.

It is possible to characterize the operator Γ_ℓ by means of local optimization.

Lemma 1.2 Let (X_2, τ_2) be topological and $g, f \in \tilde{R}^{X_2}$. We have:

(1.2)
$$\inf_{x_2 \in A_2} f(x_2) = \inf_{x_2 \in A_2} g(x_2) \qquad \forall A_2 \in A_2(X_2)$$

if and only if

(1.3)
$$\Gamma_\ell f = \Gamma_\ell g .$$

An analogous result is true for Γ_u and sup.
For functions of several variables it is possible to consider other regularizing operators: they are given by Γ-limits or hybrid limits.

Definition 1.3: Given (X_i,τ_i), $i = 1,2$, and $f \in \tilde{R}^{X_1 \times X_2}$, let us define R_ℓ and R_u by:

$$(1.4)\qquad \begin{aligned} (R_\ell f)(x_1^o,x_2^o) &= \sup_{U_2 \in \mathcal{U}_2(x_2^o)} \ \inf_{U_1 \in \mathcal{U}_1(x_1^o)} \ \sup_{x_1 \in U_1} \ \inf_{x_2 \in U_2} f(x_1,x_2) \\ (R_u f)(x_1^o,x_2^o) &= \inf_{U_1 \in \mathcal{U}_1(x_1^o)} \ \sup_{U_2 \in \mathcal{U}_2(x_2^o)} \ \inf_{x_2 \in U_2} \ \sup_{x_1 \in U_1} f(x_1,x_2). \end{aligned}$$

In what follows we list some of the main properties of the operators Γ_ℓ, cl_ℓ, R_ℓ, Γ_u, cl_u, R_u.

Property 1.4 : Let (X_i,τ_i) be topological, $i = 1,2$. For every $f, f_1, f_2 \in \tilde{R}^{X_1 \times X_2}$ we have:

1) Isotony

$$f_1 \le f_2 \Rightarrow Gf_1 \le Gf_2 \qquad G = \Gamma_\ell, cl_\ell, R_\ell, \Gamma_u, cl_u, R_u$$

2)

$$G(Gf) = Gf \qquad G = \Gamma_\ell, cl_\ell, \Gamma_u, cl_u$$

3) Locality

$$G(f/U) = Gf/U \qquad G = \Gamma_\ell, R_\ell, \Gamma_u, R_u$$

$$U = A_2, A_1 \times A_2, A_1, A_1 \times A_2$$

$$A_i \in \mathcal{A}_i(X_i) \qquad i=1,2$$

4) Invariance for increasing homomorphism

$$\tau \circ (Gf) = G(\tau \circ f) \qquad G = \Gamma_\ell, R_\ell, \Gamma_u, R_u$$

$\tau : \tilde{R} \to \tilde{R}$ increasing homomorphism

5) Invariance for continuous perturbations

$$G(f+h) = Gf + h \qquad G = \Gamma_\ell, cl_\ell, R_\ell, \Gamma_u, cl_u, R_u$$

.5) $$\forall h \in C^\circ(X_1 \times X_2, R)$$

6) Decreasing, increasing

$$Gf \le f \qquad G = \Gamma_\ell, cl_\ell$$

$$Gf \ge f \qquad G = \Gamma_u, cl_u$$

7) If x_2° is a local minimum for $f(x_1^\circ, \cdot)$, then:

$$(\Gamma_\ell f)(x_1^\circ, x_2^\circ) = f(x_1^\circ, x_2^\circ)$$

If x_2° is a global minimum for $f(x_1^\circ, \cdot)$, then:

$$(cl_\ell f)(x_1^\circ, x_2^\circ) = f(x_1^\circ, x_2^\circ)$$

If x_1° is a local maximum for $f(\cdot, x_2^\circ)$, then:

$$(\Gamma_u f)(x_1^\circ, x_2^\circ) = f(x_1^\circ, x_2^\circ)$$

If x_1° is a global maximum for $f(\cdot, x_2^\circ)$, then:

$$(cl_u)(x_1^\circ, x_2^\circ) = f(x_1^\circ, x_2^\circ)$$

If $x^\circ = (x_1^\circ, x_2^\circ)$ is a local saddle point (see §.4), then:

$$(Gf)(x_1^\circ, x_2^\circ) = f(x_1^\circ, x_2^\circ) \qquad G = \Gamma_\ell, \Gamma_u, R_\ell, R_u$$

If $x^\circ = (x_1^\circ, x_2^\circ)$ is a global saddle point for f, then:

$$(Gf)(x_1^\circ, x_2^\circ) = f(x_1^\circ, x_2^\circ) \qquad G = \Gamma_\ell, cl, R_\ell, \Gamma_u, cl_u, R_u .$$

Theorem 1.5 (Semicontinuity and comparison)
For each $f \in \tilde{R}^{X_1 \times X_2}$, the functions $\Gamma_\ell f$, $R_\ell f$ are ℓ.s.c. x_2 and the functions $\Gamma_u f$, $R_u f$ are u.s.c. x_1. Furthermore the following inequalities are true:

$$\Gamma_\ell f \le R_\ell f \le R_\ell^m f \le R_\ell^n f \le \Gamma_\ell \Gamma_u \Gamma_\ell f \le \left\{ \begin{matrix} \Gamma_\ell \Gamma_u f \\ \Gamma_u \Gamma_\ell f \end{matrix} \right\} \le \Gamma_u \Gamma_\ell \Gamma_u f \le R_u^n f \le \tag{1.6}$$
$$\le R_u^m f \le R_u f \le \Gamma_u f$$

for $n, m \in N$ and $m \le n$.

Remark The functions that appear in (1.6) are all those one can obtain from f by applications of the operators Γ_ℓ, Γ_u, R_ℓ, R_u finitely many times.
The two functions $\Gamma_u \Gamma_\ell f$ and $\Gamma_\ell \Gamma_u f$, are not generally comparable; not even in the concave-convex case as the two following examples show.

Example 1 ([13]) $f: R^2 \to R$

$$f(x_1,x_2) = \begin{cases} +\infty & x_2 > 1 \text{ or } x_2 < 0 \\ -\infty & 0 \le x_2 \le 1 \text{ and } x_1 < 0 \text{ or } x_1 > 1 \\ \alpha & x_1 = x_2 = 0 \qquad 0 \le \alpha \le 1 \\ x_1^{x_2} & \text{otherwise} \end{cases}$$

For the previous f, we have:

$$(\Gamma_u \Gamma_\ell f)(0,0) = 1 \ge \alpha = f(0,0) \ge (\Gamma_\ell \Gamma_u f)(0,0) = 0.$$

Example 2 $f: R^2 \to R$

$$f(x_1,x_2) = \begin{cases} 0 & x_1 = x_2 = 0 \\ +\infty & \{x / x_1 \cdot x_2 \ge 0\} - \{x_2 = 0,\ x_1 \le 0\} \\ -\infty & \text{otherwise} \end{cases}$$

for which we have:

$$(\Gamma_{\ell}\Gamma_{u}f)(0,0) = +\infty > 0 = (\Gamma_{u}\Gamma_{\ell}f)(0,0) \ .$$

Theorem 1.6
For each $f \in \tilde{R}^{X_1 \times X_2}$ the following equalities hold true :

(1.7)

1) $\Gamma_{\ell}\Gamma_{\ell}f = \Gamma_{\ell}f \qquad \Gamma_{u}\Gamma_{u}f = \Gamma_{u}f$

2) $(\Gamma_{\ell}\Gamma_{u})(\Gamma_{\ell}\Gamma_{u})f = (\Gamma_{\ell}\Gamma_{u})f$

$(\Gamma_{u}\Gamma_{\ell})(\Gamma_{u}\Gamma_{\ell})f = (\Gamma_{u}\Gamma_{\ell})f$

3) $\Gamma_{\ell}R_{\ell}f = R_{\ell}\Gamma_{\ell}f = R_{\ell}f$

$\Gamma_{u}R_{u}f = R_{u}\Gamma_{u}f = R_{u}f$

4) $R_{\ell}R_{u}f = \Gamma_{\ell}R_{u}f = R_{\ell}\Gamma_{u}f = \Gamma_{\ell}\Gamma_{u}f$

$R_{u}R_{\ell}f = \Gamma_{u}R_{\ell}f = R_{u}\Gamma_{\ell}f = \Gamma_{u}\Gamma_{\ell}f$

When $\Gamma_{\ell}f = f$, we have:

5) $\Gamma_{u}f = \Gamma_{u}R_{\ell}f = R_{u}f$

When $\Gamma_{u}f = f$, we have:

6) $\Gamma_{\ell}f = \Gamma_{\ell}R_{u}f = R_{\ell}f$

Remark Relations 2 have been proved in [13] ,for concave-convex functions, with the operators cl_{ℓ}, cl_{u} instead of Γ_{ℓ}, Γ_{u}.
Equalities 4 relate the iterates of operators R with the iterates of operators Γ.

For the complete proofs we send back to [7] , [9] , [11],[14], [4] . For a more detaled analysis of closure operators and duality theory see [13].

The next definition is useful to introduce definitions of convergence more stable than those known from literature.

Definition 1.7: We shall call (f,g); $f, g \in \tilde{R}^{X_1 \times X_2}$, a stable pair with respect to the operators G_ℓ, G_u if:

$$(1.8) \qquad \begin{aligned} &a) \quad f \le g \\ &b) \quad G_\ell g = f \\ &\qquad G_u f = g \end{aligned}$$

Remark When $G_\ell = \Gamma_\ell$, cl_ℓ; $G_u = \Gamma_u$, cl_u, the condition a in (1.8) is superfluous.

When (f,g) is a stable pair with respect to the operators Γ_ℓ, Γ_u; cl_ℓ, cl_u; R_ℓ, R_u, f is a ℓ.s.c. function in x_2 and g is a u.s.c. function in x_1.

Lemma 1.8 <u>If</u> (f,g) <u>is a stable pair with respect to</u> Γ_ℓ, Γ_u, <u>then</u>:

$$(1.9) \qquad \begin{aligned} &\inf_{x_2 \in U_2} f(x_1,x_2) = \inf_{x_2 \in U_2} g(x_1,x_2) \qquad \forall x_1 \in X_1,\ \forall U_2 \in A_2(X_2) \\ &\sup_{x_1 \in U_1} f(x_1,x_2) = \sup_{x_1 \in U_1} g(x_1,x_2) \qquad \forall x_2 \in X_2,\ \forall U_1 \in A_1(X_1) \\ &\sup_{x_1 \in H_1} \inf_{x_2 \in U_2} f(x_1,x_2) = \sup_{x_1 \in H_1} \inf_{x_2 \in U_2} g(x_1,x_2) \quad \forall U_2 \in A_2(X_2),\ \forall H_1 \subset X_1 \\ &\inf_{x_2 \in H_2} \sup_{x_1 \in U_1} f(x_1,x_2) = \inf_{x_2 \in H_2} \sup_{x_1 \in U_1} g(x_1,x_2) \quad \forall U_1 \subset A_1(X_1),\ \forall H_2 \subset X_2 \end{aligned}$$

Remark Particularly (1.9) will be true when we choose H_1 and H_2 open or compact. Properties analogous to (1.9) are true, with global extrema in place of local ones, for stable pairs with respect to the operators cl_ℓ, cl_u.

Lemma 1.9 Let $f, g \in \tilde{R}^{X_1 \times X_2}$; f $\ell.s.c.$ x_2, g u.s.c. x_1, $f \le g$. When we have:

(1.10) $$\Gamma_\ell g \le \Gamma_u f$$

then $(\Gamma_\ell g, \Gamma_u f)$ is a stable pair with respect to Γ_ℓ, Γ_u.

Proof: Since

$$f \le \Gamma_\ell g \le \Gamma_u f \le g$$

then

$$\Gamma_\ell g = \Gamma_\ell^2 g \le \Gamma_\ell \Gamma_u f \le \Gamma_\ell g \Rightarrow \Gamma_\ell g = \Gamma_\ell \Gamma_u f .$$

Similarly

$$\Gamma_u f \le \Gamma_u \Gamma_\ell g \le \Gamma_u^2 f = \Gamma_u f \Rightarrow \Gamma_u f = \Gamma_u \Gamma_\ell g . \qquad \#$$

Definition 1.10 We shall call $[\Gamma_\ell f, \Gamma_u f] = \{h : \Gamma_\ell f \le h \le \Gamma_u f\}$ the class of functions associated to f.
We shall say that f and g are Γ-equivalent if they have the same associated class and we shall call f a Γ-closed function when the pair$(\Gamma_\ell f, \Gamma_u f)$ is a stable pair with respect to Γ_ℓ, Γ_u.

Remark The concept of Γ-closed and closed function in the sense of Rockafellar are not equivalent. For example for

$$f(x_1,x_2) = \begin{cases} +\infty & \text{, if } x_2 > 0,\ x_1 \ge 0 \\ x_2^2 - x_1^2 & \text{, otherwise} \end{cases}$$

we have $\Gamma_\ell f = \Gamma_u f = f$, while $cl_\ell cl_u f \ne cl_\ell f = f$. It may happen that $h \in [\Gamma_\ell f, \Gamma_u f]$ without being equivalent to f. In the following example:

$$f(x_1,x_2) = \begin{cases} 1 & , x_1 \in Q, \quad x_2 \in R - Q \\ 1 & , x_1 \in R - Q, \quad x_2 \in Q \\ 0 & , \text{otherwise} \end{cases}$$

we have $\Gamma_\ell f = 0 < 1 = \Gamma_u f$ and $h = \frac{1}{2} \in [\Gamma_\ell f, \Gamma_u f]$.

From now on closedness and equivalence always will be referred to operators Γ_ℓ, Γ_u.

3. DEFINITIONS AND PROPERTIES OF Γ-LIMITS

Let (X_i, τ_i), $i = 1,2$, be topological spaces and $f_h: X_1 \times X_2 \to \tilde{R}$, $h = 1,2,\ldots$, be a sequence of functions and put:

Definition 2.1: [8] . We shall call Γ-limits of the sequence $\{f_h\}_{h=1,2,..}$ in $x^o = (x_1^o, x_2^o)$ the extended real numbers defined by:

$$\Gamma(N^{i_o}, X_{t_1}^{i_1}, X_{t_2}^{i_2}) \lim f_h(x_1^o, x_2^o) =$$

(2.1)
$$\operatorname*{ext}_{U_{t_2} \in \mathcal{U}_{t_2}(x_{t_2}^o)}^{-i_2} \ \operatorname*{ext}_{U_{t_1} \in \mathcal{U}_{t_1}(x_{t_1}^o)}^{-i_1} \ \operatorname*{ext}_{t}^{-i_o} \ \operatorname*{ext}_{h \geq t}^{i_o} \ \operatorname*{ext}_{y_{t_1} \in U_{t_1}}^{i_1} \ \operatorname*{ext}_{y_{t_2} \in U_{t_2}}^{i_2} f_h(y_1,y_2)$$

$$t_1, t_2 = 1,2 \quad t_1 \neq t_2 \ ; \ i_o, i_1, i_2 = \pm$$

In what follows we shall consider only Γ-limits of the form

$$\Gamma(N^\pm, X_1^+, X_2^-)\lim f_h, \quad \Gamma(N^\pm, X_2^-, X_1^+)\lim f_h$$

since x_1 always will be a maximization variable and x_2 a minimization variable.

The preceeding four numbers may be all different as in the example:

$$f_h(x_1,x_2) = (-1)^h \cdot 2 + \sin h(x_1+x_2) \quad x_1,x_2 \in R ,$$

R with the usual topology, for which we have:

$$\Gamma(N^-,X_1^+,X_2^-) \lim f_h = -3 , \quad \Gamma(N^+,X_1^+,X_2^-) \lim f_h = +1$$

$$\Gamma(N^-,X_2^-,X_1^+) \lim f_h = -1 , \quad \Gamma(N^+,X_2^-,X_1^+) \lim f_h = 3 .$$

We shall use the following short notations:

$$f'_\ell = \Gamma(N^-,X_1^+,X_2^-) \lim f_h \quad , \quad f''_\ell = \Gamma(N^+,X_1^+,X_2^-) \lim f_h$$

$$f'_u = \Gamma(N^-,X_2^-,X_1^+) \lim f_h \quad , \quad f''_u = \Gamma(N^+,X_2^-,X_1^+) \lim f_h$$

Let us recall now the main properties of Γ-limits; some of them will be used later.

Property 2.2: (Semicontinuity) [7]. For each sequence $f_h : X_1 \times X_2 \to \tilde{R}$, we have.

(2.2)
1) f'_ℓ , f''_ℓ are ℓ.s.c. x_2

2) f'_u , f''_u are u.s.c. x_1

Remark Other properties of semicontinuity are generally not true.

Property 2.3: (Stability for continuous perturbations) [5] . Let $g \in C^0(X_1 \times X_2, R)$ and $f_h \in \tilde{R}^{X_1 \times X_2}$, h = 1,2,..., then we have:

(2.3)

$$\Gamma(N^\pm,X_1^+,X_2^-) \lim (f_h + g) = \begin{cases} f'' + g \\ f' + g \end{cases}$$

$$\Gamma(N^\pm,X_2^-,X_1^+) \lim (f_h + g) = \begin{cases} f''_u + g \\ f'_u + g \end{cases}$$

Remark The previous property is useful in the characterizations of Γ-convergence by means of perturbation methods, like De Giorgi - Yoshida method and duality method ([4]).

Property 2.4: (Locality) [7] . Let $\{f_h\}_{h=1,2,..}$ be a sequence and $U_i \in A_i(X_i)$, $i = 1,2$, be open sets, then we have:

$$\Gamma(N^{i_o},X_{t_1}^{i_1},X_{t_2}^{i_2}) \lim f_h/_{U_1\times U_2} = (\Gamma(N^{i_o},X^{i_1},X^{i_2})\lim f_h/_{U_1\times U_2}$$

Property 2.5: (Invariance for increasing homomorphism) ([7]). Let $\omega: \tilde{R} \to R$ be continuous increasing. For each sequence $\{f_h\}_{h=1,2,..}$ we have:

$$\Gamma(N^{i_o},X_{t_1}^{i_1},X_{t_2}^{i_2}) \lim(\omega \circ f_h) = \omega \circ \Gamma(N^{i_o},X_{t_1}^{i_1},X_{t_2}^{i_2})\lim f_h$$

Remark By means of property 2.5 we can always refer to uniformly bounded sequences in the calculus of Γ-limits.

Property 2.6: (Invariance for Γ-equivalent functions). When f_h and g_h, $h = 1,2,...,$ are Γ-equivalent, we have:

$$f'_\ell = g'_\ell \quad , \quad f''_\ell = g''_\ell \quad , \quad f'_u = g'_u \quad , \quad f''_u = g''_u \ .$$

4. VARIATIONAL PROPERTIES OF Γ-LIMITS

The next inequalities, which are generalizations of those obtained in [3] , allow to characterize the Γ-limits among functions of open sets.

Theorem 3.1

Let $f_h: X_1 \times X_2 \to \tilde{R}$ be any sequence and $K_1 \subset X_1$ be compact and $U_2 \subset X_2$ be open, then we have:

$$(3.1) \qquad \inf_{\substack{A_1 \supset K_1 \\ A_1 \in A_1(X_1)}} \overline{\lim_h} \sup_{x_1 \in A_1} \inf_{x_2 \in U_2} f_h(x_1,x_2) \le \sup_{x_1 \in K_1} \inf_{x_2 \in U_2} f''_\ell(x_1,x_2)$$

Proof: Let us consider the open set function $\omega: A_1(X_1) \to \tilde{R}$ defined by:

$$\omega(A_1) = \overline{\lim_h} \sup_{x_1 \in A_1} \inf_{x_2 \in U_2} f_h(x_1,x_2)$$

For any finite family of open sets $\{A_i\}_{j=1,2,..,n}$ of X_1, we have:

$$\max_{1 \le i \le n} \omega(A_i) = \omega(\bigcup_1^n{}_i A_i) .$$

Now let us put

$$\lambda = \inf_{A_1 \supset K_1} \overline{\lim_h} \sup_{x_1 \in A_1} \inf_{x_2 \in U_2} f_h(x_1,x_2),$$

Then it results:

$$(3.2) \qquad \exists x_1^\circ \in K_1 : \forall A_1 \in \mathcal{U}_1(x_1^\circ) , \lambda \le \omega(A_1) .$$

Suppose that (3.2) is not true to obtain:

$$(3.3) \qquad \forall y_1 \in K_1, \exists A_1 \in \mathcal{U}_1(y_1) : \lambda > \omega(A_1)$$

and consider the family of open sets $\{A_{y_1}\}_{y_1 \in K_1}$, with $\{A_{y_1}\} \in \mathcal{U}_1(y_1)$ and A_{y_1} verifying (3.3).

From the open covering $\{A_{y_1}\}_{y_1 \in K_1}$ of the compact set K_1 we can extract a finite subcovering $\{A^i_{y_1}\}_{i=1,..,n}$, for which we have:

$$\lambda > \max_{1\le i\le n} \omega(A^i_{y_1}) = \omega(\bigcup_{i}^{n} A^i_{y_1}) \ge \lambda$$

which is absurd. The property (3.2) must then be verified for, at least, one $x_1^o \in K_1$. We can apply (3.2) to obtain:

$$\lambda \le \inf_{A_1\in \mathcal{U}_1(x_1^o)} \overline{\lim_h} \sup_{x_1\in A_1} \inf_{x_2\in U_2} f_h(x_1,x_2) \le$$

$$\le \sup_{V_2\in \mathcal{U}_2(x_2^o)} \inf_{A_1\in \mathcal{U}_1(x_1^o)} \overline{\lim_h} \sup_{x_1\in A_1} \inf_{x_2\in V_2} f_h(x_1,x_2) = f''_\ell(x_1^o,x_2^o)$$

$$\forall x_2^o \in U_2$$

Finally

$$\lambda \le \inf_{x_2^o\in U_2} f''_\ell(x_1^o,x_2^o) \le \sup_{x_1\in K_1} \inf_{x_2^o\in U_2} f''_\ell(x_1,x_2^o) . \qquad \#$$

Remark The analogous property obtained by substituting $\overline{\lim}$ with $\underline{\lim}$ and f''_ℓ with f'_ℓ is not true (in general) as the next example shows.

Let $A_1, A_2 \in A_1(X_1)$ and $\bar{A}_1 \cap \bar{A}_2 = \emptyset$. Put

$$f_h(x_1,x_2) = \begin{cases} \chi_{A_1}(x_1) & , h \text{ even} \\ \chi_{A_2}(x_1) & , h \text{ odd} \end{cases}$$

to obtain $f'_\ell(x_1,x_2) = 0 \quad \forall x_1 \in X_1 \quad \forall x_2 \in X_2$.

We have

$$\sup_{x_1\in K_1} f'_\ell(x_1,x_2) = 0 = \sup_{x_1\in K_1} \inf_{x_2\in U_2} f'_\ell(x_1,x_2).$$

Choose $K_1 = \{\bar{x},\bar{\bar{x}}\}$, $\bar{x} \in A_1$, $\bar{\bar{x}} \in A_2$ to find

$$\inf_{A_1 \supset K_1} \underline{\lim}_h \sup_{x_1 \in A_1} \inf_{x_2 \in U_2} f_h(x_1,x_2) = 1 > 0 = \sup_{x_1 \in K_1} \inf_{x_2 \in U_2} f'_\ell(x_1,x_2).$$

Corollary 3.2 ([3]). <u>For any</u> $\{f_h\}_{h=1,2,..}$, $K_1 \subset X_1$ <u>compact</u>, $U_2 \in A_2(X_2)$, <u>we have</u>:

$$\overline{\lim_h} \sup_{x_1 \in K_1} \inf_{x_2 \in U_2} f_h(x_1,x_2) \le \sup_{x_1 \in K_1} \inf_{x_2 \in U_2} f''_\ell(x_1,x_2) . \tag{3.4}$$

Remark Choosing in (3.4) the compact set $K_1 = \{\bar{x}_1\}$, we obtain:

$$\overline{\lim_h} \inf_{x_2 \in U_2} f_h(\bar{x}_1,x_2) \le \inf_{x_2 \in U_2} f''_\ell(\bar{x}_1,x_2)$$

which is a result known from ([7]) when X_1 carries the discrete topology.

When X_2 carries the discrete topology we have:

$$\inf_{A_1 \supset K_1} \overline{\lim_h} \sup_{x_1 \in A_1} f_h(x_1,\bar{x}_2) \le \sup_{x_1 \in K_1} f''_\ell(x_1,\bar{x}_2)$$

a result known from ([7]).

The inequality (3.4) can be strict, as the following example shows:

$$f_h(x_1,x_2) = \begin{cases} 0 & , x_1 \ne \frac{1}{h} , \frac{2}{h} \\ x_2 & , x_1 = \frac{1}{h} \\ 1-x_2 & , x_1 = \frac{2}{h} \end{cases}$$

for which we have:

$$f''_\ell(x_1,x_2) = f'_\ell(x_1,x_2) = \begin{cases} 0 & x_1 \neq 0 \\ \max\{x_2, 1-x_2\} & x_1 = 0 . \end{cases}$$

In this example we have (R^2 with the usual metric):

$$\sup_{x_1 \in X_1} \inf_{x_2 \in X_2} f''(x_1,x_2) = \frac{1}{2} > 0 = \overline{\lim_h} \sup_{x_1 \in X_1} \inf_{x_2 \in X_2} f_h(x_1,x_2).$$

Theorem 3.3

Let $f_h: X_1 \times X_2 \to \tilde{R}$ be any sequence and $K_2 \subset X_2$ a compact set, $U_1 \in A_1(X_1)$, then we have:

$$(3.5) \qquad \inf_{x_2 \in K_2} \sup_{x_1 \in U_1} f'_u(x_1,x_2) \leq \sup_{U_2 \supset K_2} \overline{\lim_h} \inf_{x_2 \in U_2} \sup_{x_1 \in U_1} f_h(x_1,x_2) .$$

Proof: The proof can be obtained by means of duality from (3.1),it will be enough to put $f_h(x_1,x_2) = -g(x_2,x_1)$. #

Corollary 3.4 ([3]). With the same hypothesis of theorem 3.3, we have:

$$(3.6) \qquad \inf_{x_2 \in K_2} \sup_{x_1 \in U_1} f'_u(x_1,x_2) \leq \underline{\lim_h} \inf_{x_2 \in K_2} \sup_{x_1 \in U_1} f_h(x_1,x_2) .$$

Remark From (3.6) and (3.5) we obtain the inequalities:

$$\sup_{x_1 \in U_1} f'_u(x_1,\bar{x}_2) \leq \underline{\lim_h} \sup_{x_1 \in U_1} f_h(x_1,\bar{x}_2)$$

$$\inf_{x_2 \in K_2} f'_u(\bar{x}_1,x_2) \leq \sup_{U_2 \supset K_2} \underline{\lim_h} \inf_{x_2 \in U_2} f_h(\bar{x}_1,x_2)$$

which are known from ([7]), when X_2 or X_1 carries the discrete topology.

We can now characterize f'_u, f'' among the functions verifying (3.1) or (3.5).

Theorem 3.5
Let $F_\infty : X_1 \times X_2 \to \tilde{R}$ verifies (3.1) $\forall K_1 \subset X_1$ compact, $\forall U_2 \subset X_2$ open ; then we have:

$$(3.7) \qquad \begin{aligned} &1) \quad F_\infty \geq f''_\ell \\ &2) \quad f''_\ell = \min \{F_\infty : F_\infty \text{ verifies } (3.1)\} \end{aligned}$$

Proof: It will be enough to verify 1 of (3.7). From (3.1) with $K_1 = \{x_1^o\}$, remembering lemma 1.2, we have:

$$\inf_{x_2 \in U_2} \Gamma_\ell F_\infty(x_1^o, x_2) = \inf_{x_2 \in U_2} F_\infty(x_1^o, x_2) \geq$$

$$\geq \inf_{A_1 \in \mathcal{U}_1(x_1^o)} \overline{\lim_h} \sup_{x_1 \in A_1} \inf_{x_2 \in U_2} f_h(x_1, x_2)$$

Taking the sup with respect to $\mathcal{U}_2$, we obtain:

$$F_\infty(x_1^o, x_2^o) \geq \Gamma_\ell F_\infty(x_1^o, x_2^o) \geq \sup_{U_2 \in \mathcal{U}_2(x_2^o)} \inf_{A_1 \in \mathcal{U}_1(x_1^o)} \overline{\lim_h}$$

$$\sup_{x_1 \in A_1} \inf_{x_2 \in U_2} f_h(x_1, x_2) = f''_\ell(x_1^o, x_2^o) \qquad \forall x_1^o, \forall x_2^o . \qquad \#$$

Theorem 3.6 Let $G_\infty : X_1 \times X_2 \to \tilde{R}$ verifies (3.5) $\forall K_2 \subset X_2$ compact, $\forall U_1 \subset X_1$ open, then we have:

$$(3.8)\quad \begin{cases} 1) & G_\infty \le f'_u \\ 2) & f'_u = \sup\{G_\infty : G_\infty \text{ \underline{verifies} } (3.5)\}. \end{cases}$$

Remark In (3.7) we may take F_∞ *ℓ.s.c.* x_2 and in (3.8) G_∞ u.s.c. x_1.

5. SOLUTION CONCEPTS

Many kinds of games are studied in the theory of games: two person games, n-person games; cooperative and individual (non cooperative) games; dinamic and differential games and so on. To each kind of game it corresponds one or more solution (or equilibrium) concepts. Extensions to n-person games of what we shall say, when possible, are easy to obtain ([61]).

Let two sets X_1, X_2 be given and a subset $\phi \neq K \subset X_1 \times X_2 = X$, furthermore let us consider two functions $J_1, J_2 : K \to \tilde{R}$ and denote with J the pair (J_1, J_2). Let $\pi_i : K \to X_i$, $i = 1,2$, be the projections on X_i.

In what follows we shall call (two person) game the pair (K,J). The set X_i is called the strategies set for the i-th player, $X = X_1 \times X_2$ is called the strategies set, while the elements of K are called admissible strategies. J_i is called the cost of the i-th player and J the vector of the costs.

Definition 4.1: An element $\hat{x} = (\hat{x}_1, \hat{x}_2) \in K$ will be called a Nash equilibrium for the game (K,J), if it verifies

$$(4.1)\quad \begin{aligned} J_1(\hat{x}) &\le J_1(x) \qquad \forall x \in K/\pi_2 x = \hat{x}_2 \\ J_2(\hat{x}) &\le J_2(x) \qquad \forall x \in K/\pi_1 x = \hat{x}_1 \end{aligned}$$

Remark In general Nash equilibria are not unique and the values of J on distinct Nash equilibria may be different.

A formal semplification in the definition of Nash equilibria can be obtained introducing the extension of a game. Let (K,J) be a given game; we shall call extension of the game (K,J), the game $(X,\hat{J})$ defined by:

$$\hat{J}_i(x) = \begin{cases} +\infty & x \notin K \\ J_i(x) & x \in K \qquad i=1,2 \end{cases}$$

Remark A game (K,J) and its extension $(X,\hat{J})$ have the same Nash equilibria in K and the same Nash equilibria with finite values.

The extension just given is different from the usual minimal and maximal extension of a zero sum game.

We shall always suppose that the set of admissible strategies is a product $X_1 \times X_2$, if this is not the case we shall consider the extension of the given game.

For extended games the definition 4.1 becomes: $\hat{x}$ is said a Nash equilibrium for (X,J) if:

$$J_1(\hat{x}) = J_1(\hat{x}_1,\hat{x}_2) = \min_{x_1 \in X_1} J_1(x_1,\hat{x}_2) \le J_1(x_1,\hat{x}_2)$$

(4.2)

$$J_2(\hat{x}) = J_2(\hat{x}_1,\hat{x}_2) = \min_{x_2 \in X_2} J_2(\hat{x}_1,x_2) \le J_2(\hat{x}_1,x_2)$$

Interpretation Nash equilibria are said individual since both players have any incentive to change their strategies unless their opponents make a change in their strategies. Two different Nash equilibria x° and $\hat{x}$ can exist and x° can be less expensive of $\hat{x}$ for both players, i.e. $J_i(x^\circ) < J_i(\hat{x})$, i = 1,2; without that the unilateral change of strategy be convenient for any player.

A game (X,J) is said a zero sum game when

$$J_1 = -J_2$$

or $J_1 + J_2 = 0$ if both costs are finite.

Definition 4.2: We shall call $x^\circ = (x_1^\circ, x_2^\circ)$ a conservative solution for the game (X,J) if:

$$(4.3)\quad \begin{aligned} J_1(x^\circ) &= J_1(x_1^\circ,x_2^\circ) = \min_{x_1} \sup_{x_2} J_1(x_1,x_2) = \sup_{x_2} J_1(x_1^\circ,x_2) \\ J_2(x^\circ) &= J_2(x_1^\circ,x_2^\circ) = \min_{x_2} \sup_{x_1} J_2(x_1,x_2) = \sup_{x_1} J_2(x_1,x_2^\circ) \end{aligned}$$

When (X,J) is a zero sum game and x° is a conservative solution, if we put $-J_1 = J_2 = f$, we have:

$$(4.4)\quad \min_{x_2} \sup_{x_1} f(x_1,x_2) = \max_{x_1} \inf_{x_2} f(x_1,x_2).$$

The following inequality, called minimax inequality, is true for all functions

$$(4.5)\quad \inf_{x_2 \in X_2} \sup_{x_1 \in X_1} f(x_1,x_2) \geq \sup_{x_1 \in X_1} \inf_{x_2 \in X_2} f(x_1,x_2) .$$

When a function satisfies the equality in (4.5) we say that f has saddle value.

Definition 4.3: We shall say that x° is a saddle point for $f: X_1 \times X_2 \to \tilde{R}$, when:

$$(4.6)\quad \begin{aligned} f(x_1^\circ,x_2^\circ) &\leq f(x_1^\circ,x_2) \quad , \quad \forall x_2 \in X_2 \\ f(x_1^\circ,x_2^\circ) &\geq f(x_1,x_2^\circ) \quad , \quad \forall x_1 \in X_1 \end{aligned}$$

Remark The equality (4.4) is necessary and sufficient for f to have a saddle point.
A function f has a saddle point if and only if the game (X,(-f,f)) has a Nash equilibrium.
In each saddle point for f, f attains the saddle value. For a zero sum game Nash equilibria and conservative solutions (or saddle points) are the same concept.

Lemma 4.4 All Γ-equivalent functions have the same saddle points(if any).
If f is Γ-closed all elements in the associated class $[\Gamma_\ell f, \Gamma_u f]$ have the same saddle points of f.

Proof: Let φ be any function equivalent to f and x° be a saddle point for f. We have:

$$f(x_1^\circ,x_2^\circ) \le \inf_{x_2} \Gamma_\ell f(x_1^\circ,x_2) = \inf_{x_2} \Gamma_\ell \varphi(x_1^\circ,x_2) =$$

$$= \inf_{x_2} \varphi(x_1^\circ,x_2) \le \varphi(x_1^\circ,x_2^\circ) \le \sup_{x_1} \varphi(x_1,x_2^\circ) =$$

$$= \sup_{x_1} \Gamma_u \varphi(x_1,x_2^\circ) = \sup_{x_1} \Gamma_u f(x_1,x_2^\circ) \le f(x_1^\circ,x_2^\circ) \qquad \#$$

6. CONVERGENCES FOR SADDLE POINT PROBLEMS AND THEIR PROPERTIES

Some convergences have been recently introduced, in mathematical literature, for sequences of saddle functions. These convergences were requested , explicitely or implicitely, to satisfy the two following fundamental properties:

(5.1)

1) When x^h is a saddle point for f_h and $x^h \to x^o$ then x^o is a saddle point for "the limit" f_o of f_h.

2) If f_h converges to f_o, the saddle values $f_h(x^h)$ converge to the saddle value $f_o(x^o)$.

All the proposed convergences satisfy other properties of stability : they will be introduced in what follows with short comments on their properties.

Definition 5.1: Given $f_h: X_1 \times X_2 \to \tilde{R}$, we shall say that:

(5.2)

i) f_h inferiorly Γ-converges to f_ℓ, if we have:

$$f'_\ell = f''_\ell = f_\ell$$

ii) f_h A-W converges to φ, when:

$$f''_\ell \leq \varphi \leq f'_u$$

and we write $\varphi \in$ A-W lim f_h

iii) f_h F-converges to φ, when:

a) (f''_ℓ, f'_u) is a Γ-stable pair

b) $f''_\ell \leq \varphi \leq f'_u$

Remark The definition i) has been suggested by De Giorgi and introduced in ([3]) to prove (5.1) as theorems. The definition ii) was proposed by [1] and (5.1) was proved again. In the definition i) the limit is unique, while in the definition ii) we have a limit class which may contain non Γ-equivalent functions. Then definition iii) has been proposed in [11] and is motivated from the desire to obtain the Γ-equivalence of all functions in the limit class (which is unique under Γ-equivalence).

In the very important case of concave-convex(quasi concave-convex) functions we can only say ([4]), in general, that f''_ℓ , f''_u are convex (quasi convex) functions, when f_h are, and f'_ℓ, f'_u are concave (quasi concave) functions, when f_h are. If f exists, it is a concave-convex function when f_h are concave-convex.

In the general case all convergences introduced are different, but they admit the common property of attenuation of duality gap.

Definition 5.2: Let $\phi \neq A_i \subset X_i$, $i = 1,2$. We shall call duality gap of f, relatively to $A_1 \times A_2$, the non negative extended real number given by:

$$(5.3) \qquad \alpha(A_1,A_2,f) = \inf_{x_2 \in A_2} \sup_{x_1 \in A_1} f(x_1,x_2) - \sup_{x_1 \in A_1} \inf_{x_2 \in A_2} f(x_1,x_2) \geq 0$$

when the difference is meaningful.

Remark If and only if f has saddle value relatively to $A_1 \times A_2$ we have $\alpha(A_1,A_2,f) = 0$.

Theorem 5.3 (<u>Attenuation of duality gap and convergence</u>).
<u>Let</u> $f_h: X_1 \times X_2 \to \tilde{R}$ <u>be a sequence and</u> $A_1 \in \mathcal{A}_1(X_1)$, $A_2 \in \mathcal{A}_2(X_2)$. <u>If there exist two non empty compact sets</u> $K_1 \subset A_1$, $K_2 \subset A_2$ <u>and two sequences</u> ε_h, $\sigma_h \geq 0$; ε_h, $\sigma_h \to 0$, <u>such that</u>:

$$(5.4) \qquad \left| \begin{array}{l} \sup\limits_{x_1 \in U_1} \inf\limits_{x_2 \in U_2} f_h(x_1,x_2) = \sup\limits_{x_1 \in K_1} \inf\limits_{x_2 \in U_2} f_h(x_1,x_2) + \varepsilon_h \\ \\ \inf\limits_{x_2 \in K_2} \sup\limits_{x_1 \in U_1} f_h(x_1,x_2) = \inf\limits_{x_2 \in U_2} \sup\limits_{x_1 \in U_1} f_h(x_1,x_2) + \sigma_h \end{array} \right.$$

<u>Then we have</u>:

$$(5.5) \qquad \underline{\lim_h} \ \alpha(U_1,U_2,f_h) \geq \alpha(U_1,U_2,\varphi)$$

for each $\varphi \in$ A-W $\lim f_h$, when both sides of (5.5) are meaningful. Furthermore if x^h is a saddle point for f_h on $U_1 \times U_2$ and $x^h \to x^\circ \in U_1 \times U_2$, then x° is a saddle point for φ and

$$(5.6) \qquad \lim_h f_h(x^h) = \varphi(x^\circ)$$

for each $\varphi \in$ A-W $\lim f_h$.

Remark When convergence i is verified the limit f is the minimum element in the class A-W $\lim f_h$.

Proof: From corollaries 3.2 and 3.4 we obtain:

$$\overline{\lim_h} \sup_{x_1 \in U_1} \inf_{x_2 \in U_2} f_h(x_1,x_2) = \overline{\lim_h} \sup_{x \in K_1} \inf_{x_2 \in U_2} f_h(x_1,x_2) \leq$$

$$\leq \sup_{x_1 \in K_1} \inf_{x_2 \in U_2} f''_\ell(x_1,x_2) \leq \sup_{x \in K_1} \inf_{x_2 \in U_2} \varphi(x_1,x_2) \leq$$

$$\leq \inf_{x_2 \in K_2} \sup_{x_1 \in U_1} \varphi(x_1,x_2) \leq \inf_{x_2 \in K_2} \sup_{x_1 \in U_1} f'_u(x_1,x_2) \leq$$

$$\leq \underline{\lim_h} \inf_{x_2 \in K_2} \sup_{x_1 \in U_1} f_h(x_1,x_2) = \underline{\lim_h} \inf_{x_2 \in U_2} \sup_{x_1 \in U_1} f_h(x_1,x_2)$$

And taking the differences we have:

$$\underline{\lim_h} \ \alpha(U_1,U_2,f_h) \geq \underline{\lim_h} \inf_{x_2 \in U_2} \sup_{x_1 \in U_1} f_h(x_1,x_2) -$$

$$- \overline{\lim_h} \sup_{x_1 \in U_1} \inf_{x_2 \in U_2} f_h(x_1,x_2) \geq \alpha(U_1,U_2,\varphi)$$

when both members of the inequality make sense.
When each f_h has saddle points in $U_1 \times U_2$, we have:

$$\alpha(U_1,U_2,\varphi) = 0 \qquad \forall\, \varphi \in \text{A-W}\lim f_h .$$

The only thing it remains to show is that each limit of a sequence of saddle points for f_h is a saddle point for φ.
Let x^h be a sequence of saddle points for f_h and $x^h \to x^\circ$ in the product topology on $X_1 \times X_2$. Let $V_1 \times V_2 \subset U_1 \times U_2$ be an open neighbourhood of x°, since $x^h \in V_1 \times V_2$ definitively, we have:

$$\inf_{V_1 \in \mathcal{U}_1(x_1^\circ)} \overline{\lim_h} \sup_{x_1 \in V_1} \inf_{x_2 \in V_2} f_h(x_1,x_2) \le \overline{\lim_h} \inf_{x_2 \in V_2} f_h(x_1^h,x_2) =$$

$$= \overline{\lim_h} f_h(x_1^h,x_2^h) \le \overline{\lim_h} \inf_{y_2 \in W_2} f_h(x_1^h,y_2) \le \overline{\lim_h} \sup_{y_1 \in W_1} \inf_{y_2 \in W_2} f_h(y_1,y_2)$$

$$\forall W_1 \in \mathcal{U}_1(x_1^\circ),\ \forall V_2 \in \mathcal{U}_2(x_2^\circ),\ \forall W_2 \in \mathcal{U}_2(\bar{x}_2)\ ;\ W_1 \subset U_1,\ V_2,W_2 \subset U_2$$

and, by saturation, we obtain:

$$f''_\ell(x_1^\circ,x_2^\circ) \le f''_\ell(x_1^\circ,\bar{x}_2) \qquad \forall \bar{x}_2 \in U_2$$

Similarly we have:

$$f'_u(x_1^\circ,x_2^\circ) \ge f'_u(\bar{x}_1,x_2^\circ) \qquad \forall \bar{x}_1 \in U_1 .$$

Now let us show that

$$(5.7) \qquad f'_u(x^\circ) = f''_\ell(x^\circ) = \lim_h f_h(x^h).$$

Since x^h is a saddle point for f_h, $\forall V_1 \in \mathcal{U}_1(x_1^\circ)$, $\forall V_2 \in \mathcal{U}_2(x_2^\circ)$, for h large enough, we have:

$$\sup_{x_1 \in V_1} \inf_{x_2 \in V_2} f_h(x_1,x_2) \ge f_h(x_1^h,x_2^h) \ge \inf_{x_2 \in V_2} \sup_{x_1 \in V_1} f_h(x_1,x_2)$$

hence

$$f''_\ell(x^\circ) \ge \overline{\lim_h} f_h(x_1^h,x_2^h) \ge \underline{\lim_h} f_h(x_1^h,x_2^h) \ge f'_u(x^\circ)$$

and from A-W convergence we deduce (5.7).

Now, for $\varphi \in$ A-W lim f_h, we have:

$$\varphi(x_1^o,\bar{x}_2) \geq f''_\ell(x_1^o,\bar{x}_2) \geq f''_\ell(x_1^o,x_2^o) = \varphi(x_1^o,x_2^o) = f'_u(x_1^o,x_2^o) \geq f'_u(\bar{x}_1,x_2^o) \geq \varphi(\bar{x}_1,x_2^o)$$

from which the result follows. #

Remark When at least one of the limits $\overline{\lim_h} \sup_{x_1 \in U_1} \inf_{x_2 \in U_2} f_h(x_1,x_2)$, $\underline{\lim_h} \inf_{x_2 \in U_2} \sup_{x_1 \in U_1} f_h(x_1,x_2)$ is finite, both members of (5.4) make sense.

We discuss some properties of the previous convergences by means of an example. Let us take $f_h: R^2 \to R$ given by $f_h(x_1,x_2) = \sin 2\pi h(x_1,x_2)$ and consider the usual metric on R^2.

We have:

$$\min_{x_2} \max_{x_1} f_h(x_1,x_2) = +1$$

$$\max_{x_1} \min_{x_2} f_h(x_1,x_2) = -1$$

so that none of the previous function has saddle points. Furthermore we get:

$$f''_\ell = f'_\ell = -1$$

$$f''_u = f'_u = +1$$

If we choose f_ℓ as limit of f_h, then it is the point of view of the minimizer that prevails in the limit game and the maximizer is obliged to pay the optimal value for the minimizer.

If we take the A-W lim f_h as the limit class, then we find in the limit class non Γ-equivalent functions and we cannot reconstruct the class if we know one limit function. Different limit functions can have different saddle values, for instance we have:

$$0 \in \text{A-W} \lim f_h = [-1,1] = [f_\ell, f_u]$$

Finally we notice that F-convergence does not take place in this example.

We observe that convergence iii) is harder to obtain that i) or ii).

We present now some compactness theorems.

Theorem 5.3

Let (X_1,τ_1), (X_2,τ_2) be second countable topological spaces. Given any sequence $f_h: X_1 \times X_2 \to \tilde{R}$ there is a subsequence that converges both in the sense i) and ii) of definition 5.1.

For a proof see [3], [1].

Theorem 5.4

Let (X_1,τ_1) be a strongly Lindelöf space and (X_2,τ_2) be a second countable space, then convergence i) is sequentially compact.

The proof depends on a recent result of R. Peirone.

As the previous example shows, compactness theorems for F-convergence are harder to prove. Here we present an elegant result on finite dimensional spaces.

Theorem 5.5

Let (X_1,τ_1) and (X_2,τ_2) be finite dimensional vector spaces. For any sequence f_h of Γ-closed quasi concave-convex functions we can pick out a subsequence that F-converges.

This theorem is due to T. Franzoni ([10]) and uses new saddle point criteria of G.H. Greco.

Let X be a topological vector space. We say that $\varphi: X \to \tilde{R}$ is superlinear when φ is proper and its conjugate φ^* is everywhere finite.

Theorem 5.6

Let X_1 and X_2 be reflexive, separable Banach spaces and f_h, φ, $\psi : X_1 \times X_2 \to \tilde{R}$ be concave-convex functions everywhere finite such that:

i) boundedness

$$\varphi \leq f_h \leq \psi$$

ii) superlinearity

$$\exists x^o = (x_1^o, x_2^o) \quad \text{such that}$$

$$\varphi(x_1^o, \cdot) \ , \ \psi(\cdot, x_2^o)$$

are superlinear.

For each sequence f_h satisfying the previous conditions we can pick out a sub-sequence convergent in the sense i), ii), iii) with respect to ω-X_1 and ω-X_2. The three convergences are equivalent and there is only one continuous concave-convex function in the limit class.

The theorem 5.6 was proved in [4] .

Before introducing our next result, we need a new definition.

Definition 5.7: We say that $f_h: X_1 \times X_2 \to \tilde{R}$ S-converges to φ if:

1) f_ℓ and f_u exist

2) $\Gamma_\ell f_u \leq \varphi \leq \Gamma_u f_\ell$.

Remark We remember that ([4]) f_ℓ and f_u are concave-convex functions when f_h are. When S-limit exists we put S-lim $f_h = [\Gamma_\ell f_u, \Gamma_u f_\ell]$. All functions in the S-limit class are Γ-equivalent, but not necessarily concave-convex (nor quasi concave-convex) also in the case in which f_h are.

We present now a theorem with practical applications.

Theorem 5.8

Let $(X_1, \| \|_1)$ and $(X_2, \| \|_2)$ be separable, reflexive, Banach spaces. Suppose that there exist two norms: $||| \; |||_1$ on X_1 and $||| \; |||_2$ on X_2 such that each closed bounded convex subset of X_i is $||| \; |||_i$, $i = 1,2$, compact. Let us consider a class $F(\psi_1, \psi_2)$ of functions such that:

1) each $f \in F$ is concave-convex

2) f is $\Gamma_\ell(||| \; |||_2)$, $\Gamma_u(||| \; |||_1)$ closed

3) $\exists\, x^\circ = (x_1^\circ, x_2^\circ)$

$\psi_1 : X_1 \to \tilde{R}$, concave, proper, $\| \|_1$ u.s.c., superlinear

$\psi_2 : X_2 \to \tilde{R}$, convex, proper, $\| \|_2$ ℓ.s.c., superlinear

such that

$$f(x_1^\circ, x_2) \geq \psi_2(x_2) \qquad \forall f \in F,\ \forall x_2 \in X_2$$

$$f(x_1, x_2^\circ) \leq \psi_1(x_1) \qquad \forall f \in F,\ \forall x_1 \in X_1$$

then for any sequence $f_h \in F$ we can select a subsequence f_{h_r} which $S(||| \; |||_1, ||| \; |||_2)$-converges to an equivalence class of Γ-closed functions; the largest and the lowest functions of the class are always convex-concave.

Remark The proof uses the abstract theorem 5.3 and the Hahn-Banach separation theorem for the stability ([5]). No result of this kind is known to the author in absence of the superlinearity condition or for quasi concave-convex functions on infinite dimensional spaces.

7. AN EXAMPLE

Let us consider the sequence of quadratic functionals

$$f_h(w) = \frac{1}{2} \int_0^1 \sum_{1\,ij}^{n} a_{ij}^h(t)\, \dot{w}_i(t)\dot{w}_j(t)dt \qquad h \in N \tag{6.1}$$

where $w = (w_1,\dots,w_n) \in [H_o^1(]0,1[)]^n$ and the coefficients of the quadratic forms are functions $a_{ij}^h \in L^\infty([0,1])$; $i,j=1,\dots,n$, $h \in N$, which verify the boundedness condition $|a_{ij}^h(t)| \le M$, $M > 0$.

Consider the matrices

$$A_h = \begin{pmatrix} A_h^1 & B^h \\ {}^T B_h & A_h^2 \end{pmatrix} \qquad h \in N$$

where A_h^1 is a symmetric positive defined $m \times m$ matrix and A_h^2 is a symmetric negative defined matrix $p \times p$, B_h is $m \times p$ and ${}^T B_h$ its transpose.

Put

$$v = (w_1,\dots,w_m) \qquad u = (w_{m+1},\dots,w_{m+p})$$

$$F_h(w) = F_h(v,u)$$

and let the matrices A_h^1, A_h^2 verify:

$$\exists \lambda_1,\Lambda_1 > 0 : -\lambda_1 |\xi^1|^2 \ge \sum_{1\,ij}^{m} a_{ij}^h(t)\xi_i^1\, \xi_j^1 \ge -\Lambda_1 |\xi^1|^2$$

$$\exists \lambda_2,\Lambda_2 > 0 : \lambda_2 |\xi^2|^2 \le \sum_{m+1\,ij}^{m+p} a_{ij}^h(t)\xi_i^2\, \xi_j^2 \le \Lambda_2 |\xi^2|^2 \tag{6.2}$$

$$(\xi^1,\xi^2) = (\xi_1^1,\dots,\xi_m^1,\xi_{m+1}^2,\dots,\xi_{m+p}^2) .$$

From (6.2) we deduce the superlinearity conditions

$$F_h(v,0) \le -\frac{1}{2}\lambda_1 c_1 \|v\|^2_{(H_1^o)^m}$$

$$F_h(0,u) \ge \frac{1}{2}\lambda_2 c_2 \|u\|^2_{(H_1^o)^p} \tag{6.3}$$

$$c_1,\ c_2 > 0$$

and the boundedness conditions

(6.4)
$$F_h(v,u) \le \sup_v F_h(v,u) \le \alpha_1(\sum_1^p \|\dot u_i\|_1)^2 + \alpha_2 \|u\|_{(H_o^1)^p} < +\infty$$
$$F_h(v,u) \ge \inf_u F_h(v,u) \ge \alpha_3(\sum_1^m \|\dot v_j\|_1)^2 + \alpha_4 \|v\|_{(H_o^1)^m} > -\infty$$

for suitable constant $\alpha_1, \alpha_2 > 0$; $\alpha_3, \alpha_4 < 0$.

We can finally apply theorem 5.6 to obtain:

Theorem 6.1

Let F_h be as in (6.1) and (6.2) be verified. Consider the topologies $w-(H_o^1)^m$, $w-(H_1^o)^p$ on $(H_1^o)^m$ and $(H_1^o)^p$ and the following convergences:

1) $F_h \to F_o$ in the sense i) of 5.1

2) $F_o = A\text{-}W \lim F_h$

3) $F_o = F\text{-}\lim F_h$

4) $F_o = S\text{-}\lim F_h$

The convergences 1, 2, 3, 4 are equivalent and furthermore each of them is equivalent to:

5) $A_h^{-1} \xrightarrow{w^*-L^\infty} A_o^{-1}$

where A_o is the matrix of the quadratic problem F_o.

For a proof see [5] . Multidimensional problems do not admit characterization 5.

8 CONVERGENCE OF NASH EQUILIBRIA

Now we introduce, directly by means of Γ-limits, N- convergence of games. Successively we give a sequential characterization of it on first countable spaces and we show that every limit point of a sequence of Nash equilibria is a Nash equilibrium for the limit game. The introduced convergence will be stable under addition of continuous perturbations.

Let (X_1,τ_1), (X_2,τ_2) be topological spaces and $f_h: X_1 \times X_2 \to \tilde{R}$ a sequence of functions. We put

$$f_o = \Gamma(N,X_1,X_2^-) \lim f_h$$

when

$$\Gamma(N^-,X_1^-,X_2^-) \lim f_h = \Gamma(N^+,X_1^+,X_2^-) \lim f_h = f_o.$$

We use similar notation with X_1 and X_2 interchanged.

Definition 7.1: Let (X,J^h) be a sequence of games on $(X_1,\tau_1)(X_2,\tau_2)$. We say that $J^h \xrightarrow{N} J^o$ (J^h N-converges to J_o) when:

$$\text{(7.1)} \quad \begin{aligned} \Gamma(N,X_1,X_2^-) \lim J_2^h &= J_2^o \\ \Gamma(N,X_2,X_1^-) \lim J_1^h &= J_1^o \end{aligned}$$

Theorem 7.2

Let (X_1,τ_1) and (X_2,τ_2) be first countable topological spaces; we have $J^h \xrightarrow{N} J^o$ if and only if the following inequalities are true:

(7.2)

1) $\forall (x_1^o,x_2^o) \in X_1 \times X_2,\ \forall u_1^h \to u_1^o,\ \forall u_2^h \to u_2^o$

$$\underline{\lim_h}\ J_i^h(u_1^h,u_2^h) \geq J_i^o(u_1^o,u_2^o) \qquad i=1,2$$

2) $\forall (x_1^o,x_2^o) \in X_1 \times X_2,\ \forall u_2^h \to u_2^o\ \ \exists u_1^h \to u_2^o$:

$$\overline{\lim_h}\ J_1^h(u_1^h,u_2^h) \leq J_1^o(u_1^o,u_2^o)$$

3) $\forall (x_1^o,x_2^o) \in X_1 \times X_2,\ \forall u_1^h \to u_1^o\ \ \exists u_2^h \to u_2^o$:

$$\overline{\lim_h}\ J_2^h(u_1^h,u_2^h) \leq J_2^o(u_1^o,u_2^o)\ .$$

Remark Some properties given for Γ- limits are true for N-convergence(see [6]). Analogous definition can be considered for n-person games ([6]).

We have the following compactness result:

Theorem 7.3
Let (X_1,τ_1) and (X_2,τ_2) be second countable spaces and $J^h : X_1 \times X_2 \to [-M,M] \times [-M,M]$, $M > 0$, a sequence of games satisfying the following equicontinuity conditions:

(7.3)
$$\forall (x_1^o,x_2^o) \in X_1 \times X_2,\ \forall \varepsilon > 0\ \ \exists A_1 \in \mathcal{U}_1(x_1^o),\ A_2 \in \mathcal{U}_2(x_2^o):$$
$$\forall x_1 \in A_1,\ \forall h \in N,\ x_2 \in A_2 \Rightarrow |J_1^h(x_1,x_2) - J_1^h(x_1,x_2)| < \varepsilon$$
$$\forall (x_1^o,x_2^o) \in X_1 \times X_2,\ \forall \varepsilon > 0\ \ \exists B_1 \in \mathcal{U}_1(x_1^o),\ B_2 \in \mathcal{U}_2(x_2^o):$$
$$\forall x_2 \in B_2,\ \forall h \in N,\ x_1 \in B_1 \Rightarrow |J_2^h(x_1,x_2) - J_2^h(x_1^o,x_2^o)| < \varepsilon$$

then there exists a N-convergent subsequence.

Remark This theorem is proved in [6] and remains true for unbounded games modifying condition (7.3).

Definition 7.4: Let (X,J^h) be a sequence of games. A sequence x^h will be said asymptotically Nash for J^h if:

(7.4)
$$\lim_h \varphi_h^{\#}(u^h) = 0$$

where

$$\varphi_h^{\#}(x) = \sup \varphi_h(x,y) =$$

$$= \sup_{y_1,y_2} [J_1^h(x_1,x_2)-J_1^h(y_1,x_2)+J_2^h(x_1,x_2)-J_2^h(x_1,y_2)].$$

When x^h is a Nash equilibrium for J^h, x^h is asymptotically Nash.

Theorem 7.5 (Convergence of Nash equilibria and values).
Let X_1 and X_2 be first countable and the sequence of games (X,J^h) verifies:

$$1)\quad J^h \xrightarrow{N} J^\circ$$

$$2)\quad x^h \text{ is asymptotically Nash for } J^h$$

$$3)\quad x^h \to x^\circ$$

then we have:

$$i)\quad x^\circ \text{ is a Nash equilibrium for } J^\circ$$

$$ii)\quad \lim_h J^h(x^h) = J^\circ(x^\circ)\ .$$

Proof: We may suppose J^h bounded. We want to show that:

$$J_1^\circ(x_1^\circ,x_2^\circ) \le J_1^\circ(x_1,x_2^\circ) \qquad \forall x_1 \in X_1$$

$$J_2^\circ(x_1^\circ,x_2^\circ) \le J_2^\circ(x_1^\circ,x_2) \qquad \forall x_2 \in X_2$$

Consider two points (x_1°,y_2) and (y_1,x_2°) and take two sequences $y_2^h \to y_2$, $y_1^h \to y_1$ verifying the inequalities 2 and 3 of theorem 7.2, that is:

$$\overline{\lim_h}\ J_1^h(y_1^h,x_2^h) \le J_1^\circ(y_1,x_2^\circ)$$

$$\overline{\lim_h}\ J_2^h(x_1^h,y_2^h) \le J_2^\circ(x_1^\circ,y_2).$$

We obtain

$$J_1^\circ(x_1^\circ,x_2^\circ)+J_2^\circ(x_1^\circ,x_2^\circ) \le \underline{\lim_h}\ J_1^h(x_1^h,x_2^h)+ \underline{\lim_h}\ J_2^h(x_1^h,x_2^h) \le \underline{\lim_h}\ [J_1^h(x_1^h,x_2^h) + $$
$$+ J_2^h(x_1^h,x_2^h)] \le \underline{\lim_h}\ [\inf J_1^h(x_1,x_2^h) + \inf_{x_2} J_2^h(x_1^h,x_2)+\varepsilon_h] \le \underline{\lim_h}[J_1^h(y_1^h,x_2^h) + $$
$$+J_2^h(x_1^h,y_2^h)+\varepsilon_h] \le \overline{\lim_h}\ J_1^h(y_1^h,x_2^h) + \overline{\lim_h}\ J_2^h(x_1^h,y_2^h) \le J_1^\circ(y_1,x_2^\circ)+J_2^\circ(x_1^\circ,y_2)\ .$$

The required result follows, assuming alternatively $y_1 = x_1^o$ and $y_2 = x_2^o$. For the convergence of values, we prove only

$$\lim_h J_1^h(x^h) = J_1^o(x^o)$$

because the second equality follows in the same way.

From the first of (7.2), we have:

$$\underline{\lim}_h J_1^h(x^h) \geq J_1^o(x^o)$$

From the second of (7.2), choosen $x_2^h \to x_2^o$ we find $y_1^h \to x_1^o$ such that

$$\overline{\lim}_h J_1^h(y_1^h, x_2^h) \leq J_1^o(x_1^o, x_2^o) \ ;$$

since x^h is a Nash equilibrium for J^h, it follows:

$$\overline{\lim}_h J_1^h(x_1^h, x_2^h) \leq \overline{\lim}_h J_1^h(y_1^h, x_2^h)$$

and the convergence of values is proved. #

Remark The previous convergence is not verified in problems with integral costs depending on all strategy derivatives ([6]).

Open problems

1) Is the theorem 5.8 true for arbitrary concave-convex proper functions? How to characterize, in this case, S-convergence by means of conjugates functions?
2) How to define a convergence for which the conclusions i) and ii) of theorem 7.5 become true also in the case of quadratic integral costs depending on all derivatives of the strategies?

REFERENCES

[1] H.Attouch, R.J.Wets. 'A convergence theory for saddle functions'. *Trans. Am. Math. Soc.* 280, 1(1983), 1-41.

[2] G.Buttazzo. 'Su una definizione generale dei Γ-limiti'. *Bull. Un. Mat. Ital.* 5, 14-B (1977), 213-220.

[3] E.Cavazzuti. 'Multiplie Γ-convergence and applications to convergence of saddle points and max-min points'. Proceed. Meet. on"Recent Methods in Non-linear Analysis and Applications". *S.A.F.A.IV*, Napoli, March 21-28 (1980), 333-340.

[4] E.Cavazzuti. 'Alcune caratterizzazioni della Γ-convergenza multipla'. *Ann. Mat. Pura Appl.* (1982) (IV) vol.XXXII, 69-112.

[5] E.Cavazzuti, N.Pacchiarotti. 'Alcune convergenze su spazi di funzioni concavo-convesse: caratterizzazioni duali ed applicazioni' . *Bull. Un. Mat. Ital.* Analisi funzionale ed applicazioni. Serie VI, vol. II c. n.1, (1983).

[6] E.Cavazzuti, N.Pacchiarotti. 'Convergence of Nash equilibria'.(To appear).

[7] E.De Giorgi, T.Franzoni. 'Su un tipo di convergenza variazionale'. *Atti Acc. Naz. Lincei*, Rend.(8) 58 (1975), 842-850.

[8] E.De Giorgi. 'Γ-convergenza e G-convergenza'. *Bull. Un. Mat.Ital.* (5), 14-A (1977), 213-220.

[9] E.De Giorgi, T.Franzoni. 'Una presentazione sintetica dei limiti generalizzati'. Preprint Pisa (1982).

[10] T.Franzoni. 'This lecture notes'.

[11] T.Franzoni, S.Francaviglia. (To appear).

[12] G.H.Greco. 'Decomposizioni di semifiltri e Γ-limiti sequenziali in reticoli completamente distributivi'. *Ann. Mat. Pura e Appl.*(1984).

[13] R.T.Rockafellar. 'Convex analysis'. Princeton University Press,New Jersey (1970).

[14] C.Sbordone .'Regularization for functions of two variables'.Colloque Optimization. Puy de Dome (1981).

Chapter 5

On Pareto's Problems

J. P. Cecconi

My intent is to report some results concerning minimization of vector functions with respect to convex cones and additive semigroups containing zero, considering both the charcterization of minimal points, some sufficient conditions for the existence of minimal points, and the possibility of reducing the research of these points to the minimization of scalar functions.

In the last decade these problems have been the subject of various researches; I shall only mention works by Yu [12], Geoffrion[10],Benson [1,2], Borwein [3,4], Hartley [8], Henig [9], Cosari-Suryanarayana [5], Corley [7] and recently Chew's work [6] to which this report is widely inspired.

1. INTRODUCTION

1. The following situation is assumed:
Y is a topological vector space on where a transitive and reflexive "≤" binary relation (called preorder) is assigned.

This occurs, for example, by fixing a subset C of Y with the following properties:

c_1) $0 \in C$

c_2) $x, y \in C \Rightarrow x + y \in C$

when $(x \leq y) \Leftrightarrow (y - x \in C)$.

Given a subset B of Y, $B \neq \emptyset$ let's consider the set of points of B which are minimal elements of B with respect to the preorder induced on Y by C.
This set, which is indicated as follows

$$g - \min_C B$$

is also called generalized Pareto set of B with respect to preorder C.
The first topic of our study is therefore the set

$$g - \min_C B$$

As it is well known, $y \in B$ is minimal in the preorder relation introduced if, and only if

$$\nexists\, z \in B\ ,\ z \neq y \quad \text{such that} \quad z \leq y\ ,\ y \nleq z\ .$$

A suitable reformulation of the above definition can be given by introducing the set

$$C^\circ = \{C \cap (-C)\}^c \cap C\ \{0\} \subset C$$

(where for every $A \subset Y$ we have indicated $A^c = Y \setminus A$) and by verifying that

$$g - \min_C B = \{y \in Y:\ B \cap (y - C^\circ)\} = \{y\} \tag{1}$$

and also that

$$g - \min_C B = \{y \in B:\ (\forall z \in B,\ z \neq y,\ z \leq y) \Rightarrow y \leq z\}$$

Note that C° is also an additive sub-semigroup of Y with a null element (just like C) as it verifies c_1) and c_2).
Another topic of our study is the set

$$\min_C B$$

given by $y \in B$ for which

$$\nexists\ z \in B \ , \quad z \neq y \quad \text{such that} \quad z \leq y \ .$$

This set is called Pareto set of points of B and is suitably characterized in the following way

$$\text{(2)} \qquad \min_C B = \{y \in Y : B \cap (y - C)\} = \{y\}$$

From definitions (1) and (2) it immediately follows that

$$\min_C B \subset g - \min_C B$$

The following example shows that this inclusion is proper

$$C = \{(x,y) \in \mathbb{R}^2 : y \geq 0\} \ , \quad Y = \mathbb{R}^2 \quad , \quad B = \{O \equiv (0,0), \ A \equiv (1,0)\}$$

Consequently

$$C^{\circ} = \{(x,y) \in \mathbb{R}^2 \ : y > 0\} \cup \{O\}$$

$$\min_C B = \emptyset$$

$$g - \min_C B = \{O,A\}$$

in fact

$$B \cap (O - C) = \{O,A\} \ , \quad B \cap (A - C) = \{O,A\}$$

$$B \cap (O - C^{\circ}) = \{O\}, \quad B \cap (A - C^{\circ}) = \{A\} \ .$$

2. In the study we shall undertake C can be a cone of Y (as well as an additive sub-semigroup with a null element). In this case the vector structure of Y shall be compatible with the algebraic structure, in the sense that $\lambda \in \mathbb{R}_+$, $x \in C$ imply $\lambda x \in C$. In this case for the property c_2), C will be a *convex cone* of Y.

Sometimes, besides having the above-mentioned property of a convex cone C will also have the "pointedness" property, i.e.

$$C \cap (-C) = \{0\}$$

When this happens we have:

1) The relation $(x \leq y) \Leftrightarrow (y - x \in C)$ will also be anti-symmetric, i.e. such that $(x \leq y,\ y \leq x) \Rightarrow (x = y)$ and therefore C induces in this case in Y a partial order.

2) $C^{\circ} = C$ and therefore

$$g - \min_C B = \min_C B$$

3. Going back to the general case, it should be observed that, for example

1) If Y is a Banach space, C is a convex cone such that $C^{\circ} \neq \{0\}$ then

$$g - \min_C B \subset \partial B$$

If $y \in g - \min_C B$ and if ad absurdum $y \notin \partial B$ then $y \in \text{int}(B)$ and therefore there is $B_Y(0,\delta)$ such that $y - B_Y(0,\delta) \subset B$. Let's assume $0 \neq c \in C^{\circ}$ and let $\lambda \in \mathbb{R}_+$ be such that $\lambda c \in B_Y(0,\delta)$.

In this way $y - \lambda c \in B$.

But C° is a cone, therefore $\lambda c \in C^{\circ}$ so that

$$y \neq y - \lambda c \in B \cap (y - C^{\circ})$$

and this is in contrast with the hypothesis $y \in g - \min_C B$.

2) In the same hypothesis for Y and C

$$g - \min_C B = g - \min_{C^{\circ}} B = g - \min_C (B + C)$$

3) If Y is a Banach space and if C is a pointed convex cone

$$\min_C B = \min_C (B + C)$$

Proofs of 2) and 3) are left to the reader.

4. The following definitions are introduced to study the Pareto sets defined above.

For every set $E \subset Y$; Y is a Banach space; indicating with Y^* the dual space of Y and with $\langle \lambda, y\rangle$ the value of $\lambda \in Y^*$ in $y \in Y$, let's assume

1) $$E^p \equiv E^p_- = \{\lambda \in Y^* : \langle \lambda, y \rangle \leq 0 \quad , \quad \forall y \in E\}$$

Evidently E^p is a closed convex cone of Y^* which is called polar cone of E.

2) $$E^- = \{\lambda \in Y^* \ : \ \langle \lambda, y \rangle < 0 \quad , \quad \forall 0 \neq y \in E\}$$

Evidently $\{0\} \cup E^-$ is a convex cone of Y^* . It is called strict polar cone of E.

3) $$E^\wedge = \{\lambda \in Y^* - \{0\} \ : \ \exists \delta \in \mathbb{R}_+ \ \langle \lambda, y \rangle \leq -\delta \|\lambda\| \ \|y\| \ , \ \forall y \in E\}$$

4) $$E^\Delta = \{\lambda \in Y^* - \{0\} \ : \ \{y \in E : \langle \lambda, y\rangle < -\delta\} \text{ is weakly compact } \forall \delta \in \mathbb{R}_+\}$$

Evidently $\{0\} \cup E^\wedge$ and $\{0\} \cup E^\Delta$ are also convex cones of Y^* .

We can state that if C *is a convex cone* then we have

(3) $$C^\Delta \subset C^\wedge \begin{array}{l} \nearrow \subset \operatorname{int}(C^p) \subset C^p \\ \searrow \subset C^- \subset C^p \end{array}$$

if, moreover, Y is reflexive

(4) $$C^{\Delta} = C^{\wedge} .$$

For the first inclusion in (3), we refer to Chew [6] . The inclusion $C^{\wedge} \subset \operatorname{int}(C^{p})$ can be proved in the following way.
Let $\lambda \in C^{\wedge}$; then $\lambda \neq 0$ and there is $\delta \in \mathbb{R}_{+}$ such that

$$< \lambda , a > \leq -\delta \|\lambda\| \|a\| , \forall a \in C$$

so that for every $\mu \in Y^{*}$ for which

$$\|\mu - \lambda\| < \delta \|\lambda\|$$

we have

$$< \mu , a > - < \lambda , a > \leq \|\mu - \lambda\| \|a\|$$

hence it follows that

$$< \mu , a > \leq < \lambda , a > + \delta \|\lambda\| \|a\|$$

and for the way in which δ is given it follows that

$$< \mu , a > \leq 0 \quad \text{for every } a \in C ;$$

this proves that $\mu \in C^{p}$ and consequently $\lambda \in \operatorname{int}(C^{p})$ and therefore

$$C^{\wedge} \subset \operatorname{int}(C^{p})$$

At this point the remaining inclusions in 3) are obvious.

For a proof of (4) we refer to the above-mentioned work by Chew.

5. Let's also introduce the following notations. If C is a *convex cone* in Banach space Y, let's say that:

1) C is *acute* if $(\operatorname{cl} C)^{-} \neq \emptyset$

where for every $A \subset Y$; with cl (A) we indicated the closure of A in Y

2) C is *strictly acute* if $C^{\wedge} \neq \emptyset$

3) C has the (C-S) *property* if $C^{\Delta} \neq \emptyset$

It should also be reminded that C is *pointed* if $C \cap (-C) = \{0\}$.

Let us now compare the properties introduced above for a *convex cone* C of a Banach space Y.

We have the following situation

(5) (prop.C-S) $\Rightarrow$(strict acuteness) $\Rightarrow$(acuteness)$\Rightarrow$(pointedness).

Moving on to proof (5), note that the first inclusion follows from (3). Let us therefore prove that if Y is a reflexive Banach space and C is a convex cone, then condition

$$C^{\wedge} \neq \emptyset \quad \text{implies} \quad (\text{cl } C)^{-} \neq \emptyset \; .$$

Let thus $\lambda \in C^{\wedge}$, then there is $\delta \in \mathbb{R}_+$ such that

$$\langle \lambda, y \rangle \le -\delta \|\lambda\| \, \|y\| \quad \text{for every} \quad y \in C \; .$$

Then let $y \in \text{cl } C$ and $y_n \in C$, $y_n \to y$. Hence

$$\langle \lambda, y_n \rangle \le -\delta \|\lambda\| \, \|y_n\| \quad \text{for every} \quad n \in N$$

and so

$$\langle \lambda, y \rangle \le -\delta \|\lambda\| \, \|y\| \quad \text{for every } y \in \text{cl } (C)$$

Therefore, for every $y \in \text{cl } (C)$, $y \neq 0$ we have

$$\langle \lambda, y \rangle \le -\delta \|\lambda\| \, \|y\| < 0$$

and this proves that

$$\emptyset \neq C^{\wedge} \subset (\text{cl } C)^{-} \; .$$

Let us then prove the final inclusion in (5) by observing that if $\lambda \in (\mathrm{cl}\ C)^-$ then for every $y \neq 0$, $y \in \mathrm{cl}\ (C)$

$$<\lambda, y> < 0$$

and therefore if $0 \neq y \in (\mathrm{cl}\ C) \cap (-\mathrm{cl}\ C)$ should exists we would then have a contradiction.

Consequently

$$C \cap (-C) \subset (\mathrm{cl}\ C) \cap (-\mathrm{cl}\ C) = \{0\}$$

(5) is thus fully proved.

At this point let us remember the notion of a bipolar of a convex cone C in a Banach space Y.

$$(C^p)_p = \{y \in Y: <\lambda, y> \leq 0 \quad , \quad \forall \lambda \in C^p\}$$

and let us state the (Bipolarity) Theorem. If C is a *closed convex cone* of a Banach space Y, then

$$C = (C^p)_p$$

Let us now prove that if to the hypothesis in which C is a convex cone of a reflexive space Y we add the hypothesis that C is closed non trivial, i.e. $C \neq \{0\}$ then we have

(6) 1) $C^\wedge = \mathrm{int}\ (C^p)$

2) $\mathrm{int}\ (C^p) \neq \emptyset$ implies $\mathrm{int}\ (C^p) = C^-$

As for 1), it suffices to prove that

$$\mathrm{int}\ (C^p) \subset C^\wedge$$

In order to prove it we observe first that if $\lambda \in \mathrm{int}\ (C^p)$ then $\lambda \neq 0$.

In fact, if $0 \in \text{int}\,(C^p)$ there exists $r \in \mathbb{R}_+$ such that $B_{Y^*}(0,z) \subset C^p$ and this implies $C^p = Y^*$.

From this bipolarity theorem and the closedness of C it follows that

$$C = (C^p)^p = \{y \in Y : \langle \lambda, y \rangle \leq 0 \quad \text{for every} \quad \lambda \in C^p = Y^*\} = \{0\}$$

which is in contrast with the hypothesis of non triviality of C.

Let $\lambda \in \text{int}\,(C^p)$ so that $\lambda \neq 0$, then there exists $\rho \in \mathbb{R}_+$ such that if $\mu \in Y^*$, $\|\mu\| = 1$ we have

$$\lambda + \rho\mu \in C^p$$

On the other hand, for every $a \in Y$, $a \neq 0$ it exists $\bar{\mu} \in Y^*$ such that

$$\langle \bar{\mu}, a \rangle = \|a\| \quad , \quad \|\bar{\mu}\| = 1 .$$

Hence, for every $a \in C$, $a \neq 0$

$$0 \geq \langle \lambda + \rho\bar{\mu}, a \rangle = \langle \lambda, a \rangle + \rho \langle \bar{\mu}, a \rangle = \langle \lambda, a \rangle + \rho \|a\| .$$

This ia also valid for a = 0. Hence

$$\langle \lambda, a \rangle \leq -\rho\|a\| = -\delta \|\lambda\| \|a\| \text{ for every } a \in C$$

if $\delta = \frac{\rho}{\|\lambda\|}$; this proves that $\lambda \in C^\wedge$.

It is thus shown that

$$\text{int}\,(C^p) \subset C^\wedge$$

As for 2), we already know that $\text{int}\,(C^p) \subset C^-$.

If $\bar{\lambda} \in C^- \setminus \text{int}\,(C^p) \subset C^p \setminus \text{int}\,(C^p)$ should exist then $\bar{\lambda} \in \partial(C^p)$ and according to separation theorems since $\text{int}\,(C^p) \neq \emptyset$, C^p being convex, there is a support hyperplane H for C in $\bar{\lambda}$ such that C^p is in only one part of H. According to the reflexiveness of Y, there is $0 \neq \bar{y} \in Y$ such that

$$\sup\{<\lambda,\bar{y}> : \lambda \in C^p\} = <\bar{\lambda},\bar{y}>$$

But, if $\lambda_o \in C^p$ then $k\lambda_o \in C^p$ for every $k \geq 0$ and therefore

$$(7) \qquad 0 \leq <\bar{\lambda},\bar{y}>$$

furthermore, for the same reason we cannot have $<\lambda_o,\bar{y}> > 0$ for $\lambda_o \in C^p$ and therefore the following is obtained

$$<\lambda,\bar{y}> \leq 0 \quad \text{for every} \quad \lambda \in C^p$$

This implies $0 \neq \bar{y} \in (C^p)_p \subset C$ and hence, since $\bar{\lambda} \in C^-$ then

$$<\bar{\lambda},\bar{y}> < 0$$

which is in contrast with (7). The statement is thus proved.

7. Note that, if $\dim Y < +\infty$ and if C is a *closed convex cone* then (5) according to what was proved by Yu [12] becomes

(8) (prop. C-S) $\Leftrightarrow$ (strict acuteness) $\Leftrightarrow$ (acuteness) $\Leftrightarrow$ (pointedness)

let's also observe that in the same hypothesis the condition: C is a convex cone is sufficient in order that

(9) (acuteness) $\Rightarrow$ int $(C^p) \neq \emptyset$

(acuteness) $\Rightarrow$ int $(C^p) = C^-$

8. Going back to the hypothesis: Y Banach space, C *closed convex cone*

according to bipolarity theorem n.5 the following characterization can be supplied for the partial order relation:

$$(x \leq y) \Leftrightarrow (<\lambda,x> \geq <\lambda,y>: \quad \forall\lambda \in C^P)$$

9. Let us now give a useful proposition to state the existence of elements of $\min_C B$.

Proposition

Let Y be a topological vector space on $\mathbb{R}$, let C be a convex cone and $\bar{\lambda} \in C^-$. If $B \subset Y$ is such that $y \in B$ exists, for which

$$<\bar{\lambda},y> \geq <\bar{\lambda},z> \quad \text{for every } z \in B$$

then $y \in \min_C B$.

Proof: Let $z \in B$, $z \neq y$ such tha $z \leq y$.

Then $y - z = c \in C$ with $0 \neq c$.

However, for the fact that $\bar{\lambda} \in C^-$ we have

$$<\bar{\lambda},c> < 0$$

and this is in contrast with the hypothesis. #

2. EXISTENCE OF PARETO POINTS

1. We can now study the problem concerning the existence of Pareto points and of generalized Pareto points in a set $B \subset Y$.

The first result in this direction is the following:

Theorem 1 (Yu [12])

Let Y be a topological vector space on $\mathbb{R}$, $\dim Y < \infty$. Let C be an *acute convex cone* of Y. Let $\emptyset \neq B \subset Y$ a compact part of Y. Then

$$g - \min_C B = \min_C B$$

Proof: From the hypotheses on Y and C and for (8) it follows that there is at least one $\bar{\lambda} \in C^-$.

When considering the continuous function

$$z \in B \to \langle \bar{\lambda}, z \rangle \in \mathbb{R}$$

it can be concluded that it is provided with a maximum by virtue of the compactness of B. Therefore $y \in B$ exists such that

$$\langle \bar{\lambda}, y \rangle \geq \langle \bar{\lambda}, z \rangle \quad , \quad \forall z \in B \; .$$

From proposition n.9 we have that

$$\min_C B \neq \emptyset$$

and from hypothesis C acute, in accordance with (8), it follows that

$$\min_C B = g - \min_C B \; . \quad \#$$

2. *Remark.* The same conclusion is reached if the B compactness hypothesis is replaced by the following: there is $y \in Y$ such that

$$B \cap (y - C) \quad \text{is compact and non empty}$$

A set of the type $B \cap (y - C)$ is called C-section of B.

3. *Remark.* The conclusion $0 \neq \min_C B$ is reached, with the same proof, if Y is a reflexive Banach space having any dimension provided C is a non trivial *closed convex cone* for which int $(C^p) \neq \emptyset$ and B admits a compact and non empty C-section; also these hypotheses imply $C^- \neq \emptyset$.

4. Other theorems on the existence of Pareto points are stated hereafter.

Theorem 2 (Hartley [8])

Let Y be a topological vector space of $\mathbb{R}$, $\dim Y < +\infty$. Let C be a *convex cone* of Y. Let $B \subset Y$ such that

$$\emptyset \neq B \cap (y - C) \text{ is a compact section.}$$

Then

$$g - \min_C B \neq \emptyset$$

Theorem 3 (Borwein [4])

Let Y be a topological vector space on $\mathbb{R}$, let C be a closed convex cone of Y and let $B \subset Y$ be such that

$$\emptyset \neq B \cap (y - C) \text{ is a compact section.}$$

Then

$$g - \min_C B \neq \emptyset$$

Theorem 4 (Corley [6])

Let Y be a topological vector space on $\mathbb{R}$, let C be a *pointed closed concex cone*. Let $\emptyset \neq B \subset Y$ be semicompact in the sense that: if

$$B \subset \bigcup_{\alpha \in A} (y_\alpha - C)^c$$

then there are $\alpha_1, \alpha_2, \ldots, \alpha_p \in A$ such that

$$B \subset \bigcup_{i=1}^{p} (y_{\alpha_i} - C)^c$$

Then

$$\min_C B = g - \min_C B$$

Theorem 5 (Cesari-Saryanarayana [5])

Let Y be a Banach space. Let C be a *closed convex cone* with property (C-S). Let $\emptyset \neq B \subset Y$ be weakly closed and lower bounded.

Then

$$\min_C B = g - \min_C B .$$

5. *Remark.* The following theorem by Borwein [4] assigns a further sufficient condition for the existence of Pareto generalized minimum in all those topological spaces Y in which, given a convex C, there exist $\lambda \in Y^*$ which is positive in the sense that:

$$<\lambda,y> \ >0 \quad \text{for every } y \in C \setminus (-C) \quad \text{and } <\lambda,y> = 0$$

if

$$y \in (C) \cap (-C).$$

Theorem

If Y and C are given as above and if $B \subset Y$ admits a

$$\emptyset \neq B \cap (y_o - C) \ \text{compact section}$$

then

$$g - \min_C B \neq \emptyset \quad .$$

When considering the problem

$$\min \{<\lambda,z> \ : \ z \in B \cap (y_o - C)\}$$

there exists $y_1 \in B \cap (y_o - C)$ such that it minimizes

$$z \to <\lambda,z> \quad \text{for} \quad z \in B \cap (y_o - C)$$

Now, if $\bar{y} \in B$ and $\bar{y} \leq y_1$ it follows that $y_1 - \bar{y} \in C$ and since $y_1 \in (y_o - C)$

also $y_o - \bar{y} \in C$.

We therefore have

$$\bar{y} \in B \cap (y_o - C)$$

and for the positivity hypothesis of λ we have $-\lambda \in C^p$ consequently

$$< -\lambda, y_1 - \bar{y} > \leq 0$$

hence

$$< \lambda, \bar{y} > \leq < \lambda, y_1 > \quad \text{and} \quad \bar{y} \in B \cap (y_o - C)$$

Based on the way in which y_1 is chosen, we have

$$< \lambda, \bar{y} > = < \lambda, y_1 >$$

and since $y_1 - y \in C$ due to the positivity of λ we'll have that

$$y_1 - \bar{y} \in -C .$$

In conclusions we therefore have

$$y_1 \leq \bar{y}$$

and this proves that

$$y \in g - \min_C B.$$

In the same work Borwein also observes that such positive functional exists in every locally convex topological vector space Y if the convex cone C has a base; i.e. there is a convex set $\Gamma \subset Y$, such that $0 \notin \mathrm{cl}\,\Gamma$ and

$$C = \{ky: k \geq 0 , y \in \Gamma\}$$

6. *Remark.* With reference to theorem 4 (Carley) it should be observed that more generally for the cone C it can only be assumed:

$$C \text{ is convex}, \quad (\mathrm{cl}\ C) \cap (-\mathrm{cl}\ C) = \{0\}$$

In order to prove it we observe that if C_1, C_2 are *pointed convex* cones such that $C_1 \subset C_2$ then

$$\min_{C_2} B \subset \min_{C_1} B$$

In fact, if

$$y \in \min_{C_2} B \qquad \text{then :} \quad B \cap (y - C_1) = \{y\}$$

and therefore

$$y \in B \cap (y - C_1) \subset B \cap (y - C_2) = \{y\}$$

Then, from the hypothesis on C expressed in the remark it follows that

$$\emptyset \neq \min_{\mathrm{cl}(C)} B \subset \min_C B$$

7. For the proof of Theorem 2 we refer to Hartley's work [8] . We just want to observe that the proof occurs for induction on the dimension of Y.

We refer to the work by Cesari-Suryanarayana [5] for the proof of theorem 3. However, as it has been pointed out by Chew [6] the hypotheses on C and B in this work imply that B is weakly *semicompact* and that C is weakly-closed, pointed, convex. This is sufficient to apply Theorem 4 by Corley.

8. As for the proofs of Theorems 3 and 4, they are based on the fact that, having introduced on Y the partial order (or the partial preorder) induced by C, it has to be demonstrated that B becomes inductively ordered (i.e. every chain $\Gamma = \{y_\alpha : \alpha \in F\} \subset B$ has a lower bound in B).

In accordance with Zorn's lemma we have a minimal element in the order (or in the preorder) induced by C, and as already seen, it becomes an element of $g - \min_C B = \min_C B$ or of $g - \min_C B$.

9. In the hypothesis of theorem 4, for example, this occurs by observing that if a chain $\Gamma = (y_\alpha : \alpha \in F) \subset B$ is not lower bounded in B, it would then be

$$(\star) \qquad \bigcap_{\alpha \in F} B \cap (y_\alpha - C) = \emptyset$$

and that on the other hand this relation leads to a contradiction.

Let us first prove that if

$$(\#) \qquad y \in \bigcap_{\alpha \in F} B \cap (y_\alpha - C)$$

then Γ is lower bounded in B. In fact from (#) we have

$$y \in B \ , \quad y \in y_\alpha - C \quad \text{for every } \alpha \in \Gamma$$

and therefore

$$y \leq y_\alpha \qquad \text{for every} \quad \alpha \in F$$

It is then sufficient to prove that $(\star)$ leads to a contradiction.

Let's assume ad absurdum $(\star)$ to be true.

Then for every $y \in B$, $\bar{\alpha} \in F$ exists such that

$$y \notin (y_{\bar{\alpha}} - C)$$

and therefore

$$y \in (y_{\bar{\alpha}} - C)^c$$

It then follows that

$$B \subset \bigcup_{\alpha \in F} (y_\alpha - C)^c$$

and hence for the hypothesis of semicompactness of B we deduce that there are $\alpha_1, \alpha_2, \ldots, \alpha_p \in \Gamma$ such that

$$(10) \qquad B \subset \bigcup_{i=1}^{p} (y_{\alpha_i} - C)^c$$

But $\{y_{\alpha_1}, y_{\alpha_2},\dots,y_{\alpha_p}\}$ is totally ordered, $\bar{I} \in \{1,2,..,p\}$ exists such that

$$y_{\alpha_{\bar{i}}} \le y_{\alpha_i} \quad ; \quad i = 1, 2, \dots, p ;$$

so that

$$y_{\alpha_{\bar{i}}} - c \in y_{\alpha_i} - C \quad \text{for every} \quad i = 1,2,..,p$$

and for every $c \in C$, consequently

$$(y_{\alpha_{\bar{i}}} - C)^c \supset (y_{\alpha_i} - C)^c \text{ for every } i = 1,2,..,p.$$

From (10) it then follows that

$$B \subset (y_{\alpha_{\bar{i}}} - C)^c$$

and this is a contradiction since $y_{\alpha_{\bar{i}}} \in B$ and $y_{\alpha_{\bar{i}}} \notin (y_{\alpha_{\bar{i}}} - C)^c$.

The proof of Theorem 3 is obtained is a similar way.

3. PROPERTIES OF PARETO POINTS: SUPPORT POINTS, EFFICIENCY

As we observed in the proof of Yu theorem there are some points in Pareto set $\min_C B$ that can be obtained as maximum points of restrictions to B of functionals $\lambda \in C^p$.

This leads us to introduce the notion of lower support of B with respect to C in relation to a part Γ of polar C^p of C in the following way:
if $\Gamma \subset C^p$ let's pose

$$\Gamma - \text{supp}_C\, B = \{y \in B:\ \exists \lambda \in \Gamma \text{ such that } \langle \lambda, z\rangle \le \langle \lambda, y\rangle\ \forall z \in B\}$$

From the proposition in the premise to Yu's existence theorem we have the following:

Theorem 6

If C is a convex cone of a topological spaee on $\mathbb{R}$ then

$$(C^-)\ \mathrm{supp}_C\ B \subset \min_C B$$

Furthermore

Theorem 7 (Hartley [8])

If Y is a topological vector space, if dim Y < ∞, if C is a *convex cone* of Y, different from a subspace and if B is a C-convex in the sense that (B + C) is convex, then

$$g - \min_C B \subset (C^p) - \mathrm{supp}_C\ B$$

Furthermore, if C is a pointed closed cone and if (B + C) is convex, closed, then

$$(C^-)\ \mathrm{supp}_C\ B \subset \min_C B = g - \min_C B \subset \mathrm{cl}\{(C^-)\mathrm{supp}_C\ B\}.$$

We refer to the already mentioned work by Hartley for the proof of this theorem. We limit ourselves to notify also the following result.

In the same hypothesis with regard to Y, if C is a convex cone such that C\{0} is open and if (B + C) is convex, then

$$g - \min_C B = (C^p) - \mathrm{supp}_C\ B$$

2. In order to characterize the set (C^-)-supp_C B, limited to the case in which Y is a Banach space and C a convex cone, and with the intent to analyse the gap between this set and the sets $\min_C$ B, g - $\min_C$ B let us introduce the following sets consisting in *proper Pareto points* with the meaning stated hereafter:

$$p_{loc} - \min_C B = \{y \in \min_C B: \mathrm{cl}\{T(B + C, y)\} \cap (-\mathrm{cl}\ C) = \{0\}\}$$

$$p_{glob} - \min_C B = \{y \in \min_C B : \mathrm{cl}\{P(B + C - y)\} \cap (-\mathrm{cl}\, C) = \{0\}\}$$

$$p_H - \min_C B = \{y \in g - \min_C B : \exists M \in R_+ \text{ such that if } \exists\lambda \in C^p,\ \|\lambda\| = 1,\ x \in B \text{ for which } \langle\lambda, y\rangle < \langle\lambda, x\rangle \text{ then } \exists\mu \in \bar{C}, \|\mu\| = 1, \text{ for which } \langle\lambda, x - y\rangle \le M\langle\mu, y - x\rangle\}$$

where for $E \subset Y$ the following has been set

$$T(E,y) = \{z \in y : \exists y_i \in Y,\ \lambda_i \in \mathbb{R}_+,\ i \in N, \text{ such that } \lambda_i(y_i - y) \to z\}$$

so that $T(E,y)$ is the cone tangent to E in y; and having also set

$$P(E) = \{z \in Y : \exists\lambda \in \mathbb{R}_+,\ y \in E \text{ such that } z = \lambda y\}$$

so that $P(E)$ is the cone projecting E from the origin 0 of Y.

The set $P_H - \min_C B$ was introduced by Geoffrion [10] in the case $Y = \mathbb{R}^n$, $C = \mathbb{R}^n_+$ to restrict Pareto set $\min_C B$ so as to eliminate from it those points $y_0 \in \min_C B$ defined as "improper Pareto points" for which it occurs that irrespective of how large $M \in \mathbb{R}_+$ is set, there is $y \in B$ such that

$$y^{(\bar{i})} \le y_0^{(\bar{i})}$$

for at least one $\bar{i} = 1, 2, \ldots, n$ and

$$\frac{y_0^{(\bar{i})} - y^{(\bar{i})}}{y^{(i)} - y_0^{i}} > M$$

for every $i \ne \bar{i}$ for which $y^{(i)} > y_0^{(i)}$.

The definition, adopted above, $p_H - \min_C B$ is Hartley's [8] and is the direct extension of Geoffrion's idea referred to the general case.

The definition of sets $P_{glob} - \min_C B$ and $P_{loc} - \min_C B$ were introduced by Benson [1],[2] and Borwein [3] respectively, referred to the case $Y = \mathbb{R}^n$, C closed convex cone, to eliminate from $\min_C B$ the elements with the undesirable behaviour described above.

3. In order to clarify the meaning of the sets which have just been introduced, let us consider some examples.

Example 1 (Henig [9])

$Y = \mathbb{R}^1$, $C = \mathbb{R}^2_+$

$$B = \{z \equiv (x,y) \in \mathbb{R}^2 : \|z\| \leq 1,\ y \leq -x\}$$

Then

$$\min_C B = g - \min_C B = \{(\cos\theta,\ \sin\theta)\ ;\ \pi \leq \theta \leq \frac{3}{2}\pi\}$$

On the other hand, having set for every $n \in N$: $z_n = (-1+\frac{1}{n}, \frac{-1}{n^2})$ the following is obtained

$$z_n \in B\ ,\ z_n \to (-1,0) \in \min_C B\ ,\ \frac{x_n+1}{y_n} \to +\infty \quad \text{as } n \to \infty$$

such that (1-,0) is not a proper Pareto point according to Geoffrion, furthermore the following can be easily pointed out

$$(-1,0) \notin (p_{loc} - \min_C B) \cup (p_{glob} - \min_C B) \cup (p_h - \min_C B)$$

Example 2 (Henig [9])

$Y = \mathbb{R}^2$, $C = \mathbb{R}^2_+$

$$A = \{(x,y) \in \mathbb{R}^2 ;\ x \leq 0,\ y \leq 0\} \cap \{x + y \leq -1\}$$

$$B = \mathbb{R}^2 \setminus A\ .$$

Then

$$\min_C B = g - \min_C B = \{(-a, a-1) \in \mathbb{R}^2 : 0 < a < 1\}$$

On the other hand, having set for every $0 < a < 1$ and for every

$$(x_n, y_n) = (-a + \frac{1}{n}, (a - 1) - \frac{1}{n^2})$$

we have

$$\frac{(-a + \frac{1}{n}) - (-a)}{(a - 1) - (a - 1 - \frac{1}{n^2})} \to +\infty \quad \text{as } n \to \infty$$

so that $z_a = (-a, a - 1)$ is not a proper Pareto point, according to Geoffrion, for every $0 < a < 1$.

Furthermore, the following can be easily pointed out

$$\mathrm{cl}\{T(B + C, z_a)\} \cap (-\mathrm{cl}\ C) = \{0\}$$

$$\mathrm{cl}\{P(B + C - z_a)\} \cap (-\mathrm{cl}\ C) =$$

$$= \{(x,y) \in \mathbb{R}^2 : x = -a, y \le a-1\} \cup \{(x,y) \in \mathbb{R}^2 : x \le -a, y = a-1\}$$

such that

$$z_a \in p_{loc} - \min_C B$$

$$z_a \notin (p_{glob} - \min_C B) \cup (p_H - \min_C B)$$

for every $0 < a < 1$.

4. *Remark.* The following should be observed in relation to the sets $T(E,y)$; $y \in E$; $P(E)$ which have just been introduced

1) $T(E,y) \subset \mathrm{cl}\ \{P(E,y)\}$

2) $T(E,y) = \mathrm{cl}\ \{P(E,y)\}$ if E is convex

we refer, for instance, to Varaya [11].

5. Moving on to the comparison between the notions which have just been introduced and limited to the case $Y = \mathbb{R}^n$ we have the situation described in the following

Theorem 8 (Borwein [3], Benson [1,2], Hartley [8], Henig [9], Chew[6]) If C is a closed convex cone we have the following inclusions

1) $(C^-)\ \mathrm{supp}_C\ B \subset p_{glob} - \min_C B \subset p_{loc} - \min_C B$

2) $(C^-)\ \mathrm{supp}_C\ B \subset p_H - \min_C B$ if, furthermore, C is pointed

3) if, furthermore, B is C-convex and C is pointed, then

$(C^-)\ \mathrm{supp}_C\ B = p_H - \min_C B = p_{glob} - \min_C B = p_{loc} - \min_C B.$

Proof: As for the inclusion in 1) let

$$y_o \in (C^-)\ \mathrm{supp}_C\ B \text{ and } \bar{\lambda} \in C^- \text{ be such that}$$

$$<\bar{\lambda}, y> \le <\bar{\lambda}, y_o> \text{ for every } y \in B.$$

Let afterwards $z \in \mathrm{cl}\{P(B + C - y_o)\} \cap (-C)$.

Then for every $n \in N$ there is $z_n \in P(B + C - y_o) \cap (-C)$ such that $z_n \to z$ and also $k_n \in \mathbb{R}_+$, $y_n \in B$, c_n, $c'_n \in C$, $\forall n \in N$ such that

$$z_n = k_n(y_n + c_n - y_o) = -c_n .$$

It then follows that

$$<\bar{\lambda}, y_n> \le <\bar{\lambda}, y_o>$$

and therefore from the fact that

$$k\{<\bar{\lambda}, y_n - y_o> + <\bar{\lambda}, c_n>\} \to <\bar{\lambda}, z>$$

and furthermore

$$k\{<\bar{\lambda}, y_n - y_o> + <\bar{\lambda}, c_n>\} \leq 0$$

for every $n \in N$, hence

$$<\bar{\lambda}, z> \leq 0 .$$

On the other hand, if $z \neq 0$ for the fact that $\bar{\lambda} \in C^-$ we have

$$<\bar{\lambda}, z> < 0$$

It is thus proved that $z = 0$, hence

$$cl\{P(B + C - y_o)\} \cap (-C) = \{0\}$$

and therefore that $y \in p_{glob} \min_C B$.

Let us now prove the inclusion of (2).

Like above let $\bar{\lambda} \in C^-$ and $y_o \in B$ be such that

$$<\bar{\lambda}, y> \leq <\bar{\lambda}, y_o>$$

From (9) of part 1 we have $\bar{\lambda} \in C^\wedge = int\,(C^p)$. There exists $\delta \in \mathbb{R}_+$ such that

$$<\bar{\lambda}, y> \leq -\delta \|\bar{\lambda}\| \cdot \|y\| \quad , \quad \forall y \in C$$

and we can assume $0 < \delta < 1$.

Let

$$d = \{\inf \|y - c\| : y \in S \; , \; c \in C\}$$

where

$$S = \{y \in Y: \|y\| = 1, <\bar{\lambda}, y> \geq 0\}$$

If $d = 0$, then for every $n \in N$ there are $y_n \in S$, $c_n \in C$ such that

$$\|y_n - c_n\| \to 0 \quad \text{for } n \to \infty$$

and therefore

$$<\bar{\lambda}, y_n - c_n> \to 0 \quad \text{for } n \to \infty.$$

For a sufficiently large n we have

$$\delta > \|y_n - c_n\| \geq \|y_n\| - \|c_n\| = 1 - \|c_n\|$$

hence

$$\|c_n\| > 1 - \delta$$

and consequently

$$<\bar{\lambda}, y_n - c_n> \;\geq\; <-\bar{\lambda}, c_n> \;\geq\; \delta\, \|\bar{\lambda}\| \cdot \|c_n\| > \delta(1-\delta)\, \|\bar{\lambda}\|$$

This is in contrast with the fact that $<\bar{\lambda}, y_n - c_n> \to 0$.

It must therefore be $d > 0$ and we can consider

$$M = \frac{\|\bar{\lambda}\| + 1}{d}$$

Let's assume $<\lambda, \bar{y}> \; > \; <\lambda, y_o>$ for at least one $\lambda \in C^p$ and one $\bar{y} \in B$, and let's fix

$$z = (y_o - \bar{y})\, (\|y_o - \bar{y}\|)^{-1}$$

so that $z \in S$. Let $w \in C$ be such that

$$\|w - z\| = \inf\{\|c - z\| : c \in C\}$$

so that

$$\|w - z\| \geq d\,.$$

For every $c \in C$ and for every $\alpha \in [0,1]$

$$\alpha(c - w) + w \in C$$

and consequently

$$\|w - z\|^2 \leq \|\alpha(c - w) + w - z\|^2 = \alpha^2\|c - w\|^2 + 2a(c - w, w - z) + \|w - z\|^2$$

Hence

$$0 \leq \alpha^2 \|c - w\|^2 + 2\alpha(c - w, w - z) \quad \text{for every } c \in C$$

and therefore

$$< c - w, w - z > \ \geq 0 \quad \text{for every } c \in C .$$

In particular, if $c = 0$ we obtain

$$- < w, w - z > \ \geq 0$$

and if $c = 2w$ we obtain

$$< w, w - z > \ \geq 0$$

hence

$$< w, w - z > \ = 0$$

and therefore

$$\|w - z\|^2 = < w - z, w - z > \ = - < z, w - z >$$

i.e.

$$\|w - z\| = - < z, w - z > (\|w - z\|)^{-1}.$$

Let

$$\mu = \bar{\lambda} + (w - z)\ (\|w - z\|)^{-1}$$

then

$$\|\mu\| \leq \|\bar{\lambda}\| + 1$$

and

$$\|\mu\| \cdot <\lambda,\bar{y}> \cdot (\|\lambda\| \; \|\bar{y} - y_o\|)^{-1} \leq (M + 1) = Md \leq M\|w - z\| =$$

$$= M < -z, \; w - z> (\|w - z\|)^{-1} = M < -z, \; \bar{\lambda} - \mu> =$$

$$= M < z,\mu> \;\; \leq M < \mu, \; y_o - \bar{y} > (\|y_o - \bar{y}\|)^{-1} .$$

Hence

$$\|\mu\| < \lambda, \bar{y} - y_o > \;\; \leq M \; \|\lambda\| < \mu, \; y_o - \bar{y} >$$

and since we can assume $\|\bar{\lambda}\| \neq 1, \;\; \|\lambda\| = 1$ it results

$$< \lambda, \bar{y} - y_o > \;\; \leq M < \frac{\mu}{\|\mu\|} \; , \; y_o - \bar{y} > .$$

At this point we only have to prove that μ and therefore $\frac{\mu}{\|\mu\|}$ belong to C^-.

For this purpose let $c \in C$, $c \neq 0$ we then have

$$< \mu, c> \; = \; < \bar{\lambda}, c > - \; < \frac{w - z}{\|w - z\|} \; , \; c > \; =$$

$$< \lambda, c > - \; < \frac{w - z}{\|w - z\|} \; , \; c - w > - < \frac{w - z}{w - z} \; , \; w >$$

$$= \; < \bar{\lambda}, c > - \; < \frac{w - z}{\|w - z\|} \; , \; c - w > \; < 0$$

Let us move on to the proof of relation 3), in Theorem 8.

Based upon well known properties of convex sets, it follows that if (B+C) is convex, then for every $y \in B$

$$cl\{P(B + C - y)\} = cl\{T(B + C, y)\}$$

and therefore

$$p_{glob} - \min_C B = p_{loc} - \min_C B .$$

We already know that

$$(C^-) \text{ supp}_C B \subset p_{loc} - \min_C B$$

On the other hand if $y_o \in p_{loc} - \min_C B$, according to what has been previously pointed out, we have

$$B + C - y_o \subset P(B + C - y_o) = T(B + C, y_o)$$

and this set is convex.

However, since

$$y_o \in p_{loc} - \min_C B$$

then

$$T(B + C, y_o) \cap (-C) = \{0\}$$

and therefore for every $c \in C$, for every $\lambda \in C^-$ and for every $y \in B$ and in consistency with well known separation theorems (see Borwein [3]) we have

$$<\bar{\lambda}, y - y_o + c> \leq 0$$

hence, for $c = 0$

$$<\bar{\lambda}, y> \leq <\bar{\lambda}, y_o>$$

and therefore

$$y_o \in (C^-) - \text{supp}_C B$$

It still has to be proved that $P_H - \min_C B \subset (C^-)\text{-supp}_C B$: for this we refer to Hartley' work [8]. #

6. Finally, for the comparison between the sets defined in n.2 and set $(C^-)\text{-supp}_C B$, with Y as a Banach space of any dimension, we simply point out the inclusions described by the following

Theorem 9 (Borwein [3], Chew [6])
Let Y be a Banach space, let C be a non trivial closed convex cone of Y. Let B be a part of Y. Then

1) (C^-) $\text{supp}_C\ B \subset P_H - \min_C B$ if X is a Hilbert space and C is strictly acute

2) (C^-) $\text{supp}_C\ B \subset P_{glob} - \min_C B \subset p_{loc} - \min_C B$

3) Furthermore, if C is strictly acute B is C-convex,

i.e. B + C is convex, then

$$(C^-) - \text{supp}_C\ B = p_{loc} - \min_C B \ .$$

For the proofs of these theorems, which are similar to previous ones , we refer to the above-mentioned works.

4. EXISTENCE OF PARETO MINIMA FOR VECTOR FUNCTIONS

1. We shall now briefly survey a few results regarding the optimization of vector functions.

Let X be a topological vector space (e.g. locally convex), let A be a part of X, let Y be a Banach space. Let's fix

$$f \ : \ A \to Y$$

Let's assume, for example, that a convex cone C is assigned in Y.

We shall then consider the following sets

$$\min_C f(A) \quad , \quad g - \min_C f(A)$$

and the elements $f(x_o)$, $x_o \in A$ belonging to these sets.

In this case we can say for example that x_o is Pareto minimum point (or generalized minimum point) of $f : A \to Y$ with respect to the positivity cone C of Y.

We shall now briefly indicate how, through the introduction of an adequate lower semicontinuity property of f with respect to C, we can obtain Pareto minima theorems (also in the generalized sense) of $f:A \to Y$.

Given $f : A \to Y$ and the convex cone C of Y, we say that f is C-lower semicontinuous (in short C-lsc) if for every Y the set

$$f^{-1}(y - \text{cl } C) \quad \text{will be closed.}$$

If $\dim Y < \infty$ and C is also closed and finitely generated, then C-lsc of f is equivalent to lsc of the vector function

$$\lambda \circ f : A \to \mathbb{R}$$

for every $\lambda \in C^p$.

We shall only indicate the following version of Corley's existence theorem for C-lsc functions.

Theorem 10 (Corley [7])

Let X be a topological vector space, $A \subset X$, Y a Banach space. If $f : A \to Y$ is C-lsc, with respect to a convex cone C of Y for which we have

$$(\text{cl } C) \cap (-\text{cl } C) = \{0\},$$

if A is σ-compact, then it exists $x_o \in A$ such that

$$f(x_o) \in \min_C f(A) = g - \min_C f(A) \quad .$$

Under these conditions C-lsc of f and the compactness of A imply the semicompactness of f(A).

In order to extend relation 3) of theorem 8 the C-convexity concept of function $f : A \to Y$ was introduced in the following way: f is C-convex if A is convex and for every $x,y \in A$, $t \in [0,1]$ we have

$$t\, f(x) + (1 - t)f(y) \in C - f(tx + (1 - t)y) .$$

We can then enunciate, for example,

Theorem 11

If $\dim Y < \infty$, $f : A \to Y$ is C-convex if C is a pointed closed convex cone, then

$$(C^-)\text{-}\mathrm{supp}_C\, B = p_{glob} - \min_C B = p_{loc} - \min_C B = p_H - \min_C B$$

The hypothesis that f is C-convex implies that $f(A)$ is a C-convex set of Y.

REFERENCES

[1] H.Benson. *Journal of Math. Anal. and Appl.*, 71, (1974).
[2] H.Benson. *Journal of Math. Anal. and Appl.*, 93, (1983).
[3] J.Borwein. *SIAM Journal of Control and Opt.*,15, (1977).
[4] J.Borwein. *Mathem. of Optim. Research*, 8, (1983).
[5] L.Cesari, M.Suryanarayana. *Trans. of Am. Math. Soc.*, 244, (1978).
[6] K.L.Chew. *Journal of Opt. Theory and Appl.*, 44, (1984).
[7] H.Corley. *Journal of Opt. Theory and Appl.*, 31, (1980).
[8] R.Hartley. *SIAM Journal Appl. Math.*, 34, (1978).
[9] J.M.Henig. *Journal of Opt. Theory and Appl.*, 36, (1982).
[10] A.Geoffrion. *Journal Math. Anal. and Appl.*, 22, (1968).
[11] P.Varaya. *SIAM Journal of Appl. Math.*, 15, (1967).
[12] P.Yu. *Journal of Opt. Theory and Appl.*, 14, (1974).

Chapter 6

Tonelli's Regurarity Theory in the Calculus of Variations: Recent Progress

F. H. Clarke

Abstract. Three of the fundamental issues in the calculus of variations are existence, necessary conditions, and regularity of the solution. The name of Tonelli is usually associated with the first of these, but in fact he contributed to all three. This article reviews the interplay amongst these issues and Tonelli's results concerning them, and also goes on to describe new discoveries which deepen and complete our understanding of the basic problem in the calculus of variations.

1. INTRODUCTION: THE NAIVE THEORY

This article is about a venerable problem in mathematics known as the *basic problem* in the calculus of variations. It has played a central role in the development of mathematics (and physics!) for some three centuries, yet manages to remain relevant and useful today. It continues to generate deeply interesting questions and still admits progress on fundamental issues. In its simplest form, the basic problem, denoted (P), is that of minimizing a functional J defined by

$$J(x) := \int_a^b L(t, x(t), \dot{x}(t))dt$$

over some class X of functions $x(\cdot)$ mapping the interval [a,b] to R^n, and subject to the boundary conditions $x(a) = A$, $x(b) = B$. Here the points A, B in R^n and the function $L(t,x,v)$ mapping $[a,b]\ R^n \times R^n$ to R are all given, and $\dot{x}$ is synonymous with dx/dt. To recapitulate, our central character is the problem

$$(P) \qquad \text{minimize } J(x) : x \in X, \quad x(a) = A, \quad x(b) = B.$$

The choice of X, the class of competing functions, is critical.The early history of the calculus of variations takes X to be, for example,the class C^1 of continuously differentiable (smooth) functions. The existence question, whether there is an element x at which J attains a minimum, is not directly adressed (hence "naive" theory). Instead, attention is focused upon *necessary conditions*, conditions which any potential solution x must satisfy.

The principal necessary condition is the *Euler equation* (1744):

$$\frac{d}{dt} L_v(t,x,\dot{x}) = L_x(t,x,\dot{x}) \ ,$$

a relationship that is easy to derive if we assume that x is a solution to (P) for $X = C^1$, and that L is continuously differentiable.

It became evident very early that there was interest in admitting to the theory functions x with "corners" (i.e., points of nondifferentiability). A commonly used class of such functions for which most of the classical theory goes through readily is PWS, a certain class of piecewise-smooth functions defined as follows: corresponding to any x in PWS is a finite partition $c_o, c_1, \ldots, c_k$ ($c_o = a$, $c_k = b$) of $[a,b]$ such that x is smooth on each subinterval (c_i, c_{i+1}) and admits finite one-sided derivatives at each partition point. With X now taken to be PWS, it remains an easy matter to derive the Euler equation as a necessary condition (except at corners, for which $\dot{x}$ is undefined). But the Euler equation is not as precise a condition in PWS as it is in C^1, until it is supplemented by the additional fact that $t \to L_v(t,x,\dot{x})$ has only removable discontinuities.

This fact is known as the *first Erdmann condition*; it is an outgrowth of the work of Weierstrass.

Only a few years after this work, duBois-Reymond (by means of the well-known lemma named after him) found a relatively straightforward way to subsume the Euler equation and the first Erdmann condition (when X = PWS) in a single relationship:

$$L_v(t,x,\dot{x}) = \text{constant} + \int_a^t L_x(s,x,\dot{x})ds \ .$$

We refer to this as the *integral form* of the Euler equation.

The Erdmann condition, or the integral form of the Eulèr equation, will imply under certain circumstances that $\dot{x}$ itself is continuous (i.e., that x is smooth). An assertion that a solution x must belong to a smaller class than was at first supposed (e.g., C^1 instead of PWS) is called a *regularity theorem*. An early example due to Hilbert asserts that when L is twice continuously differentiable and L_{vv} is globally positive definite, then x is C^2. The proof of this uses the necessary conditions, and illustrates the interplay between necessary conditions and regularity theorems. The former frequently imply the latter; on the other hand, an a priori assumption of regularity (e.g., that x lies in C^1 or PWS) plays a critical role in the derivation of the necessary conditions.

There is a serious weakness in the naive theory, and it is this: even if the necessary conditions uniquely define one element x of X, it does not follow that x solves (P). The difficulty, of course, is not knowing to begin with that (P) actually has a solution. The classical theory had only indirect ways around this: the use of so-called *sufficient conditions* to verify that a given function x is indeed a solution. The issue of existence was first broached in general terms by Leonida Tonelli in 1911.

2. TONELLI'S EXISTENCE THEORY

Tonelli's great contribution was to recognize both a choice of X and attendant conditions on L which would guarantee the existence of a solution to (P). He specified for X the class AC of *absolutely continuous* functions: x lies in AC if there is a Lebesque integrable function v such that

$$x(t) = x(a) + \int_a^t v(s)ds$$

for all t in [a,b]. It follows that $\dot{x}(t) = v(t)$ almost everywhere (a.e.) in [a,b] .

Besides an underlying twice continuous differentiability, the following conditions were imposed on L:

(i) *Coercivity:* for certain constants β, λ with $\beta > 0$, one has

$$L(t,x,v) \geq \beta|v|^2 + \lambda , \quad \text{for all } (t,x,v) .$$

(ii) *Convexity:* for each (t,x), the function $v \to L(t,x,v)$ is convex.

The convexity condition is equivalent to positing that $L_{vv}(t,x,v)$ is everywhere positive semidefinite ($L_{vv} \geq 0$). We see foreshadowed in these conditions the elements of weak convergence and semicontinuity and their connection to convexity.

Tonelli's existence theorem [20] is that J attains a minimum in AC (subject to given boundary conditions) when the coercivity and convexity conditions hold.

This is a satisfying result, but it does engender some very natural questions, the central one being "how different is the problem with X = AC from the problem with X = C^1 or X = PWS?". Specifically, one may ask:

(Q1) Does J admit the same infimum over AC (subject to the boundary conditions) as over PWS?

(Q2) Can reasonable problems really have solutions in AC not lying in PWS? What does "reasonable" mean in this context?

(Q3) Will a solution to the problem in AC satisfy the necessary conditions, for example the Euler equation, as did a solution in PWS?

To facilitate the discussion, let us introduce another subclass Lps of AC consisting of all those x satisfying a Lipschitz condition on [a,b]: for some constant K, for all s and t in [a,b] , one has

$$|x(s) - x(t)| \leq K|s - t| .$$

This is equivalent to the requirement that $\dot{x}(\cdot)$ be essentially bounded. It is not hard to show that (subject as always to the given boundary conditions) one has

$$\inf\{J(x): x \in \text{Lps}\} = \inf\{J(x): x \in X\}$$

provided X is a subclass of Lps containing all polynomials (for example). In particular then, Lps and PWS generate the same infimum. In view of the multitude of ways in which elements of AC can be approximated by those of Lps, it is a truly surprising fact that the following can be the case:

$$\inf\{J(x): x \in \text{Lps}\} > \inf\{J(x): x \in \text{AC}\}.$$

This situation is referred to as the *Lavrentiev phenomenon*, after its discoverer (1926); see Cesari [4, Section 18.5] . Thus the answer to (Q1) is "no". (Incidentally, it would be interesting to know whether there is a general class between AC and Lps which can generate an intermediate value of the infimum).

Let us note the relevance of the Lavrentiev phenomenon to a well-known approximation procedure called the *Rayleigh-Ritz method.* The method considers the subproblem of minimizing J(x) among those x expressible in the form

$$x(t) = x_o(t) + \sum_1^N \lambda_i \phi_i(t) ,$$

where x_o is a given smooth (typically) function satisfying the boundary conditions, and where $\phi_1, \phi_2, \ldots$ is a given complete family of functions vanishing at a and b. The word "complete" refers here to properties of approximation, as in the Stone-Weierstrass Theorem; a typical example is the family defined by

$$\phi_i(t) := (t-a)(t-b)t^{i-1} .$$

For N fixed, the minimization problem consists of finding the optimal values of $\lambda_1, \lambda_2, \ldots, \lambda_N$; i.e., it is a problem in R^N, and therefore one that can be solved in principle by conventional methods. The idea is that for N sufficiently large, the infimum of the subproblem will approximate that of (P) with X = AC, the only version of the problem for which a solution is known to exist. Since only elements of Lps are considered in the subproblem, however, this hope is doomed in general since the Lavrentiev phenomenon may occur, even when the Lagrangian L is a polynomial.

Let us turn now to question (Q2), which requires some elaboration, since it is easy to construct versions of (P) whose solution fails to lie in Lps, and hence all the more in PWS. An example would be

$$\min \int_0^1 (x^3 - t^2)^2 , \quad x(0) = 0 , \quad x(1) = 1 ,$$

whose unique solution in AC is $x(t) = t^{2/3}$. A better question then, and one that must have seemed crucial to Tonelli, is

(Q2') Under the hypotheses of the Tonelli existence theorem, can the solution x fail to lie in Lps?

Tonelli conjectured the affirmative. An example to confirm this has only recently been found; we pause now to discuss it.

3. THE BALL-MIZEL PROBLEM

Consider the version of (P) consisting of minimizing

$$\int_0^1 \{r\dot{x}^2 + (x^3 - t^2)^2 \dot{x}^{14}\}dt$$

for some $r > 0$, and where the dimension n is 1. Note that the hypotheses of the Tonelli existence theorem are satisfied. This problem was proposed by Ball and Mizel [2] , who made the significant observation that the function $\hat{x}(t) = kt^{2/3}$ satisfies the Euler equation, provided that r and k are linked by

$$r = (2k/3)^{12}(1 - k^3)(13k^3 - 7) .$$

If then we choose the boundary conditions $x(0) = 0$, $x(1) = k$, we have circumstantial evidence for believing that the solution to (P) might be the non-Lipschitz function $\hat{x}$.

The fact that this is so is confirmed by Clarke and Vinter in [12], using the Hamilton-Jacobi verification technique (see [8, Section 3.7] for a discussion of this topic) extended to a nonsmooth setting. The argument, which is elementary, depends on exhibiting a function with certain properties which turn out to confirm the fact that $\hat{x}$ is the unique solution to (P).

The function in question is nondifferentiable, and constitutes the first example of the role to be played by nonsmooth analysis in the developments described in this article.

Note that the function $\hat{x}$ does not satisfy the classical necessary conditions. In particular, the first Erdmann condition fails since $L_v(t,\hat{x},\dot{\hat{x}})$ becomes infinite at $t = 0$; the integral form of the Euler equation fails because $L_x(t,\hat{x},\dot{\hat{x}})$ is not integrable. We now turn to question (Q3): what necessary conditions hold under merely those hypotheses guaranteeing existence?

4. TONELLI'S NECESSARY CONDITION AND REGULARITY THEOREM

In 1915, as a companion result to his existence theorem, Tonelli studied the situation in which the dimension n is equal to 1, L is twice continuously differentiable and coercive, and L_{vv} is everywhere positive (the last being a sufficient but not necessary condition in order for the function $v \to L(t,x,v)$ to be strictly convex). He proved that under these conditions there corresponds to any solution x in AC of (P) a closed subset S of [a,b] of measure zero with the property that in the complement of S, x is C^2 and satisfies the Euler equation.

The example of Section 3 lends interest to this theorem by showing (for the first time) that S may be necessarily nonempty.

In [11] Clarke and Vinter extend (and stratify) the theorem to $n > 1$ dimensions and to a much less restrictive class of Lagrangian L. We shall now state a simplified version of the result. (Henceforth we shall always take X = AC in the statement of (P)).

Theorem 1

Let L be locally Lipschitz and satisfy the coercivity and convexity conditions. Then there corresponds to any solution x of (P) a closed subset S of [a,b] of zero measure such that in the complement of S, x is locally Lipschitz and satisfies the Euler equation (in a form appropriate to nonsmooth Lagrangians). If in addition for each t the function $v \to L(t,x(t),v)$ is strictly convex, then x is C^1 in the complement of S.

If in addition L is C^r $(r \geq 2)$ and L_{vv} is positive definite, then x is C^r in the complement of S.

The proof of the theorem is inspired in part by Tonelli's but is also fundamentally different due to the fact that, roughly speaking, for n = 1 there are but two directions, while for n > 1 there are a continuum. This comes into play in constructing a certain family of auxiliary Lagrangians, a goal achieved by appealing to the methods of nonsmooth analysis [8], which are intrinsically involved. In particular, the necessary conditions for nondifferentiable problems which are alluded to in the theorem play a central role in the proof.

5. CLASSES OF WELL - BEHAVED PROBLEMS

It is natural to yearn for simple hypotheses guaranteeing that the set S of Theorem 1 is empty, for then the solution x will lie in Lps and all the classical necessary conditions will apply, in particular the integral form of the Euler equation. Further, the Lavrentiev phenomenon would be excluded, and so Rayleigh-Ritz approximation justified. The following new result, excerpted from Clarke and Vinter [11], postulates the hypotheses of Theorem 1, which plays a central role in the proof.

For ease of exposition we shall assume in addition that L is differentiable.

Theorem 2

Suppose that for certain nonnegative constants c and k one has

$$|L_t(t,x,v)| \leq c|L(t,x,v)| + k$$

for all (t,x,v). Then any solution x to (P) belongs to Lps.

The problem (P) is called *autonomous* when L has no explicit dependence on t. The following new result illustrates that basic facts about

(P) are still be found:

Corollary When (P) is autonomous, its solutions lie in Lps.

There is another kind of growth condition which implies good behavior of the solution, one which does have a history (see for example [8]): $|L_x| + |L_v| \le c|L| + k$. The following result [11] transposes one of the terms in this requirements, as well as proving its sufficiency (in a general multidimensional setting) for solutions to lie in Lps.

Theorem 3
Suppose that for certain nonnegative constants c_1, c_2 and k one has

$$|L_x(t,x,v)| \le c_1|L_v(t,x,v)| + c_2|L(t,x,v)| + k .$$

Then any solution x to (P) belongs to Lps.

6. CONDITIONS ON THE BERNSTEIN FUNCTION

In addition to the Tonelli existence hypotheses, we suppose for this section that L is twice continuously differentiable and that L_{vv} is positive definite. Under these conditions, when the Euler equation holds it can be written

$$\ddot{x} = F(t,x,\dot{x}) ,$$

where F is the *Bernstein function* $L_{vv}^{-1}(L_x - L_{vt} - L_{vx}v)$.
Bernstein [3] was the first to exploit this fact in a qualitative analysis of extremals. It can also be used for a different purpose, in conjunction with Theorem 1, to identify another class of well-behaved problems, as shown in [11] :

Theorem 4

Suppose that F satisfies

$$|F(t,x,v)| \leq c(1 + |v|^3)$$

for some constant c. Then all solutions to (P) belong to Lps.

A different approach assumes only that F is polynomially bounded:

$$|F(t,x,v)| \leq c_1(|v| + c_2)^r ,$$

where, as is the case with all such hypotheses, the condition is only required to hold for x in bounded subsets of R^n. There then exist determinable numbers δ and M such that:

Theorem 5 11

Let F be polynomially bounded. Then, whenever the problem (P) is such that $|b - a| < \delta$ and $|B - A| \leq M(b - a)$, any solution to (P) is twice continuously differentiable.

7. POLYNOMIAL LAGRANGIANS

Even when the Lagrangian L is a polynomial, it is possible for the solution x of (P) to have "non-Lipschitz points" (i.e., for the set S of Theorem 1 to be necessarily nonempty), as demonstrated by the Ball-Mizel example in Section 3. Nonetheless one may a priori further limit S both in size and location when L is a polynomial (in (t,x,v)) and $n=1$. We continue to assume that L_{vv} is everywhere positive, and we write L in the form

$$L(t,x,v) = a_o(t,x)v^p + a_1(t,x)^{p-1} + \ldots + a_p(t,x) .$$

The polynomial a_o has a certain number of irreducible factors $b_j(t,x)$, and each b_j is expressible in the form

$$c_{jo}(t)x^{p_j} + c_{j1}(t)^{p-1} \dots + c_{jp_j}(t) .$$

Define

$$T_j := \{s \in [a,b]: c_{jo}(s) = 0 \text{ or } \Delta_j(s) = 0\},$$

where Δ_j is the discriminant of $b_j(t,x)$ treated as a polynomial in x (see [13, Section 4]). Define also T by $T = \cup_j T_j$.

The set T contains at most finitely many points. It is proved in [13] that any point of accumulation of S must lie in T. In consequence, S is at most countable with finitely many points of accumulation.

Note that the calculation of the set T, which serves as an a priori upper bound on the set of points where a solution can behave extremely badly, depends only upon the boundary conditions.

8. LOCAL EXISTENCE

In his Fondamenti [21, Volume II, Theorem 109], a work which unaccountably seems never to have been translated, Tonelli improved upon his 1915 regularity theorem (see Section 4) by obtaining the same conclusion without the coercivity hypothesis. Because in the absence of this hypothesis existence is no longer assured, Tonelli used a fundamentally different approach, a byproduct of which was a type of existence theorem in the small ("in piccolo"). He showed through an ingenious and very complicated argument that when the dimension n is equal to 1, when L is twice continuously differentiable and satisfies $L_{vv}>0$ (globally), then whenever b - a is sufficiently small and the ratio $|B - A|/(b - a)$ satisfies a certain a priori bound, the problem (P) admits at least one local solution, and any such is smooth and satisfies the Euler equation. This conclusion easily implies the existence of the set S of Tonelli's 1915 regularity theorem.

A counterpart of this local existence result, stated formally below, was proven by Clarke and Vinter [14] for the vector setting ($n \geq 1$), under the sole hypothesis that $L(t,x,v)$ is a locally Lipschitz function strictly convex in the v variable. An essential role is played here by the strict convexity, a condition present in both Tonelli's versions of his regularity theorem (through the stronger hypothesis $L_{vv} > 0$), but absent in the new Theorem 1 of Section 4. Thus in the extended setting, in contrast with Tonelli's, the two different approaches lead to truly different results.

Theorem 6

Let a point (t_o, x_o) in $R \times R^n$ be given, and any $M > 0$. Then there exist positive ε and γ such that whenever the data a, b, A, B are such that $b > a$, with a, b within ε of t_o, and A, B within ε of x_o, and with $|B-A| \leq M(b - a)$, then the problem of minimizing $J(x)$ over all absolutely continuous functions x satisfying

$$x(a) = A \ , \quad x(b) = B \ , \quad |x(t) - x_o| < \gamma \quad (t \in [a,b])$$

has at least one solution x, and all solutions are continuously differentiable on $[a,b]$.

We remark that an immediate consequence of the theorem is the existence of a closed set S of zero measure in the complement of which x is continuously differentiable (see [14, Corollary 1]).

9. LOCAL EXISTENCE AND THE DUAL ACTION

A classical and very interesting boundary-value problem in mathematical physics is concerned with periodic solutions $x(t)$, $p(t)$ of Hamilton's equations:

$$-\dot{p} = H_x(x,p), \quad \dot{x} = H_p(x,p) \ .$$

We refer to [15] [18] [19] [9] for surveys of the long history and active present of this problem, and for further references.

When the Hamiltonian H is a convex function, a certain *dual action principle* due to Clarke has proven useful. It involves the following function G:

$$G(u,v) := \sup_{x,p} \; (u,v)\cdot(x,p) - H(x,p) \;,$$

where the supremum is taken over all (x,p) in $R^n \times R^n$. The dual action is the following functional J_T in the calculus of variations:

$$J_T(x,p) := \int_0^T \{<\dot{p}(t),\, x(t)> + G(-\dot{p}(t),\, \dot{x}(t))\}dt \;.$$

Its utility is predicated on the (originally surprising but now well-known) fact that when H is convex, and under homogeneous boundary conditions, extremals of J_T correspond to solutions of Hamilton's equations of period T. (Here the term "extremal" means a solution of the Euler equation).

A central question then becomes: how to produce extremals of the dual action J_T? Originally this was done by imposing hypotheses on H, of one type or another, which would guarantee the existence of a global minimum of J_T (see for example [5] [6] [7] [10]), and hence an extremal by the application of necessary conditions. Later, critical point theory was applied to J_T to derive the existence of certain extremals(see for example [1] [16] [17] [18]).

A third approach has been initiated by Clarke [9] , who derived under certain circumstances the existence of *local* minima of J_T (and hence of extremals). The approach is based in part upon the type of arguments used to prove Theorem 6, which in turn drew upon the line of reasoning employed by Tonelli in his Fondamenti. We assume that H is a smooth nonnegative convex function vanishing only at the origin, and satisfying the growth condition (superlinearity):

$$\lim_{|(x,p)|\to\infty} \frac{H(x,p)}{|(x,p)|} = \infty.$$

The problem P_T^γ refers to the minimization of $J_T(x,p)$ over all smooth functions (x,p) mapping $[0,T]$ to $R^n \times R^n$ which satisfy the boundary conditions

$$(x,p)(0) = (x,p)(T) = (0,0)$$

as well as the constraint

$$|x(t)| < \gamma \qquad \text{for all } t \text{ in } [0,T] .$$

The number T_γ is defined by

$$T_\gamma := \sup_{\alpha>0} \inf_{x,p} \{\gamma\alpha/H(x,p) : |x| \le \gamma,\ |p| \le \alpha\}.$$

Theorem 7 [9]

If γ is any positive number, and if T is any number in $(0,T_\gamma)$, then the problem P_T^γ admits a solution.

When combined with hypotheses assuring that the solution to P_T^γ is nontrivial (i.e., not identically zero) and with necessary conditions , the theorem leads to various types of conclusions regarding Hamilton 's equations, as shown in [9] . The following is an illustration.(H is said to be *radially increasing* in p if one has $\langle p, H_p(x,p)\rangle > 0$ for all (x,p) with $p \neq 0$).

Theorem 8

(i) Suppose that H satisfies the condition

(C_o) $\qquad |(x,p)| < \delta$ implies $H(x,p) \ge a|x|^{1+r} + b|p|^{1+s}$,

where δ, a, b, r and s are positive constants with $rs < 1$. Then for all

T sufficiently small there is a solution of Hamilton's equations having period T. If H is radially increasing in p, there is a solution of minimal period T.

(ii) Suppose that H satisfies the condition

$$(C_\infty) \qquad |(x,p)| > \Delta \quad \text{implies} \quad H(x,p) \le A|x|^{1+R} + B|p|^{1+S},$$

where Δ, A, B, R and S are positive constants with $RS < 1$. Then for all T sufficiently large there is a solution of Hamilton's equations having minimal period T.

(iii) Suppose that H satisfies both (C_0) and (C_∞). Then for any $T > 0$ there is a solution of Hamilton's equations having minimal period T.

REFERENCES

[1] A.Ambrosetti and G.Mancini. 'Solutions of minimal period for a class of convex Hamiltonian systems'. *Math. Ann.* 255(1981) 405-421.

[2] J.Ball and V.Mizel. 'Singular minimizers for regular one-dimensional problems in the calculus of variations'. *Bull. Amer. Math. Soc.* 11(1984) 143-146.

[3] S.Bernstein. 'Sur les équations du calcul des variations'.*Annales Sci.* de l'Ecole Norm. Sup (III) 29(1912) 431-485.

[4] L.Cesari. 'Optimization - Theory and Applications'. *Springer-Verlag*, New York, (1983).

[5] F.H.Clarke. 'Solutions périodiques des équations hamiltoniennes' . *Comptes Rendus Acad. Sci.*, Paris 287(1978) 951-952.

[6] F.H.Clarke. 'A classical variational principle for periodic Hamiltonian trajectories'. *Proc. Amer. Math. Soc.*, 76(1978) 186-188.

[7] F.H.Clarke. 'Periodic solutions of Hamiltonian inclusions'. *J.Differential Equations*, 40(1981) 1-6.

[8] F.H.Clarke. 'Optimization and nonsmooth analysis'. *Wiley Interscience*, New York, (1983).

[9] F.H.Clarke. 'Periodic solutions of Hamilton's equations and local minima of the dual action'.*Trans.Amer.Math.Soc.*287(1985) 239-251.

[10] F.H.Clarke and I.Ekeland. 'Hamiltonian trajectories having prescribed minimal period'. *Comm. Pure and Appl. Math.*, 33(1980) 103-116.

[11] F.H.Clarke and R.B.Vinter. 'Regularity properties of solutions to the basic problem in the calculus of variations'. *Trans. Amer.Math Soc.* 289 (1985) 73-98.

[12] F.H.Clarke and R.B.Vinter. 'On the conditions under which the Euler equation or the maximum principle hold'. *Appl. Math. Optimization*, 12 (1984) 73-79.
[13] F.H.Clarke and R.B.Vinter. 'Regularity of solutions to variational problems with polynomial Lagrangians'. *Bull.Polish Acad. Sci.*, to appear.
[14] F.H.Clarke and R.B.Vinter. 'Existence and regularity in the small in the calculus of variations'. *J. Diff. Eq.*, to appear.
[15] N.Desolneux-Moulis. 'Orbites périodiques des systèmes hamiltoniens autonomes'. *Sémin. Boubaki*, 32(1979), n°552.
[16] I.Ekeland. 'Periodic Hamiltonian trajectorics and a theorem of P '. Rabinowitz, *J. Differential Equations* 34(1979) 523-534.
[17] I.Ekeland and J.M.Lasry. 'On the number of periodic trajectories for a Hamiltonian flow on a convex energy surface'. *Ann. Math.*, 112 (1980) 283-319.
[18] H.Mancini. 'Periodic solutions of Hamiltonian systems having prescribed minimal period'. in '*Advances in Hamiltonian Systems*'. Birkhauser, Boston (1983).
[19] P.H.Rabinowitz. 'Periodic solutions of Hamiltonian systems:a survey'. *SIAM J. Math. Anal.*, 13(1982) 343-352.
[20] L.Tonelli. 'Sur une méthode directe du calcul des variations'.*Rend. Circ.*, Palermo 39(1915) 233-264.
[21] L.Tonelli. 'Fondamenti di calcolo delle variazioni' (two volumes), Zanichelli, Bologna, (1921-1923).

Chapter 7

Convergence of Unilateral Convex Sets

G. Dal Maso

We shall consider minimum problems of the following form:

$$(1) \qquad \min_{\substack{u=0 \text{ on } \partial\Omega \\ u\geq f \text{ on } \Omega}} \left[\int_\Omega |Du|^2 dx + \int_\Omega gudx\right] ,$$

where Ω is a bounded open subset of $\mathbb{R}^n$, $g \in L^2(\Omega)$, and the obstacle f is an arbitrary function from Ω into $\overline{\mathbb{R}}$.

The function space in which we study problem (1) is the Sobolev space $H^1(\Omega)$, defined as the space of all functions in $L^2(\Omega)$ whose first order distribution derivatives are in $L^2(\Omega)$, so that the condition "$u = 0$ on $\partial\Omega$" becomes "$u \in H^1_o(\Omega)$", where $H^1_o(\Omega)$ is the closure of $C^1_o(\Omega)$ in $H^1(\Omega)$.

Since every element u of $H^1_o(\Omega)$ is only a class of equivalent functions with respect to almost everywhere equality in Ω, and we do not assume that f is regular, we have to define precisely the meaning of the second condition "$u \geq f$ on Ω".

In order to study problems with thin obstacles, i.e. obstacles f for which the set $\{x \in \Omega: f(x) > -\infty\}$ (which is the set where the constraint $u \geq f$ is actually imposed) has Lebesgue measure zero, we introduce the concept of capacity.

For every set $E \subseteq \Omega$ we define cap(E) as the infimum of all numbers

$\int_\Omega |Du|^2 dx$ corresponding to those functions $u \in H^1_0(\Omega)$ for which u=1 a.e. in an open neighbourhood of E.

We define also Cap(E) as the infimum of all numbers $\int_\Omega |Du|^2 dx + \int_\Omega |u|^2 dx$ corresponding to those functions $u \in H^1_0(\Omega)$ for which u=1 a.e. in an open neighbourhood of E.

By Poincaré's inequality, there exists a constant $K = K(\Omega)$ such that

$$\text{(2)} \qquad \mathrm{cap}(E) \le \mathrm{Cap}(E) \le K\,\mathrm{cap}(E)$$

for every $E \subseteq \Omega$, thus the capacities cap and Cap are equivalent in many problems. For instance we have

$$\mathrm{cap}(E) = 0 \iff \mathrm{Cap}(E) = 0 ,$$

so the capacities cap and Cap are equivalent in every situation where only sets of capacity zero are taken into account. But in other problems the capacities cap and Cap can lead to different results, as we shall see later.

In order to define the meaning of the inequality "$u \ge f$ on Ω " we need the concept of Cap-quasi continuous function.

Given an increasing set function $\alpha: P(\Omega) \to [0,+\infty]$ and a function $\phi: \Omega \to \overline{\mathbb{R}}$, we say that ϕ is α-quasi continuous (or quasi continuous with respect to α) if for every $\varepsilon > 0$ there exists an open set $A \subseteq \Omega$, with $\alpha(A) < \varepsilon$, such that the restriction $\phi|_{\Omega - A}$ is continuous on $\Omega - A$.

By Lusin's theorem, every measurable function on Ω is quasi continuous with respect to Lebesgue measure.

In the case of the capacity Cap, not every function in Cap -quasi continuous, but it is known that for every function $u \in H^1(\Omega)$ there exists a Cap-quasi continuous function $\tilde{u}$ such that $\tilde{u} = u$ a.e. on Ω (see [6], proposition 7.7). This function $\tilde{u}$ is said to be a Cap-quasi continuous representative of u. The Cap-quasi continuous representative of a function $u \in H^1(\Omega)$ is essentially unique (see [6], corollary 7.4), in the sense that any two Cap-quasi continuous represetatives of u coincide Cap-a.e. on Ω (in general we say that a property $A(x)$ holds Cap-a.e. on Ω if $\text{Cap}(\{x \in \Omega: A(x) \text{ is false}\}) = 0$).

Using the inequalities (2) it is easy to see that a function is Cap-quasi continuous if and only if it is cap-quasi continuous.

We are now in a position to give a rigorous formulation to problem (1).

For every function $f: \Omega \to \overline{\mathbb{R}}$ we define

$$K(f) = \{u \in H_0^1(\Omega): \tilde{u} \geq f \quad \text{Cap - a.e. on } \Omega\}.$$

Then the rigorous form of problem (1) is:

$$\min_{u \in K(f)} \left[\int_\Omega |Du|^2 dx + \int_\Omega gu dx\right] \tag{3}$$

It is easy to see that the set $K(f)$ is convex; using some properties of the Cap-quasi continuous representatives, it is easy to prove that $K(f)$ is closed in $H_0^1(\Omega)$ (see [5], theorem 1.9). Therefore the existence of a solution to problem (3) can be obtained by a straightforward application of the direct methods of the calculus of variations, using the fact that the functional

$$\int_\Omega |Du|^2 dx + \int_\Omega gu dx \tag{4}$$

is weakly lower semicontinuous and coercive on $H_0^1(\Omega)$. If $K(f) \neq \emptyset$, the uniqueness of the solution to problem (3) follows from the strict convexity of the functional (4) on $H_0^1(\Omega)$.

For every obstacle f with $K(f) \neq \emptyset$ and for every $g \in L^2(\Omega)$ we denote by $u(f,g)$ the unique solution of problem (3) and by $m(f,g)$ the minimum value of problem (3).

We are interested in the behaviour of $u(f,g)$ and $m(f,g)$ as f varies. In particular, given a sequence (f_n) of obstacles and an obstacle f, we look for conditions on (f_h) and f which ensure the strong convergence in $H_o^1(\Omega)$ of $u(f_h,g)$ to $u(f,g)$ for every $g \in L^2(\Omega)$.

A first result in this direction was obtained by U.Mosco, who introduced a convergence for sequences of convex sets which is useful in the study of variational inequalities.

Let us recall Mosco's definitions (see [7]). Let (K_h) be a sequence of convex subsets of a Banach space V. We define the strong lower limit

$$s - \liminf_{h\to\infty} K_h$$

as the set of all $u \in V$ for which there exists a sequence (u_h) strongly converging to u with $u_h \in K_h$ for all large h. We define the weak upper limit

$$w - \limsup_{h\to\infty} K_h$$

as the set of all $u \in V$ for which there exists a sequence (u_h) weakly converging to u with $u_h \in K_h$ for infinitely many h. If

$$K = s\text{-}\liminf_{h\to\infty} K_h = w\text{-}\limsup_{h\to\infty} K_h$$

we say that (K_h) converges to K and write $K_h \to K$.

Let (f_h) be a sequence of functions from Ω into $\overline{\mathbb{R}}$ and let f be a function from Ω into $\overline{\mathbb{R}}$. Assume that $K(f_h) \neq \emptyset$ for every $h \in \mathbb{N}$ and that $K(f) \neq \emptyset$. From Mosco's papers [7] and [8] we obtain the following theorem.

Theorem 1

The following conditions (a), (b), (c) are equivalent:

(a) $K(f_h) \to K(f)$ (in $H_o^1(\Omega)$),

(b) $u(f_h,g) \to u(f,g)$ strongly in $H_o^1(\Omega)$ for every $g \in L^2(\Omega)$,

(c) $m(f_h,g) \to m(f,g)$ for every $g \in L^2(\Omega)$.

Since the direct verification of the condition "$K(f_h) \to K(f)$" is not trivial, it is useful to find conditions on (f_h) and f which imply the convergence of $K(f_h)$ to $K(f)$.

In order to avoid the unessential problems originated by the fact that f_h can be unbounded from above in Ω or positive near $\partial\Omega$, in contrast with the condition "$u = 0$ on $\partial\Omega$", we shall always assume that there exist a constant M and a compact set $H \subseteq \Omega$ such that

$$f_h(x) \le M \qquad \forall h \in \mathbb{N} \qquad \forall x \in \Omega,$$

$$f_h(x) \le 0 \qquad \forall h \in \mathbb{N} \qquad \forall x \in \Omega - H.$$

Under these assumptions, there are several sufficient conditions for the convergence of $K(f_h)$ to $K(f)$. For instance, if $f_h \to f$ uniformly, then $K(f_h) \to K(f)$ (see [2], [3]). If $f_h \to f$ strongly in $H_o^1(\Omega)$, then $K(f_h) \to K(f)$ (see [7]). But the condition $f_h \to f$ weakly in $H_o^1(\Omega)$ is not sufficient for the converge of $K(f_h)$ to $K(f)$ (see [2]). Nevertheless, if $f_h \to f$ weakly in $W_o^{1,p}(\Omega)$ for some $p > 2$, then $K(f_h)$ converges to $K(f)$ in $H_o^1(\Omega)$ (see [2]).

All these conditions are far from being neceaary for the convergence of $K(f_h)$ to $K(f)$, as the following example shows.

For every $E \subseteq \Omega$ let 1_E be the characteristic function of E, defined by $1_E(x) = 1$ if $x \in E$ and $1_E(x) = 0$ if $x \in \Omega - E$. For every $x \in \mathbb{R}^n$ and every $r > 0$ we set $B(x,r) = \{y \in \mathbb{R}^n : |y-x| < r\}$.

Example 2

Let $\Omega = B(0,2)$, $E = B(0,1)$, $E_h = B(0,1-\frac{1}{h})$, $F_h = B(0,1+\frac{1}{h})$. Then $K(1_{E_h})\to K(1_E)$ and $K(1_{F_h})\to K(1_E)$, but the sequences (1_{E_h}) and (1_{F_h}) do not converge to 1_E uniformly, nor weakly in $H^1_0(\Omega)$.

The following conditions, due to H.Attouch and C.Picard (see [1]), do not involve the regularity of (f_h) and f, and are considerably weaker than uniform convergence.

Theorem 3

Assume that:

(a) $$\lim_{h\to\infty} \int_0^{+\infty} \mathrm{cap}(\{f_h - f > t\})\, t\, dt = 0$$

(b) $$\lim_{h\to\infty} \mathrm{cap}(\{f - f_h > t\}) = 0 \quad \underline{\text{for every}}\ t > 0$$

Then $K(f_h)\to K(f)$.

Conditions (a) and (b) are not necessary for the convergence of $K(f_h)$ to $K(f)$, as we can see in example 2. Indeed the sequence (1_{F_h}) and the function 1_E of example 2 do not satisfy condition (a), whereas the sequence (1_{E_h}) and the function 1_E do not satisfy condition (b).

In order to give a necessary and sufficient condition for the convergence of $K(f_h)$ to $K(f)$, it is convenient to consider first the case in which $f_h = 1_{E_h}$ and $f = 1_E$, where E_h and E are subsets of a fixed compact subset H of Ω. For every $F \subseteq H$ we denote by w_F the solution of the minimum problem

$$\min_{u\in K(1_F)} \left[\int_\Omega |Du|^2 dx + \int_\Omega u^2 dx\right]$$

Then we can prove the following theorem.

Theorem 4

The following conditions (a), (b), (c) are equivalent:

(a) $K(1_{E_h}) \to K(1_E)$

(b) $\mathrm{Cap}(E_h) \to \mathrm{Cap}(E)$ and $\mathrm{Cap}(E \cup E_h) \to \mathrm{Cap}(E)$

(c) $w_{E_h} \to w_E$ strongly in $H^1_o(\Omega)$.

The implication (a) $\Rightarrow$ (b) follows from the variational properties of the convergence of convex sets in the sense of Mosco and from the variational definition of Cap. The implication (b) $\Rightarrow$ (c) follows from the remark that $w_{E\cup E_h} \to w_E$ strongly by (b), and that $w_{E_h} - w_{E\cup E_h} \to 0$ by (b) and by the parallelogram identity. The implication (c) $\Rightarrow$(a) is the most difficult one, and requires a nontrivial approximation argument (see [4]).

The following example shows that theorem 4 does not hold if we use the capacity cap instead of Cap in condition (b).

Example 5

Let $\Omega = B(0,2)$, $E = \partial B(0,1)$, $F = \overline{B(0,1)}$. Then $E \subseteq F$, $\mathrm{cap}(E)=\mathrm{cap}(F)$,and $K(1_E) \neq K(1_F)$. If we take $E_h = F$ for every $h \in \mathbb{N}$, we obtain that

$$\mathrm{cap}(E_h) = \mathrm{cap}(F) = \mathrm{cap}(E)$$

$$\mathrm{cap}(E_h \cup E) = \mathrm{cap}(F) = \mathrm{cap}(E),$$

hence conditon (b) of theorem 4 is satisfied by E_h and E with cap instead of Cap, but

$$K(1_{E_h}) \to K(1_F) \neq K(1_E).$$

The general necessary and sufficient conditions for the convergence of $K(f_h)$ to $K(f)$ are given by the following theorem, in which we suppose, for the sake of simplicity, that there exist a constant M and a compact set $H \subseteq \Omega$ such that

$$f_h(x) \leq M \qquad \forall h \in \mathbb{N} \qquad \forall x \in \Omega ,$$

$$f_h(x) = -\infty \qquad \forall h \in \mathbb{N} \qquad \forall x \in \Omega - H.$$

A necessary and sufficient condition can be obtained also withuot this additional hypothesis, but the statement of the theorem would be more involved (see [4]).

Theorem 6

The following conditions (a), (b), (c) are equivalent:

(a) $K(f_h) \to K(f)$;

(b) there exists a dense subset D of $\mathbb{R}$ such that

$$K(1_{\{f_h>t\}}) \to K(1_{\{f>t\}})$$

for every $t \in D$;

(c) there exists a dense subset D of $\mathbb{R}$ such that

$$\mathrm{Cap}(\{f_h > t\}) \to \mathrm{Cap}(\{f > t\})$$

$$\mathrm{Cap}(\{f \vee f_h > t\}) \to \mathrm{Cap}(\{f > t\})$$

for every $t \in D$.

If (a) is satisfied, then we can prove that a particular set D for which (b) and (c) hold is the set of continuity points of the decreasing function

$$t \to \mathrm{Cap}(\{f > t\});$$

note that $\mathbb{R} - D$ is countable in this case. In general (b) and (c) do not hold with $D = \mathbb{R}$, as the following example shows.

Example 7

Let H be a compact subset of Ω with $\mathrm{Cap}(H) > 0$. Let f_h, f be defined by

$$f_h(x) = \begin{cases} \frac{1}{h} & \text{if} \quad x \in H, \\ -\infty & \text{if} \quad x \in \Omega - H \end{cases}$$

$$f(x) = \begin{cases} 0 & \text{if} \quad x \in H, \\ -\infty & \text{if} \quad x \in \Omega - H. \end{cases}$$

Then $K(f_h) \to K(f)$, and it is easy to prove that conditions (b) and (c)of theorem 6 hold for $D = \mathbb{R} - \{0\}$, but not for $D = \mathbb{R}$.

REFERENCES

[1] H.Attouch, C.Picard. 'Inéquations variationnelles avex obstacles et espaces fonctionnels en théorie du potentiel'. *Applicable Anal.* 12 (1981), 287-306.

[2] L.Boccardo, F.Murat. 'Nouveaux résultats de convergence dans des problèmes unilatéraux'. In "Nonlinear partial differential equations and their applications. Collège de France Seminar. Volume II", 64-85, ed. by H.Brezis and J.L.Lions. *Research Notes in Mathematics*, Pitman, London (1982).

[3] H.Brezis. 'Problémes unilatéraux'. *J.Math. Pures Appl.* 51(1972),1-68.

[4] G.Dal Maso. 'Some necessary and sufficient conditions for the convergence of sequences of unilateral convex sets'.*J.Funct.Anal.* 62, (1985) 119-159.

[5] G.Dal Maso, P.Longo. 'Γ-limits of obstacles'.*Ann. Mat. Pura Appl.*, 128 (1980) 1-50.

[6] P.A.Fowler. 'Capacity theory in Banach spaces'. *Pacific J.Math.* 48 (1973), 365-385.

[7] U.Mosco. 'Convergence of convex sets and of solutions of variational inequalities'. *Advances in Math.* 3(1969), 510-585.

[8] U.Mosco. 'On the continuity of the Young-Fenchel transform'. *J.Math. Anal. Appl.* 35(1971), 518-535.

Chapter 8

Continuity of Bilinear and Non-Bilinear Polarities

S. Dolecki

1. INTRODUCTION

Polarities (and more generally, Galois connections) constitute an essential aspect of numerous mathematical objects and relations (let us mention duality in the theory of topological vector spaces, duality in optimization, the Stone transform ...). We present here continuity properties of general polarities with respect to several basic convergences on spaces of sets. A more detailed discussion will be done in [9].

A polarity is a mapping from the space of subsets of a set X to the space of subsets of another set Y that converts the unions into intersections. Every polarity is determined by a subset of $X \times Y$, say, R. The R-polar of a subset A of X is the greatest subset B of Y such that

(1.1) $$A \times B \subset R$$

Accordingly, a couple A, B of sets such that one is the R-polar of another is maximal with respect to (1.1). Therefore, the continuity considerations for polarities aim at an understanding of the dynamics of the equilibrium (1.1) in which the growth of one set forces the shrinking of the other.

We shall consider polarities determined by some level sets, epigraphs and diepigraphs of extended-real-valued functions on $X \times Y$. In the particular bilinear case this yields dualities between polar cones, closed convex sets and their support functions, closed convex functions and their Fenchel conjugates. As regards the bilinear polarities Wijsman [26] established the homeomorphism through the conjugation between upper and lower variational convergences in euclidean space. A considerable progress was then due to Mosco [22] who extended these results to reflexive Banach spaces and to Joly who presented a beautiful theory of continuity of conjugation in locally convex spaces [17] . The works of Matzeu [19] and Attouch [2] refine some and give some other results in reflexive Banach spaces. As for locally convex spaces a recent work of Back [3] rediscovers, in a different way, some results of Joly.

On the other hand, the questions of relationship between semicontinuity properties of value functions and of constraint multifunctions constitute another example of continuity-of-polarities problems for generally non bilinear coupling functions. Here we quote the important theorems of Valadier and Castaing ([23],[4]) concerning weak semicontinuity of value functions.

Another situation in which non bilinear polarities occur is that of conjugation by level sets, in particular, the quasi-convex duality . Continuity results for these polarities have been recently obtained by Volle [24].

We present several general sufficient conditions and characterizations for continuity of arbitrary polarities with respect to the fundamental convergences on spaces of sets: adherence, lower and upper Vietoris topologies and to their variants. These convergence notions, in general non topological, have their functional formulations in terms of Γ-limits of De Giorgi and Franzoni [8] (the special cases of which have been used already by Wijsman and, more or less implicitly by Joly and Mosco).

We show how our general results on continuity of polarities and on convergence recover the above listed facts from several seemingly different domains.

2. POLARITIES

Let X, Y be sets and let R be a relation in $X \times Y$. The *polar* of a subset A of X is, by definition.

$$P_R(A) = \bigcap_{x \in A} Rx \quad . \tag{2.1}$$

Any mapping of 2^X to 2^Y that transforms the unions into intersections is of the above form and is called a *polarity*. The *conjugate polarity* of P_R is P_{R^-}, where R^- stands for the inverse relation to R. Every mapping on 2^X that is isotone, extensive and idempotent is called a *hull* operator. For every relation R on $X \times Y$,

$$cl_R = P_{R^-} P_R \quad \text{and} \quad cl_{R^-} = P_R P_{R^-}$$

are hulls. A set A for which $cl_R A = A$ is called R-*closed* (similarly one defines R^--closed sets).

The image by an R-polarity consists precisely of R^--closed sets and the image its conjugate polarity of R-closed sets. The restriction to R-closed sets of a polarity P_R is an isomorphism onto R^--closed sets, P_{R^-} being the inverse isomorphism.

If M is a subset of X and N a subset of Y, then the polarity with respect to $R \cap M \times N$ maps the subsets A of M into $N \cap P_R A$ and other subsets of X into the empty set. We shall call it the restriction of R to $M \times N$.

We shall consider certain polarities defined with the aid of extended-real-valued functions $\langle \, , \, \rangle$ on $X \times Y$.

The *level* polarity with respect to $\langle \, , \, \rangle$ is defined by the relation $\{(x,y): \langle x,y \rangle \le 0\}$ (we shall consider here only 0-level relations). When $\langle \, , \, \rangle$ is a bilinear coupling function, then the corresponding level polarity maps each set A on its dual cone A^* .

The *epigraphic* polarities are those polarities from $X \times \mathbb{R}$ to Y (or from X to $Y \times R$), where R stands for the real line, that are generated by the relation epi $\langle\, ,\rangle = \{(x,r;y) : \langle x,y\rangle \le r\}$. This polarity maps every subset A of Y onto the epigraph of its *support* function $\sup_A$ on X defined by

$$\sup_A x = \sup_{y \in A} \langle x,y\rangle$$

The polarly closed subsets of $X \times R$ are, in this case, the epigraphs of support functions, while the polarly closed subsets of Y are $\langle\, ,\rangle$-*convexoid* sets, i.e., the sets B of the form

$$B = \bigcap_{(x,r)\in A} \{ y : \langle x,y\rangle \le r \}$$

Clearly, in the case of bilinear coupling functions, these are precisely convex $\sigma(Y,X)$ closed sets.

Finally, the *diepigraphic* polarities with respect to $\langle\, ,\rangle$ map the epigraphs of (extended-real-valued) functions on X to the epigraphs of functions on Y. They are determined by the relation on $(X \times R) \times (Y \times R)$

$$\text{diepi} \langle\, ,\rangle = \{(x,r;y,s) : \langle x,y\rangle \le r+s\} .$$

Given an (arbitrary) function $\langle\, ,\rangle$ on $X \times Y$ and a function f on X (g on Y) the *conjugate* function $f^{\langle,\rangle}$ is defined on Y (respectively, $g^{\langle,\rangle}$ on X) through

$$f^{\langle,\rangle}(y) = \sup_{x\in X} [\langle x,y\rangle \dot{+} -f(x)]$$

(analogously for $g^{\langle,\rangle}$) where $\dot{+}$ is the lower extension of the addition (Moreau [20]). We observe that

Proposition 2.1

The diepigraphic polar of the epigraph of a function f is equal to the epigraph of the conjugate of f.

Diepigraphic polarities with respect to $\langle \, , \, \rangle$ may be seen as epigraphic polarities with respect to $[\, , \,]$ (defined by $[x,r;y] = \langle x,y \rangle - r$) and similarly the latter may be seen as appropriate level polarities . Moreover, we may see, say, an epigraphic polarity as a restriction of a level polarity. Let $\langle \, , \, \rangle$ be a function on $X \times Y$. Consider the level polarity corresponding to the following coupling function $[(x,r),(y,s)] = \langle x,y \rangle + rs$. Then the polarity $P_{\mathrm{epi}\langle , \rangle}$ from $2^{X \times R}$ to 2^{Y} is the restriction of the level polarity of $[\, , \,]$ to $(X \times R) \times (Y \times \{-1\})$. One procedes similarly in the case of diepigraphic polarities (see, for example, Walkup and Wets [25]).

3. CONVERGENCES

3.a Generalities

A convergence on a set Z is a rule of assigning to every filter F on Z a subset $\lim F$ of Z. A convergence t is finer than a convergence s $(t \geq s)$, whenever, for each filter F, $\lim_t F \subseteq \lim_s F$. Consider a convergence on a set Z. A subset A of Z is *compactoid* (*compact*) if, for every ultrafilter U that contains A, $\lim U$ is non empty (intersects A). A mapping f from a convergence space (X,t) to a convergence space (Y,s) is said to be continuous, if, for each filter F on X, $f(\lim_t F) \subset \lim_s f(F)$. A convergence is called *centered*, whenever it is coarser than the *discrete* convergence i, i.e., the convergence in which for each point z there is precisely one convergent filter: the principal ultrafilter of z. A convergence is said to be isotone if $\lim F \subset \lim G$ whenever G is finer than F. A convergence is called a pseudotopology if it is isotone and if z belongs to $\lim F$, whenever z is in $\lim U$, for each ultrafilter U of F . Every convergence determines its closure: let t be a convergence on X; a point x belongs to the closure $\mathrm{cl}_t A$ of the set A, whenever there is a filter F t-convergent to x, every element of which intersects A.

Each topology defines the convergence in which a filter converges to a point z if and only if it is finer than the neighborhood filter of z. We consider convergences, not just topologies, for several reasons.

Firstly, we shall deal with sequential convergences, that is those in which the only filters that converge (i.e. have non empty limits) are the filters generated by sequences. These convergences are not topologies except for trivial cases.

Secondly, given a convergence t on Z and a family M of subsets of Z, it is often very useful to consider the convergence $t|M$ in which a filter F converges to z, if it does for t and if it contains a set from M. If t is a topology, then a filter F converges to z in $t|M$, if only if there is M in M such that F is finer than $N_t(z) \vee M$ (the trace on M of the neighborhood filter of z). This convergence is not, in general, a topology; for instance, if $K(t)$ is the family of compact sets of a regular topology t, then $t|K(t)$ (denoted also $t|K$) is a topology, if and only if t is locally compact. However $t|M$ is a pseudotopology, provided that t is a pseudotopology and that M is closed under finite unions.

Finally, continuous convergences even if constructed with the aid of topologies are not, in general, topologies. In particular, this is the case of several classical convergences of sets and functions. Let X,Z be sets, Y a subset of Z^X. Let t be a convergence on X and r a convergence on Z. The *continuous convergence* r^t is the coarsest convergence s on Y for which the coupling function $\langle\ ,\ \rangle : X \times Y \to Z$, defined by $\langle x,y \rangle = y(x)$, is continuous from $t \times s$ to r.

The pointwise convergence of elements of Z^X is a particular continuous convergence, namely, r^i, i being, as usual the discrete topology on X. Its restriction to a set Y will be denoted here by $\sigma(Y,X)$; in the case where $\langle\ ,\ \rangle$ is a bilinear coupling function and n is the usual topology of the real line, this notation coincides with the classical one used for weak topologies. It t is centered, the pointwise convergence r^i is coarser than the continuous convergence r^t. A set on which both these convergences coincide is called t-*evenly continuous*; the collection of t-evenly continuous sets will be denoted by $\mathcal{E}(t)$. Clearly, if E is t-evenly continuous, then < , > is continuous on $X \times E$ equipped with $t \times \sigma(Y,X)$. If < , > is valued in extended reals, then evenly continuous sets become equi lower semicontinuous or equicontinuous depending on whether one considers the "upper half" topology or the usual topology of the reals.

3.b Convergences of functions

Given filters $\mathcal{F}$ on I, $\mathcal{G}$ on X and a function < , > on the product of I and X, the following expressions:

$$\lim_{\Gamma(\mathcal{F}^-,\mathcal{G}^-)} \langle\, , \rangle = \sup_{G\in\mathcal{G}} \sup_{F\in\mathcal{F}} \inf_{i\in F} \inf_{x\in G} \langle i,x \rangle$$

(3.1)

$$\lim_{\Gamma(\mathcal{F}^+,\mathcal{G}^-)} \langle\, , \rangle = \sup_{G\in\mathcal{G}} \inf_{F\in\mathcal{F}} \sup_{i\in F} \inf_{x\in G} \langle i,x \rangle$$

are particular Γ-limits of De Giorgi and Franzoni ([8]).They may be used to define some convergences of families of extended-real-valued functions. Let t be an isotone convergence on X and let $\{f_i\}$ be a family of functions on X filtered by $\mathcal{F}$. This family converges to a function f in $\Gamma(^-,t^-)$ and in $\Gamma(^+,t^-)$ respectively, if

$$\left(\lim_{\Gamma(\mathcal{F}^-,t^-)} f_i\right)(x) = \inf_{\mathcal{G}} \lim_{\Gamma(\mathcal{F}^-,\mathcal{G}^-)} f_i \geq f(x)$$

(3.2)

$$\left(\lim_{\Gamma(\mathcal{F}^+,t^-)} f_i\right)(x) = \inf_{\mathcal{G}} \lim_{\Gamma(\mathcal{F}^+,\mathcal{G}^-)} f_i \leq f(x)$$

where $\mathcal{G}$ ranges over the filters convergent to x in t. Let t be a convergence on X and let n_- stand for the upper half topology of the real line. The first convergence in (3.2) is the continuous convergence n_-^t .

Let t be a topology on X and let $\mathcal{M}$ be a family of subsets of X. A family of functions $\{f_i\}$ converges to a function f in $\Gamma(-,t|M^-)$ (respectively in $\Gamma(+,t|M^-)$, if and only if $\{f_i+\psi_M\}$ $\Gamma(-,t^-)$-converges (respectively $\Gamma(+,t^-)$-converges to f, where ψ_M is the indicator function of M. We prove in [9] that in the case of Hausdorff topology and for the family of all compact sets, the above $\Gamma(-,-)$-convergence amounts to the lower convergence of $\{\inf_K f_i\}$ to $\inf_K f$ for each compact set K, which is nothing else but the t-compact n_--open topology. Dino Dal Maso gives another characterization of this compact open topology in [6].

The sequential convergences $\Gamma_{seq}(-,t^-)$ and $\Gamma_{seq}(+,t^-)$ are, by definition, the convergences $\Gamma(-,tve^-)$ and $\Gamma(+,tve^-)$, where tve is the restriction of t to sequences. It follows from [15] or [21] that in the case of sequences of functions on topological spaces our definition coincides with the classical definitions of sequential Γ-convergences.

We say that a convergence is *semi-angelic* if, for every compactoid set A each point x in cl A, there is a sequence on A convergent to x. Every Fréchet topology (in particular every first-countable topology) is semi-angelic as is the weak topology of a metrizable locally convex topology.

Theorem 3.1

If t is a semi-angelic topology, then the $\Gamma(-,t|K^-)$ and $\Gamma(-,tve^-)$ limits of sequences of functions coincide. The same is true for the corresponding $\Gamma(+,-)$ limits.

It follows that in the case of first-countable topologies all the three $\Gamma(-,-)$ convergences coincide for sequences.

The following approximation theorem of the Moreau-Yosida type may be found implicitely in the paper of Joly [17] . Let t be a locally convex topology on X. Extending the definition of [17] , we say that a subset C of the extended-real-valued functions on X is t-*correct*, if there exist a continuous seminorm p, points $x_1,\ldots,x_n$ and reals $r_1,\ldots,r_n$ such that, for each f in C there is $i=1,\ldots,n$ for which $f \geq r_i - p(.-x_i)$.

Theorem 3.2

Let P be the family of continuous seminorms of a locally convex topology t. If a family $\{f_i\}$ is t-correct, then

$$\lim_{\Gamma(-,t^-)} f_i = \sup_{p\in P} \lim_{\Gamma(-,t^-)} f_i \nabla p$$

$$\lim_{\Gamma(+,t^-)} f_i = \sup_{p\in P} \lim_{\Gamma(+,t^-)} f_i \nabla p$$

where ∇ stands for the infimal convolution (see [20]).

Note that the Γ-limits on the right-hand side are equal to the analogous Γ-limits with respect to the discrete topology, as the involved functions are equi lower semicontinuous.

3.c Convergences of Sets

One may define some classical limits of families of sets with the aid of the already introduced notions of limits applied to the families of the corresponding indicator functions [7] , namely, a family $\{A_i\}$ converges to a set A in $(-,t^-)$ or *adheres to* A, if the corresponding indicator functions $\Gamma(-,t^-)$-converge; it converges to A in $(+,t^-)$ or in the *persistence convergence*, if the corresponding indicator functions $\Gamma(+,t^-)$-converge. In the case where t is a topology on X, the latter convergence is the *lower Vietoris topology* on 2^X. It is useful to define the limits for families of sets corresponding through the indicator functions to those in (3.2):

$$\text{Lim}_{(-,t^-)} A_i \quad \text{and} \quad \text{Lim}_{(+,t^-)} A_i \tag{3.3}$$

Conversely, the above limits with respect to the convergence $t \times n$ applied to the epigraphs of $\{f_i\}$ are the epigraphs of the corresponding limits defined in (3.2).

We shall also consider the *cocompact convergence* $\chi(t)$ in which a (filtered by F) family $\{A_i\}$ converges to A if, for each compact set C disjoint from A ,there is F in $\mathcal{F}$ such that, for i in F, A_i is disjoint from C. Note that $\{A_i\}$ converges to A in the cocompact convergence with respect to t, if and only if $\{A_i \cap K\}$ converges to A(equivalently to $A \cap K$) in t. This convergence is precisely the adherence with respect to $t|K$. Similarly, the convergence of $\{A_i\}$ to A in $(+, t|K)$ amounts to $(+, t^-)$ convergence of $\{A_i \cap K\}$ to A. Similarly we define the *coclosed convergence* $\mathcal{A}(t)$ by replacing in the preceding definition "compact" by "closed". If t is a topology on X the above becomes the *upper Vietoris topology* on 2^X.

A sequence (A_n) sequentially converges to A, if for each subsequence (n_k) and every sequence $x_k \in A_{n_k}$ convergent to a point x, x belongs to A. This convergence amounts to the sequential $\Gamma(-,-)$-convergence of indicator functions. Therefore, we have

Theorem 3.3

On semi-angelic spaces the cocompact convergence and the sequential adherence coincide for sequences.

3.d Convergences of Polarly Closed Sets and Functions

We shall discuss the situations in which a convergence of a family of sets implies the convergence (of the same type) of the family of the corresponding hulls. As well, we shall see that it is possible for some convergences of polarly closed sets to replace the original topology by a coareser topology (related to the polarity) without afflicting the convergence.

Let t be a topology on X and let C be a hull operator on x. C is called *algebraic*, if, for every set A and each x in CA, there is a finite subset H of A such tha x belongs to CH. This algebraic hull operator is called *lower semicontinuous*, if, for each natural n, the relation:

$$(x_1, \ldots, x_n) \to C\{x_1, \ldots, x_n\}$$

is lower semicontinuous from t^n to t. Denote by C_t the supremum of C and of the operator cl_t.

Theorem 3.4

Let C be a t-lower semicontinuous algebraic hull operator. If $\{A_i\}$ converges to A in $(+,t^-)$, then $\{CA_i\}$ converges to CA and $\{C_tA_i\}$ converges to C_tA in $(+,t^-)$.

Well-known examples of *ℓ*.s.c. algebraic hull operators are the convex, absolutely convex, linear, affine, conical and convex conical hulls on a linear topological space. In the case of bilinear couplings and locally convex topologies compatible with them, the level, epigraphic and diepigraphic polarities give rise to hull operators which are the suprema of the topological closure operators and of some of the above listed operators. The above theorem is given in [14] for the special case of convex hull.

Let R be a relation from X to Y. A family $\mathcal{A}$ of subsets of X is called R-*separated* from a family $\mathcal{B}$ of subsets of X, if for each A in $\mathcal{A}$ disjoint from a B in $\mathcal{B}$, there is y in Y such that $A \subset R^-y$ and $B \cap R^-y = \emptyset$. A family $\mathcal{C}$ of subsets of Y is (*dually*) R-*separated* from $\mathcal{B}$, if it is R-separated from the family of R^c-polars of the sets of $\mathcal{B}$. A family $\mathcal{A}$ is R-separated from a topology t, if it is R-separated from a base of t. $\mathcal{A}$ is dually R-separated from t, if it is from a base of t.

A topology t on X is said to be R-*adapted*, if it admits a base for open sets composed of R^c-closed sets. The finest R-adapted topology coarser than t is called the R-*adaptation* of t and is denoted by $R^c t$.

Theorem 3.5

If R-closed sets are R-separated from t, then the limits (3.3) of R-closed sets with respect to t and to the R-adaptation of t are the same.

Consider a bilinear coupling $\langle\ ,\ \rangle$ on $X \times Y$ and a locally convex topology t compatible with $\langle\ ,\ \rangle$. By the Hahn-Banach theorem the adaptation of t with respect to the level polarity may be described as follows: a set Q is a neighborhood of a non zero point x, if it includes a t-open convex cone with vertex at 0. Topologies with bases composed of cones (with vertex at 0) will be called *angular*. In euclidean spaces such topologies were considered by Rockafellar and Wets under the name of "cosmic topologies"; similar idea led to the notion of "gap" between linear subspaces [18] . The quotient space of an angular topology is a Hausdorff topology on the set of rays from 0.

Lemma 3.6 Let t be a locally convex topology on X and let y be a non zero linear continuous functional on X. The restriction of t to $\{x: \langle x,y\rangle = 1\}$ is homeomorphic to the quotient of the angularization of t restricted to the rays that intersect $\{x: \langle x,y\rangle = 1\}$.

As a consequence of the above theorem, we have that

Theorem 3.7
If a topology t is compatible with the coupling of X and Y, then the $\{\langle\, ,\, \rangle \le 0\}^c$-polar of each open set (with respect to the angularization of t) is compact with respect to the angularization of $\sigma(Y,X)$.

4. CONTINUITY OF POLARITIES WITH RESPECT TO(−, +) AND $\dashv$

Let t be a topology on X and let s be a topology on Y. Consider a relation R from X to Y. A subset A of Y is called R-*fitted* from t to s, if for every open set W that includes A there exist elements $x_1,\dots,x_n$ of X and their neighborhoods $Q_1,\dots,Q_n$ such that

$$A \subset \bigcap_{k=1}^{n} Rx_k \subset \bigcap_{k=1}^{n} RQ_k \subset W \tag{4.1}$$

A subset of Y is R-fitted to s, it if is R-fitted from the discrete topology i to s.

Theorem 4.1

If a family of subsets of X converges to A in $(+,t^-)$, then the family of the corresponding R-polars converges to $P_R A$ in $\mathcal{A}(s)$, if and only if $P_R A$ is R-fitted from t to s.

For bilinear couplings < , > and adaptation of a theorem of Castaing and Valadier (see for instance [4]) implies that every convex $\sigma(Y,X)$-compact set is epi < , > -fitted to $\sigma(Y,X)$.

Theorem 4.2

A polarity P_{R^-} is continuous from $\mathcal{A}(s)$ to $(+,t^-)$, provided that

(i) R^- is lower semicontinuous,

(ii) R-polars in Y are dually R-separated from t,

(iii) t is R-adapted.

Condition (i) holds the discrete topology (and thus for every topology) on X for epigraphs (from $X \times R$ to Y) and diepigraphs if, in particular, the elements of X are upper semicontinuous in s. In the case of bilinear coupling functions and a compatible topology on Y the corresponding level relation also satisfies (i) outside zero, hence it does for the angularizations.

(ii) As we have seen (Theorem 3.7) each locally convex topology t on X admits a base which polars (with respect to $\{< , > \leq 0\}^c$ and to epi$<,>^c$ from $X \times R$ to Y) are convex sets compact with respect to the corresponding adaptations of the weak topology $\sigma(X',X)$.

On the other hand, polarly closed sets with respect to $\{ < , > \leq 0\}$, epi < , > are $\sigma(X',X)$-closed and convex. Therefore, they are polarly dually separated from t and from its polar adaptation.

(iii) Clearly, the R^c-hulls of t are R-adapted. In the discussed case of bilinear coupling and a compatible topology t, the persistence convergence with respect to t and to its adaptation coincide by virtue of Theorem 3.4.

All in all, the polarities corresponding to epi < , > , < , > being bilinear, are always continuous from $(-,\sigma(X',X)^{-})$ to $(+,t\times n^{-})$ for every locally convex topology t and similarly the level polarities with respect to angularizations. One may observe that coclosed convergence of epigraphs with respect to usual topologies is very restrictive by the theorem on compactness of the active boundary [13] , [12] .

In order to obtain a stronger result for arbitrary (!) couplings, we observe that the R^c-polars of singletons from X are of the form $R^c x$, so that the R-polars in Y are always R-separated from the discrete topology i as well as from its R-adaptation. Accordingly, we obtain the continuity of the epigraph polarity from the coclosed topology of s for which the elements of X are s-upper semicontinuous to the persistence convergence of i, as it is again possible to apply Theorem 3.4. This result is particularly useful and we shall provide its simple proof.

Theorem 4.3

If for a coupling function < , > on $X\times Y$, the elements of X are s-upper semicontinuous, then the epigraphic polarity is continuous from $\mathcal{A}(s)$ to $(+,i\times n^{-})$.

Proof: If $r>\sup_A x$, then A is included in the s-open set $\{y:\langle x,y\rangle<r\}$ thus for a family $\{A_i\}$ filtered by $\mathcal{F}$ convergent in $\mathcal{A}(s)$ to A, there is F in $\mathcal{F}$ such that, for i in F, A_i is included in $\{y:\langle x,y\rangle<r\}$, equivalently $r>\sup_{A_i} x$. #

5. CONTINUITY OF POLARITIES WITH RESPECT TO (+, −) AND (−, −)

Let t be a topology on X. Consider a relation R from X to Y. It is easy to describe the topology on 2^Y homeomorphic to $(+,t^{-})$ by the polarity P_R. We constrict a subbase $\mathcal{B}$ for open sets: let Q be a t-open set; A belongs to an element of $\mathcal{B}$ corresponding to Q, if there is x in Q such that $A\subset Rx$. We shall call this topology R-*flexible* topology of t.

Let s be a topology on Y.

Theorem 5.1
The polarity P_R is continuous from $(+,t^-)$ to $(-,s^-)$, if and only if R is $t\times s$-closed.

Proof: (Sufficiency). Let y be out of an R^- closed set A. Then there is an x such that $y \notin Rx \supset A$. By closedness, there are neighborhood Q of x and W of y such that W is disjoint from RQ. A set A' is in the neighborhood of A corresponding to Q, if there is x' in Q such that $A' \subset Rx'$, hence A' is disjoint from W.

(Necessity). Suppose that (x,y) is in $cl_{t\times s}R \setminus R$. The family of singletons $\{\{x'\}\}$ filtered by $N_t(x)$ converges to $\{x\}$ in $(+,t^-)$. On the other hand, $\mathrm{Lim}_{(N(x)^-,s^-)}Rx'$ contains y, thus $\{Rx'\}$ does not adhere to Rx for s. #

If R is a level set, an epigraph or a diepigraph of an arbitrary extended-real-valued function $\langle\, ,\, \rangle$ on $X \times Y$, then R is closed if $\langle\, ,\, \rangle$ is (jointly) lower semicontinuous; in the latter cases this is also a necessary condition. Therefore, given an arbitrary topology t on X, the coupling function is ℓ.s.c. for the product of t and $\sigma(Y,X)|E(t)$, where $E(t)$ is the family of t-evenly lower semicontinuous subsets of Y. Since the $(-,t^-)$ convergence of a family $\{A_i \cap E\}$ for every E in $E(t)$ amounts to the $(-,t|E(t)^-)$ convergence of $\{A_i\}$, we have

Corollary 5.2 Given a topology t on X, the level, epigraphic and diepigraphic polarities are continuous from the persistence topology for t to the adherence of $\sigma(Y,X)|E(t)$.

It is important to note that the above polarity is still continuous, if we replace $E(t)$ by any its subfamily. On the other hand, in the case of bilinear couplings, if t is a locally convex topology compatible with the coupling, then every equicontinuous set is $\sigma(Y,X)$ compactoid. A partial converse is true (i.e., each $\sigma(Y,X)$ compact absolutely convex set is equicontinuous), if t is a Mackey topology. Moreover, if $\sigma(Y,X)$ is quasi-complete, then $K(\sigma(Y,X)) \subset E(t)$,

Corollary 5.3 If t is a Mackey topology, then the level, epigraphic and diepigraphic polarities are continuous from the persistence topology of t to the adherence of $\sigma(Y,X)|K_c$ (where K_c denotes the family of all compact, asbolutely convex sets).

We have an analogous situation to that of Corollary 5.2, if we take an arbitrary topology r on Y and $\sigma(X,Y)$ on X. Again the convergence of $\{A_i\}$ in $(+,s|E(r)^-)$ amounts to the convergence of $\{A_i \cap E\}$ in $(+,s^-)$, for every E in $E(r)$. Hence,

Corollary 5.4 Given a topology r on Y, the level, epigraphic and diepigraphic polarities are continuous from the persistence convergence relative to $\sigma(X,Y)|E(r)$ to the adherence of r.

Let t be a topology on X and s a topology on Y. Consider a relation R from X to Y.

Theorem 5.5
The polarity P_{R^-} is continuous from the adherence of s to the persistence topology of t, provided that

(i) R^- is lower semicontinuous,

(ii) R-closed sets in X are R-separated from t,

(iii) R^c-polars of t-open sets are s-compactoid

We consider now Theorem 5.5 in the particular case of bilinear coupling functions.

Condition (i) of Theorem 5.5 (as observed in Section 4) holds for the epigraphs from $X \times R$ to Y and diepigraphs, as well as for the level relations outside zero if t is arbitrary and s is finer than $\sigma(Y,X)$. Condition (ii) is satisfied for all the discussed types of polarities provided that t is a locally convex topology on X compatible with the coupling, because then polarly closed sets are t-closed and convex. Condition (iii) is satisfied for level polarities for s the angularization of $\sigma(Y,X)$ by virtue of Theorem 3.7. The same holds for the epigraphic polarity with respect to X and $Y \times \mathbb{R}$. Hence,

Theorem 5.6

If t on X is compatible, the level and the epigraphic (in $X\times(Y\times\mathbb{R})$) polarities restricted to polarly closed sets are continuous from the adherence of $\sigma(Y,X)$ to the persistence topology of t.

If we consider now the restriction of the above bilinear coupling to $X \times K$, where K is a $\sigma(Y,X)$-compact subset of Y, then the conditions of Theorem 5.5 are satisfied for the diepigraphs and epigraphs from $X\times R$ to Y with respect to the discrete topology i of X. Thus, we recover once again Theorem 4.3. Here the bilinear case offers the opportunity of using the approximation theorem (Theorem 3.2) in order to a (partial but considerable) reinforcement of Theorem 5.6.

Indeed in this case we have, for each function f on Y valued in $(-\infty,+\infty]$

$$(f + \psi_E)^* = f^* \underline{\nabla} \sup_E , \tag{5.1}$$

where $\underline{\nabla}$ stands for the t-closed infimal convolution. If for a locally convex topology t, $E_c(t)$ is the family of t-equicontinuous absolutely convex sets, then the corresponding support functions are precisely t-continuous seminorms. As is well-known, equicontinuous sets are always $\sigma(Y,X)$-compactoid, while $\sigma(Y,X)$ compact absolutely convex sets are equicontinuous in the case of the Mackey topology $\tau(X,Y)$, thus in particular, for barreled and bornological topologies.

A subset of the set of functions on Y is called t-*proper*, if its diepigraphic polar is t-correct. Similarly, a family of subsets of Y is called t-proper, whenever its epigraphic polar is t-correct. The notion of proper sets of convex functions is due to Joly [17] , as is that of correct sets of convex functions.

Corollary 5.7 A t-proper family of convex $\sigma(X',X)$-closed subsets of Y adheres to a set A in $\sigma(X',X)|E(t)$ (respectively in $\sigma(X',X)|E_c(t)$) if and only if their support functions converge to the support function of A in $(+,t^-)$.

We recover now the following fundamental theorem of Joly [17] :

Corollary 5.8 A t-proper family of $\sigma(X',X)$-closed convex functions converges in $(-,\sigma(X',X)|E(t)^-)$ (respectively in $(-,\sigma(X',X)|E_c(t'))$, if and only if the family of corresponding Fenchel conjugates converges in $(+,t^-)$.

Proof: Consider a t-proper family adhering in $\sigma(X',X)|E(t)$. Since every equicontinuous set is weakly compact we may use Theorem 4.3 and (5.1) and, finally, Theorem 3.2 to get the $(+,t^-)$ convergence of the corresponding t-correct family of conjugate (or support) functions.The converses follow immediately from Theorem 5.1. #

6. SEMICONTINUITY OF VALUE FUNCTIONS

Consider a coupling function $<,>$ on $X \times Y$. If we equip X with the discrete topology, then the $\Gamma(+,-)$ and $\Gamma(-,-)$ convergences of support functions of subsets of Y amount to the upper and lower convergences of the suprema of the functions from X on those sets. In particular, if X reduces to a singleton {x} , then the above convergences correspond to semicontinuity properties of marginal (value) functions of parametrioc optimization problems. In this context a special case of polarly flexible topologies has been already recognized in [13] .

As a result of Theorem 4.1, we recover the following theorem of Castaing and Valadier ([4, Thm. II.20], [23]).

Theorem 6.1
Suppose that A is a convex $\sigma(Y,X)$-compact and $\{A_i\}$ a family of subsets of Y filtered by F. If, for every x in X.

$$\inf_{F\in F} \sup_{i\in F} \sup_{A_i} x \le \sup_A x \quad , \tag{6.1}$$

then $\{A_i\}$ converges to A in the coclosed topology of $\sigma(Y,X)$.

Translating the above result to the language of multifunctions,one has, for instance ,

Corollary 6.2 Let G be a relation from a topological space T to a linear topological space Y such that Gt_o is convex and weakly compact. If for every continuous linear functional x on Y, the marginal function $\inf_G x$ is weakly lower semicontinuous at t_o, then G is weakly upper semicontinuous at t_o.

The converse of the above result follows from our Theorem 4.3 on continuity of polarities from coclosed to encounter convergences which may be restated as follows

Theorem 6.3
If $\{A_i\}$ converges to A coclosedly and if x is an upper semicontinuous function, then (6.1) holds.

Again, an interpretation in the language of multifunctions yields the classical result that the upper semicontinuity of a muntifunction G and the lower semicontinuity of a function x entail the lower semicontinuity of the marginal function $\inf_G x$.

Let Y be equipped with the discrete topology i and X with a topology s. Then the whole X is equi lower semicontinuous whenever $\langle\cdot,y\rangle$ is ℓs.c. for each y in Y (in other words, s if finer than $\sigma(X,Y)$). Therefore as an immediate consequence of Theorem 5.1, one has the classical

Theorem 6.4

If $\{A_i\}$ filtered by $\mathcal{F}$ converges to A in $(+,t^-)$ and if y is ℓ.s.c. in t, then

$$(6.2) \qquad \sup_{F \in \mathcal{F}} \ \inf_{i \in F} \ \sup_{A_i} y \geq \sup_A y$$

In the language of multifunctions this means that the lower semicontinuity of a relation G and the upper semicontinuity of a function y imply the upper semicontinuity of the marginal function $\inf_G y$.

The above and the next results yield [4, Thm. II.21] of Castaing and Valadier and of Godet-Thobie [16] .

Theorem 6.5

Let X be a locally convex linear topological space and let K be its totally bounded subset. If for a family of convex closed subsets A_i and for a convex closed subset A of K (6.2) holds for every continuous linear functional y on X, then A_i converges to A in the persistence topology.

As we shall see the above theorem follows from our Theorem 5.5 on continuity of polarities from adherence to persistence topology.

Proof: We have already observed that all the asumptions of Theorem 5.5 are satisfied for the epigraphic polarity considered in $X \times (X' \times \mathbb{R})$ with $\sigma(X',X)$ on X'. More even than that, $(\text{epi} \langle\ ,\ \rangle)^c$-polars of open subsets of X are compact for the angularization of the topology s of uniform convergence on totally bounded subsets of X. Since all A_i and A are subsets of a totally bounded set K, $\Gamma(-,s^-)$ and $\Gamma(-,i)$ coincide for $\{\sup_{A_i}\}$. #

7. CONTINUITY OF CONJUGATION

In this section we shall comment on how the results on continuity of conjugation of Joly [17] and Mosco [22] relate to our Section 5.Only at this meeting in Erice I became acquainted with the papers of Matzeu [19] and Back [3] as well as with the results of Attouch presented at his talk and contained in his forthcoming book [2]. In [2] and [19] the setting is that of reflexive Banach spaces, in the latter case,separable.

Corollary 5.8 is due to Joly. It is essentially Proposition 8 of [17] together with Proposition 1 and 6.

If X' admits a metrizable locally convex topology compatible with the coupling, then, thanks to Theorem 3.1, the $\sigma(X',X)$ sequential adhering of this sequence is equivalent to the adhering in $\sigma(X',X)|K$ and thus implies the adhering with respect to $\sigma(X',X)|K_c$.

If, besides X' is quasi-complete in $\sigma(X',X)$ (in particular, in X is semireflexive), then all the weakly compact sets are equicontinuous, hence we may use Theorem 5.1 to have continuity from the persistence topology of t to the weak sequential adherence.

Recall that a sequence of functions (f_n) on a locally convex space (X,t) converges to f in the sense of *Mosco*,if it $\Gamma(+,t^-)$ and $\Gamma_{seq}(-,\sigma(X,X')^-)$ converges to f. If f is not identically $+\infty$, then (f_n) convergent to f in the sense of Mosco is proper for $\tau(X',X)$ ([17]). Thus one deduces from Corollary 5.8 the Mosco-Joly theorem ([22], [17]):

Theorem 7.1

On a reflexive Fréchet space a sequence (f_n) of proper convex ℓ.s.c. functions converges to a proper convex ℓ.s.c. function f (not equal identically to $-\infty$ not to $+\infty$) in the sense of Mosco, if and only if (f_n^*) converges to f^* in the sense of Mosco.

Theorems 1 and 2 of Back [3] may be derived from the presented theory. Take the first one, for example. The involved functions are convex, ℓ.s.c. and proper ($\not\equiv +\infty, -\infty$).

Theorem 7.2

Let t be a metrizable topology on X. Suppose that (f_n) converges to f in $\Gamma_{seq}(-,\sigma(X,X')^-)$ and in $\Gamma(+,t^-)$. Then (f_n^*) converges to f^* in $\Gamma(+,\tau(X',X)^-)$ and in $\Gamma(-,\sigma(X',X)|K_c^-)$.

Proof: By Theorem 5.1, the convergence in $\Gamma(+,t^-)$ implies the convergence of the conjugates in $\Gamma(-,\sigma(X',X)|K_c^-)$ since t is a Mackey topology; it implies also that (f_n) is $\tau(X',X)$ proper. Therefore, $\sigma(X,X')$ being semi-angelic, the convergence in $\Gamma(+,\tau(X',X)^-)$ follows from Corollary 5.8. #

We observe that Corollary 5.4 together with Theorem 3.3 imply the following result of Matzeu (in reflexive Banach spaces) which is of different type than those discussed by Joly in [17] .

Theorem 7.3

(i) If a sequence of (proper convex ℓ.s.c.) functions weakly sequentially $\Gamma(+,-)$ converges, then the sequence of the conjugates strongly $\Gamma(-,-)$ converges.

REFERENCES

[1] R.Arens, J.Dugundji, 'Topologies for function spaces'. *Pacific J. Math.*, 1(1951) 5-31.

[2] H.Attouch, 'Variational Convergences for Functions and Operators'. Pitman Ed., London (1984).

[3] K.Back, 'Continuity of the Fenchel transform of convex functions'. To appear.

[4] C.Castaing, M.Valadier, 'Convex Analysis and Measurable Multifunctions'. *Lect. Notes in Math.*, Springer Verlag, Berlin,Heidelberg,N. York (1977).

[5] G.Choquet, 'Convergences'. *Annales Univ. Grenoble*,23(1947-48),55-112.

[6] D.Dal Maso, 'Questioni di topologia legate alla Γ-convergenza'.*Quaderni Matematici*,Univ. Trieste, 65(1983).

[7] E.De Giorgi, 'Convergence problems for functionals and operators'. In 'Recent Methods in Nonlinear Analysis'. E.De Giorgi, E.Magenes and U.Mosco (Editors), Pitagora, Bologna (1979) 131-188.

[8] E.De Giorgi, T.Franzoni, 'Su un tipo di convergenza variazionale'. *Atto Acc. Naz. Lincei*, 58(8) (1975),842-850.

[9] S.Dolecki, 'Continuité des polarités'. In preparation.

[10] S.Dolecki, 'Introduzione alla teoria delle convergenze'.*Lect.Notes*, Univ. Trento (1983/84).

[11] S.Dolecki, G.H.Greco, 'Cyrtologies of convergences,I'.*Math.Nachr.*, to appear (probably 123 (1985)).

[12] S.Dolecki, G.H.Greco, A.Lechicki, 'Compactoid and compact filters'. *Pacific J. Math.*, 117(1985), 69-98.

[13] S.Dolecki, S.Rolewicz, 'Semicontinuity preserving muntifunctions'. *J. Math. Anal. Appl.* 65(1978), 26-31.

[14] S.Dolecki, G.Salinetti, R.Wets, 'Convergence of functions : equi semicontinuity'. *Trans. Amer. Math. Soc.* 276(1983), 409-429.

[15] G.H.Greco, 'Decomposizioni di semifiltri e Γ-limiti sequenziali in reticoli completamente distributivi'. *Annali Mat.Pura e Appl.*, 137 (1984), 61-82.

[16] C.Godet-Thobie, 'Multimesures et multimesures de transition'.Thése Univ. Montpellier (1975).

[17] J.L.Joly, 'Une famille de topologies sur l ensemble de fonctions convexes pour lesquelles la polarité est bicontinue'.*J.Math.Pures Appl.*, 52(1973), 421-441.

[18] T.Kato, 'Perturbation Theory for Linear Operators'.Springer Verlag, Berlin, Heidelberg, N.York (1976).

[19] M.Matzeu, 'Su un tipo di continuità dell'operatore subdifferenziale'. *Boll. U.M.I.*,14-B(1977), 480-490.

[20] J.J.Moreau, 'Fonctionnelles convexes'. Collège de France (1966).

[21] G.Moscariello, 'Γ-limiti in spazi con convergenza'. *Ric. Mat.* 28 (1979), 301-321.

[22] U.Mosco, 'Convergence of convex sets and of solutions of variational inequalities'.*Adv. in Math.* 3(1969), 510-585.

[23] M.Valadier, 'Contribution à l'analyse convexe'. Thèse, Univ. Montpellier (1970).

[24] M.Volle, 'Quelques resultats relatifs à l'approche par les tranches de l'épiconvergence'. Manuscript.

[25] D.Walkup, R.Wets, 'Continuity of some convex-cone valued mappings'. *Proc. Amer. Math. Soc.*, 18(1967), 229-235.

[26] R.A.Wijsman, 'Convergences of sequences of convex sets, cones and functions,II'. *Trans. Amer. Math. Soc.*, 123(1966), 32-45.

Chapter 9

Two Results in Convex Analysis

I. Ekeland

1. INTRODUCTION

We present two results of convex analysis. The first, which is well-known, is the theorem of Brønsted-Rockafellar ([1],[5]); the second, which is more recent, is an extension of a result of Toland ([6]). The two parts are independent, having in common only the fact that they both use the following result by the present writer ([3],[4]):

Theorem 1

Let X be a Banach space and $F: X \to \mathbb{R} \cup \{+\infty\}$ a ℓ.s.c.(1) function bounded from below. For each point $\bar{x} \in X$ and each $\lambda > 0$ there exists a point $x_\lambda \in X$ such that:

(1.1) $$F(x_\lambda) \leq F(\bar{x}) \quad (2)$$

(1) lower semi-continuous

(2) this estimate can be evaluated more precisely as:

$$F(x_\lambda) \leq F(\bar{x}) - \lambda^{-1}(F(\bar{x}) - \inf F)\|\bar{x} - x_\lambda\|$$

$$(1.2)\qquad \|\bar{x} - x_\lambda\| \le \lambda$$

$$(1.3)\qquad \forall x \neq x_\lambda\ ,\quad F(x) > F(x_\lambda) - \lambda^{-1}(F(\bar{x}) - \inf F)\|x - x_\lambda\|.$$

If we specify besides that F is convex, we obtain the following corollary, where the topological dual of X is indicated by X^*.

Corollary 1 Let X be a Banach space and $F: X \to \mathbb{R} \cup \{+\infty\}$ a convex ℓ.s.c. function bounded from below. There exist two sequences $x_n \in X$ and $x_n^* \in X^*$ such that:

$$(1.4)\qquad F(x_n) \to \inf F$$

$$(1.5)\qquad \forall n \in N\ ,\ x_n^* \in \partial F(x_n)$$

$$(1.6)\qquad x_n^* \to 0$$

$$(1.7)\qquad F(x_n) - \langle x_n^*, x_n\rangle \to \inf F\ .$$

Proof: One first constructs a sequence of points $\bar{x}_n \in X$ such that $F(\bar{x}_n) \le \inf F + n^{-1}$; this is always possible. For each fixed n, one chooses λ_n large enough so that:

$$(1.8)\qquad \lambda_n^{-1}(F(\bar{x}_n) - \inf F)\max(\|\bar{x}_n\|, 1) \le n^{-1}\ .$$

One then applies theorem 1 with $\bar{x} = \bar{x}_n$ and $\lambda = \lambda_n$. It gives us a sequence $x_n \in X$ such that

$$(1.9)\qquad F(x_n) \le F(\bar{x}_n) \le \inf F + n^{-1}$$

$$(1.10)\qquad \|\bar{x}_n - x_n\| \le \lambda_n$$

$$(1.11)\qquad \forall x \neq x_n,\ F(x) > F(x_n) + \lambda_n^{-1}(F(\bar{x}_n) - \inf F)\ \|x - x_n\|\ .\#$$

Introduce the convex cone C_n of $X \times \mathbb{R}$, defined as:

(1.12) $$C_n = \{(x, \ ) \mid \alpha - F(x_n) + \lambda_n^{-1}(F(\bar{x}_n) - \inf F)\ \|x - x_n\| < 0 .$$

The formula (1.11) implies that the epigraph of F and the convex C_n are disjoint. Since the latter is clearly open, one can apply the Hahn-Banach theorem; there exists in $X \times \mathbb{R}$ an affine and closed hyperplane H separating epi F and C_n. We easily verify that it contains the couple $(x_n, F(x_n))$, and that it cannot be vertical. It is therefore the graph of a continuous and affine function:

(1.13) $$\exists\ x_n^* \in X \ : \ (\alpha, x) \in H \Leftrightarrow \alpha = \langle x, x_n^* \rangle - \gamma_n$$

where γ_n is given by the equation $F(x_n) = \langle x_n, x_n \rangle - \gamma_n$. Since H is located below epi F, the function $x_n^* - \gamma_n$ is less than F on the entire X, and coincides with F at point x_n. This means exactly that x_n^* belongs to the sub-differential of F in x_n (which is not empty):

(1.14) $$x_n^* \in \partial F(x_n)$$

It remains only to establish formulas (1.6) and (1.7). As to the first, it is enough to write that H is located above C_n:

(1.15) $$\forall x \in X, \langle x, x_n^* \rangle - \gamma_n \geq F(x_n) - \lambda_n^{-1}(F(\bar{x}_n) - \inf F)\|x - x_n\|$$

Replacing γ_n by its value, we obtain:

(1.16) $$\forall x \in X, \langle x - x_n, x_n^* \rangle \geq - \lambda_n^{-1}(F(\bar{x}_n) - \inf F)\|x - x_n\|$$

(1.7) $$\|x_n\| \leq \lambda_n^{-1}(F(\bar{x}_n) - \inf F)$$

Applying the inequality (1.8), we obtain at once $\|x_n^*\| \to 0$. Moreover, we have:

$$(1.18)\qquad |F(x_n) - \langle x_n^*, x_n\rangle - \inf F| \le (F(x_n) - \inf F) + |\langle x_n^*, x_n\rangle|$$

Applying the inequalities (1.9) and (1.17), this gives:

$$(1.19)\qquad |F(x_n) - \langle x_n^*, x_n\rangle - \inf F| \le n^{-1} + \lambda_n^{-1}(F(\bar{x}_n) - \inf F)\|x_n\|$$

But we have $\|x_n\| \le \|\bar{x}_n\| + \lambda_n$, thanks to estimate (1.10), and we have only to use once again the inequality (1.8) to obtain the result:

$$(1.20)\qquad |F(x_n) - \langle x_n^*, x_n\rangle - \inf F| \le 3n^{-1}.$$

If one retains only formulas (1.4) and (1.5) of Corollary 1, one can state the result by saying that F admits a minimizing series made up of points where it is sub-differentiable.

2. THEOREM OF BRØNSTED-ROCKAFELLAR

In their article ([1]) Brønsted and Rockafellar studied the idea of ε-subgradient, and have drawn the following result:

Theorem 2

Let X be a Banach space and $F: X \to \mathbb{R} \cup \{+\infty\}$ a ℓ.s.c. convex function. The set of points F is sub-differentiable is dense on dom F[1].

(1) $\operatorname{dom} F = \{x \mid F(x) \ne +\infty\}$

At the price of a stronger hypothesis on X, we now get a more precise conclusion, equally due to Brønsted and Rockafellar.

Theorem 2 bis

Let X be the dual of a Banach space Y, and $F: X \to \mathbb{R} \cup \{+\infty\}$ a ℓ.s.c. convex function. For every point $x \in \text{dom } F$, there exists a sequence $x_n \in \text{dom } F$ such that:

(2.1) $$\forall n, \quad \partial F(x_n) \cap Y \neq 0$$

(2.2) $$x_n \to x \quad \text{when} \quad n \to \infty$$

(2.3) $$F(x_n) \to F(x) \quad \text{when} \quad n \to \infty$$

Proof: Let us consider the convex function $G: Y \to \mathbb{R} \cup \{+\infty\}$, conjugate of F in the duality (X,Y), defined as:

(2.4) $$G(y) = \sup \{< x,y > - F(x) \mid x \in X\}$$

It is evidently lower semi-continuous on Y. From the preceding formula we deduce at once that $G(y) - < x,y > \geq -F(x)$. If $x \in \text{dom } F$, function $G - x$ is therefore bounded below on Y, and we can apply corollary 1 to it. There exist then a sequence $y_n \in Y$ and a sequence $\xi_n \in \partial(G - x)(y_n)$ such that:

(2.5) $$\|\xi_n\| \to 0$$

(2.6) $$G(y_n) - < x,y_n > - < \xi_n, y_n > \to \inf\{G(y) - < x,y > \mid y \in Y\}$$

But the second member of (2.6) is none other than $-G^*(x)$, that is, $-F(x)$. On the other hand, the condition $\xi_n \in \partial(G - x)(y_n)$ is conveniently expressed:

$$G(y_n) - \langle x,y_n \rangle + (G - x)^* (\xi_n) = \langle \xi_n,y_n \rangle \tag{2.7}$$

$$G(y_n) - \langle x,y_n \rangle + F(\xi_n + x) = \langle \xi_n,y_n \rangle \tag{2.8}$$

$$G(y_n) + F(\xi_n + x) = \langle \xi_n + x,y_n \rangle \tag{2.9}$$

This means precisely that $y_n \in \partial F(\xi_n + x)$. Hence formulas (2.1) and (2.2), providing we put $x_n = x + \xi_n$ and take into account (2.5) . And finally, by taking (2.9) into (2.6), we obtain exactly (2.3), and the result is established. #

Corollary 2 Let X be a reflexive Banach and $F: X \to \mathbb{R} \cup \{+\infty\}$ an ℓ.s.c. convex function. For every point $x \in \text{dom } F$, there exists a sequence x_n of points where F is sub-differentiable such that:

$$x_n \to x \tag{2.10}$$

$$F(x_n) \to F(x) \; . \tag{2.11}$$

If we keep only formula (2.10), we recover Theorem 2, in the most restrained frame-work of reflexive spaces. Formula (2.11) carries a supplementary accuracy. It by no means proceeds from the previous one, except if x is a point of continuity of F, in which case theorems 2 and 2 bis become trivial (choose the constant sequence $x = x_n$).

In any case, theorem 2 and corollary 2 are interesting only if domF has an empty Ω interior. Indeed, we know that if $\Omega \neq \emptyset$, then $\text{dom } F \subset \bar{\Omega}$ and F is continuous, therefore sub-differentiable, on Ω ([4],[5]). It is enough then to approach x by a sequence $x_n \in \Omega$, to obtain the conclusion of theorem 2. As for corollary 2, it is sufficient to choose a point $x_o \in \Omega$ and to approach x by a sequence of points x_n belonging to the segment $[x_0,x]$, which is entirely contained in Ω (except possibly the extremum x): one is led thus to the dimension 1.

3. THEOREM OF TOLAND

In a very elegant article ([6]) Toland proves, among other things, the following results. Let X and X^* be two vector spaces in separating duality; let $F: X \to \mathbb{R} \cup \{+\infty\}$ and $G: X \to \mathbb{R} \cup \{+\infty\}$ be two ℓ.s.c. convex functions, $F^*: X^* \to \mathbb{R} \cup \{+\infty\}$ and $G^*: X^* \to \mathbb{R} \cup \{+\infty\}$ be their conjugate. We consider the functionals $J: X \to \overline{\mathbb{R}}$ and $J^*: X \to \overline{\mathbb{R}}$ to be defined as:

$$J(x) = G(x) - F(x) \tag{3.1}$$

$$J^*(x^*) = F^*(x^*) - G^*(x^*) \tag{3.2}$$

with the convention that $(+\infty) - (+\infty) = +\infty$. One can say that $x \in X$ is a critical point of J if

$$\partial G(x) \cap \partial F(x) \neq \emptyset \tag{3.3}$$

with an analogous definition for J^* . We then have:

Theorem 3

(a) the functionals J and J^* have the same lower bound : $\inf J = \inf J^*$

(b) if x is a critical point of J, then every $x^* \in \partial G(x) \cap F(x)$ is a critical point of J^* . In addition:

$$J(x) = J^*(x^*). \tag{3.4}$$

Proof: ([6]) We have $G - F \geq \inf J$ by definition, which is re-written $G \geq F + \inf J$, from which comes $G^* \leq F^* - \inf J$ by taking the conjugates, whence $J^* \leq \inf J$ and thus $\inf J^* \geq \inf J$. To obtain the inverse inequality, we start from $F^* \geq G^* + \inf J^*$, and we get $F \leq G - \inf J^*$ by means of conjugation, from which $\inf J \geq \inf J^*$.

Let $x \in X$ be a critical point J and $x^* \in \partial G(x) \cap \partial F(x)$ we know that the relation $x^* \in \partial G(x)$ is equal to $x \in \partial G^*(x^*)$, and that $G(x) + G^*(x^*) = \langle x, x^* \rangle$. The same is true for F. Finally, we have in fact $x \in \partial G^*(x^*) \cap F^*(x^*)$ and $J(x) - J^*(x^*) = 0$. #

It will be noted that, if F is everywhere finite, a natural hypothesis that we will make from now on to eliminate the value $-\infty$, every minimal point for J is a critical point. In fact, F is then continuous everywhere, and thus sub-differentiable ([4],[5]). If $J(\bar{x}) = \inf J$ for a certain $\bar{x} \in X$, one can choose y^* into $F(\bar{x})$, and the inequality $J(x) > J(\bar{x})$ becomes:

$$G(x) - G(\bar{x}) > F(x) - F(\bar{x}) > \langle y^*, x - \bar{x} \rangle, \tag{3.5}$$

thus y^* belongs also to $\partial G(\bar{x})$. If then J reaches its lower boundary, the stated (b) covers the stated (a); this is no longer true if the lower boundary is not reached.

We now want to make a statement which covers both (a) and (b); we shall unfortunately reach it only by means of hypotheses more restrictive than those of Toland. For that reason, we introduce the following idea:

Definition 1 We suppose that X is a Banach and that X^* is its dual. We can say that $a \in \mathbb{R}$ is a critical value of J if there exist two sequences $x_n \in X$ and $x_n^* \in X^*$ related by $x_n^* \in \partial G(x_n) - \partial F(x_n)$ such that:

$$\|x_n\| \to 0 \tag{3.6}$$

$$J(x_n) \to a \tag{3.7}$$

$$\langle x_n, x_n^* \rangle \to 0 \tag{3.8}$$

and that three sequences $y_n \in X$, $z_n \in X$ and $y_n^* \in X$, related by $y_n^* \in \partial G(y_n) \cap \partial F(z_n)$, and such that

(3.9) $$\|y_n - z_n\| \to 0$$

(3.10) $$G(y_n) - F(z_n) \to a$$

(3.11) $$< y_n - z_n, y_n^* > \to 0.$$

These conditions are best understood in terms of envelopes. The first three express the fact that the tangent hyperplanes in x_n to J, with equation $\alpha = < x - x_n, x_n^* > + J(x_n)$, converge toward the horizontal hyperplane with ordinate a. The last three are dual of the preceding ones.

If $\bar{x}$ is a critical point, $J(\bar{x})$ is a critical value (take the constant sequences $x_n = y_n = z_n = x$, $x_n = 0$). Conversely, if a is a critical value, one can say that it is *strict* if there exist a critical point $\bar{x}$ such that $J(\bar{x}) = A$. This is by no means always true, except when one makes a hypothesis of the type condition (C) of Palais- Smale ([2]). We shall see, for example, that if inf J is finite, it is a critical value: according to proposition 1, it is strict only if this lower bound is reached.

We will denote by {ext J} the set of critical values of J.

Proposition 1

We suppose that X is a Banach and that G and F are convex (and) ℓ.s.c., the latter being everywhere finite. If inf $J > -\infty$, there exist two sequences $x_n \in X$, $x_n^* \in \partial G(x_n) - \partial F(x_n)$, verifying conditions (3.6), (3.7), (3.8) with a = inf J.

Proof: We have seen that F was everywhere continuous and sub-differentiable. Function J is thus ℓ.s.c. and bounded from below. One can therefore apply theorem 1 to it. Let $\bar{x}_n$ be a sequence of points such that $J(\bar{x}_n) \leq \inf J + n^{-1}$. For each fixed n, set $\lambda_n = \max (1, \|\bar{x}_n\|)$. The associated sequence x_n verifies:

$$\begin{cases} J(x_n) \leq \inf J + n^{-1} \\ \|x_n - \bar{x}_n\| \leq \lambda_n \\ \forall x \neq x_n,\ J(x) \quad J(x_n) - n^{-1}\lambda_n^{-1}\|x - x_n\| \end{cases}$$

Replacing J by G - F in this last formula, we obtain:

$$(3.12) \qquad \forall x \neq x_n,\ G(x) - G(x_n) \quad F(x) - F(x_n) - x^{-1}\,{}_{n}^{-1}\|x - x_n\|$$

We know that $\partial F(x_n)$ is nonempty, and we can thus choose a point y_n. By definition:

$$(3.13) \qquad \forall x \in X,\ F(x) - F(x_n) \geq \langle y_n^*, x - x_n \rangle.$$

Taking into account the inequality (3.13), we obtain:

$$(3.14) \qquad \forall x \neq x_n,\ G(x) - G(x_n) - \langle y_n^*, x - x_n \rangle \geq - n^{-1}\lambda_n^{-1}\|x - x_n\|$$

As usual, we separate in $X \times \mathbb{R}$ the epigraph of the ℓ.s.c. convex function $G - y_n^* + \langle y_n^*, x_n \rangle - G(x_n)$ from the open convex cone of the equation $\alpha < - n^{-1}\lambda_n^{-1}\|x - x_n\|$. The (closed and affine) separation hyperplane H is non-vertical and contains point $(x_n, 0)$. It is thus the graph of a refined continuous function $x \to \langle x^*, x - x_n \rangle$.

We have $\langle x^*, x - x_n \rangle \geq - n^{-1}\lambda_n^{-1}\|x - x_n\|$ for every $x \in X$, thus $\|x_n\|^* \leq n^{-1}\lambda_n^{-1} \to 0$. In the same way:

$$(3.15)\qquad |\langle x_n, x_n^*\rangle| \le \|x_n^*\|(\|\bar{x}_n\| + \lambda_n) \le 2n^{-1} \to 0$$

At last, we have in like manner:

$$(3.16)\qquad \forall x \in X,\ G(x) - G(x_n) - \langle y_n^*, x - x_n\rangle \ge \langle x_n^*, x - x_n\rangle$$

$$(3.17)\qquad \forall x \in X,\ G(x) - G(x_n) \ge \langle y_n^* + x_n^*, x - x_n\rangle$$

which proves that $z_n^* = x_n^* + y_n^* \in \partial G(x_n)$. We then have $x_n^* \in \partial G(x_n) - \partial F(x_n)$, and the sequences x_n and x_n^* have the desired properties. ⧺

Corollary 3 Let us suppose that X is a reflexive Banach, of dual X^*, that F and G are convex ℓ.s.c. functions, F and G^* being everywhere finite, and that the lower bound of $J = G - F$ is finite. Then it is the smallest of the critical values:

$$(3.18)\qquad \inf J \in \{\text{ext } J\}$$

Proof: It is enough to show that inf J is a critical value, it will necessarily be the smallest. We have already found sequences x_n and x_n^* verifying the conditions of definition 1. Sequences y_n and z_n remain to be found.

To that end we again apply proposition 1, but this time to J^*. There exist then three sequences $y_n^* \in X^*$, $y_n \in \partial G^*(y_n^*)$, $z_n \in \partial F^*(y_n^*)$, such that:

$$(3.19)\qquad \|y_n - z_n\| \to 0$$

$$(3.20)\qquad F^*(y_n^*) - G^*(y_n^*) \to \inf J^*$$

$$(3.21)\qquad \langle y_n^*, y_n - z_n\rangle \to 0$$

But coming back to the definition of sub-gradients, we have:

$$F^*(y_n^*) + F(z_n) = \langle y_n^*, z_n \rangle \tag{3.22}$$

$$G^*(y_n^*) + G(y_n) = \langle y_n^*, y_n \rangle \tag{3.23}$$

Subtracting on both sides , we obtain:

$$F^*(y_n^*) - G^*(y_n^*) + F(z_n) - G(y_n) + \langle y_n^*, z_n - y_n \rangle = 0 \tag{3.24}$$

Hence, taking account of (3.20) and (3.21):

$$G(y_n) - F(z_n) \to \inf J^* . \tag{3.25}$$

Now we know that $\inf J^* = \inf J$, according to theorem 3. Hence the result. #

It will be noted that the hypothesis that X be reflexive is indispensable. Indeed if we take for F the null function, we obtain that every convex ℓ.s.c. function $G: X \to \mathbb{R} \cup \{+\infty\}$ such that G^* be everywhere finite achieves its minimum.

Here is, finally, the predicted variant of Toland's result:

Theorem 3 bis

Let X be a reflexive Banach, of dual X^*; F and G convex ℓ.s.c. function on X, with F and G^* everywhere finite. Then J and J^* have the same critical values, and the same strict critical values:

$$\{\text{ext } J\} = \{\text{ext } J^*\}. \tag{3.26}$$

Proof: Let a be a strict critical value, and let $x_n \in X$, $x_n^* \in X^*$ be two sequences verifying the three first conditions of definition 1. We write $x_n^* = y_n^* - z_n^*$, with $y_n^* \in \partial G(x_n)$ and $z_n^* \in \partial F(x_n)$, namely

(3.27) $$G(x_n) + G^*(y_n^*) = \langle x_n, y_n^* \rangle$$

(3.28) $$F(x_n) + F^*(z_n^*) = \langle x_n, z_n^* \rangle$$

We have, in succession:

(3.29) $$\| y_n^* - z_n^* \| = \| x_n^* \| \to 0$$

(3.30) $$F^*(z_n^*) - G^*(y_n^*) = G(x_n) - F(x_n) + \langle x, z_n^* - y_n^* \rangle \to \alpha$$

(3.31) $$\langle z_n^* - y_n^*, x_n \rangle \to 0$$

Taking into account the relation $x_n^* \in \partial G^*(y_n^*) \cap \partial F^*(z_n^*)$, we see that the three sequences $y_n^* \in X^*$, $z_n^* \in X^*$, and $x_n \in X$ verify the three last conditions of definition 1 applied to J^*. Conditions (3.6), (3.7), (3.8) and (3.9), (3.10), (3.11) are thus in duality, hence the result. #

Inclosing, we call attention to the fact that in [2] there will be found a theory of non-convex duality, in finite dimensión and within C^∞, containing theorem 3 as a particular case.

REFERENCES

[1] Brønsted and Rockafellar. 'On the subdifferentiability of convex functions'. *Proc. AMS*, 16(1965), 605-611.

[2] Ekeland. 'Duality in non-convex optimization and calculus of variations'. *SIAM J. Optimization Control* 15, (1977) 905-934.

[3] Ekeland. 'On the variational principle'. *J. Math. An. Appl.* , 47 (1974), 324-353.
[4] Ekeland and Teman. 'Convex analysis and variational problems'.North-Holland-Elsevier, (1976).
[5] Moreau. 'Fonctionnelles convexes'. Séminaire Leray,Collège de France, (1967).
[6] Toland. 'A duality principle for nonconvex optimization and the calculus of variations'. *Arch.Rat.Mech.An.* 71, (1979) 41-61.

Chapter 10

Abstract Γ-Convergence

T. Franzoni

Let X be a topological space and (f_n) a sequence of functions from X into $\overline{\mathbb{R}}$, the extended real line. Given a point x in X, denoting by $\mathcal{I}(x)$ the set of the neighbourhoods of x, we write

$$\Gamma(X^-) \liminf_{n\to\infty} f_n(x) = \sup_{U\in\mathcal{I}(x)} \liminf_{n\to\infty} \inf_{\xi\in U} f_n(\xi),$$

$$\Gamma(X^-) \limsup_{n\to\infty} f_n(x) = \sup_{U\in\mathcal{I}(x)} \limsup_{n\to\infty} \inf_{\xi\in U} f_n(\xi).$$

So the following functions from X into $\overline{\mathbb{R}}$ are defined:

$$\Gamma(X^-) \liminf_{n\to\infty} f_n \quad \text{and} \quad \Gamma(X^-) \limsup_{n\to\infty} f_n .$$

We now say that $f_n(x)$ $\Gamma(X^-)$-converges in x to $\ell \in \overline{\mathbb{R}}$ if

$$\ell = \Gamma(X^-) \liminf_{n\to\infty} f_n(x) = \Gamma(X^-) \limsup_{n\to\infty} f_n(x) ,$$

and in this case we write $\ell = \Gamma(X^-) \lim_{n\to\infty} f_n(x)$.

Moreover, if the function f from X into $\overline{\mathbb{R}}$ satisfies the equality

$f(x) = \Gamma(X^-) \lim_{n\to\infty} f_n(x)$ for all x in X, then we say that (f_n) $\Gamma(X^-)$- converges to f, and we write $f = \Gamma(X^-) \lim_{n\to\infty} f_n$.

Analogously, the $\Gamma(X^+)$ convergence is introduced by setting

$$\Gamma(X^+) \liminf_{n\to\infty} f_n(x) = \inf_{U\in I(x)} \liminf_{n\to\infty} \sup_{\xi\in U} f_n(\xi),$$

$$\Gamma(X^+) \limsup_{n\to\infty} f_n(x) = \sup_{U\in I(x)} \limsup_{n\to\infty} \sup_{\xi\in U} f_n(\xi),$$

and so on.

The Γ-convergence is related to the Kuratowski convergence by the following results.

Let (Z_n) be a sequence of subsets of X, and let (χ_n) be the sequence of their characteristic functions; then

$$\Gamma(X^+) \liminf_{n\to\infty} \chi_n = \chi'_\infty \quad , \quad \Gamma(X^+) \limsup_{n\to\infty} \chi_n = \chi''_\infty \quad ,$$

where χ'_∞, χ''_∞ are the characteristic functions of the inferior and superior limits, with respect to the Kuratowski convergence, of (Z_n). Moreover,

$$\Gamma(X^-) \liminf_{n\to\infty} f_n = f'_\infty \quad , \quad \Gamma(X^-) \limsup_{n\to\infty} f_n = f''_\infty$$

if, and only if, $T(f'_\infty)$ and $T(f''_\infty)$ are the superior and the inferior limits, with respect to the Kuratowski convergence, of $(T(f_n))$ (we denote by $T(f)$ the epigraph of the function f: $T(f) = \{(x,t) \in X \times \overline{\mathbb{R}} : f(x) \le t\}$).

With the previous notations, we pose

$$f'_\infty = \Gamma(X^-) \liminf_{n\to\infty} f_n \quad \text{and} \quad f''_\infty = \Gamma(X^-) \limsup_{n\to\infty} f_n.$$

We set the following results:

a) f'_∞ and f''_∞ are lower semicontinuous functions from X into $\overline{\mathbb{R}}$;

b) $f'_\infty = \Gamma(X^-) \liminf_{n\to\infty} sc^-(f_n)$, $f''_\infty = \Gamma(X^-) \limsup_{n\to\infty} sc^-(f_n)$, where $sc^-(f)$ denotes the greatest lower semicontinuous function which is less than or equal to f;

c) if $f_n = f \; \forall \; n$, then $\Gamma(X^-) \lim_{n\to\infty} f_n = sc^-(f)$;

d) if (f_n) is a sequence of equisemicontinuous functions at the point x, then $f'_\infty(x) = \liminf_{n\to\infty} f_n(x)$, $f''_\infty(x) = \limsup_{n\to\infty} f_n(x)$;

e) if (f_n) is an increasing or a uniformly convergent sequence of lower semicontinuous functions, then (f_n) $\Gamma(X^-)$- converges to $\lim_{n\to\infty} f_n$;

f) if X is a locally convex topological vector space, and (f_n) is a $\Gamma(X^-)$ convergent sequence of convex functions, or of semidefinite quadratic positive forms (which may assume the value $+\infty$), then the function $\Gamma(X^-) \lim_{n\to\infty} f_n$ is of the same type.

Let $\phi : \overline{\mathbb{R}} \to \overline{\mathbb{R}}$ be a continuous non-decreasing function, and (g_n) a sequence of functions from X into $\overline{\mathbb{R}}$, uniformly convergent to g_∞. Then we have

$$\Gamma(X^-) \liminf_{n\to\infty} (\phi \circ f_n) = \phi \circ f'_\infty ,$$

$$\Gamma(X^-) \limsup_{n\to\infty} (\phi \circ f_n) = \phi \circ f''_\infty ,$$

$$\Gamma(X^-) \liminf_{n\to\infty} (f_n + g_n) = f'_\infty + g_\infty ,$$

$$\Gamma(X^-) \limsup_{n\to\infty} (f_n + g_n) = f''_\infty + g_\infty .$$

In view of preceding results, it is possible to see the $\Gamma(X^-)$ convergence as a convergence on the set of lower semicontinuous functions from X into $\overline{\mathbb{R}}$; for this notion of convergence we have the following results:

a) if $x \in X$ satisfies the first axiom of denumerability, then there exists a sequence $(\tilde{f}_n)$, extracted from (f_n), which $\Gamma(X^-)$ - converges at the point x, such that $\Gamma(X^-) \lim_{n\to\infty} \tilde{f}_n(x) = f'_\infty(x)$;

b) $\forall\, t \in \overline{\mathbb{R}}$ such that $t < f''(x)$ there exists a sequence $(\tilde{f})$, extracted from (f_n), with the property

$$t < \Gamma(X^-) \lim_{n\to\infty} \tilde{f}_n(x) \le f''_\infty(x) \;;$$

c) if X satisfies the second axiom of denumerability, then there exists a sequence $(\tilde{f}_n)$, extracted from (f_n), which is $\Gamma(X^-)$ convergent.

The $\Gamma(X^-)$ convergence is related to the calculus of variations by the following results:

a) if A is an open subset of X, then

$$\inf_{x\in A} f'_\infty(x) \ge \liminf_{n\to\infty} (\inf_{x\in A} f_n(x)),$$

$$\inf_{x\in A} f''_\infty(x) \ge \limsup_{n\to\infty} (\inf_{x\in A} f_n(x));$$

b) if K is a compact subset of X, then

$$\min_{x\in K} f'_\infty(x) \le \liminf_{n\to\infty} (\inf_{x\in K} f_n(x));$$

c) if (f_n) $\Gamma(X^-)$ converges to f_∞, and if (x_n) is a sequence of points of X which converges to the point x_∞, and such that

$$\lim_{n\to\infty} f_n(x_n) = \lim_{n\to\infty} (\inf_{x\in K} f_n(x)),$$

then

$$\lim_{n\to\infty} f_n(x_n) = f_\infty(x_\infty) = \min_{x\in X} f_\infty(x).$$

The preceding results characterize the $\Gamma(X^-)$-convergence, in the sense that the following statement hold.

Let (f_n) be as previously, and let f be a function from X into $\overline{\mathbb{R}}$. Then $f = \Gamma(X^-) \lim_{n\to\infty} f_n$ if, and only if:

i) for all open subsets A of X we have

$$\inf_{x\in A} f(x) \geq \limsup_{n\to\infty} (\inf_{x\in A} f_n(x)),$$

ii) for all compact subsets K of X we have

$$\inf_{x\in K} f(x) \leq \sup_{A} \liminf_{n\to\infty} (\inf_{x\in A} f_n(x)) ,$$

where A varies on the open subsets of X which contain K.

If X is a metric space, and $\lambda \in \overline{\mathbb{R}}$ is greater than 0, then we define the Yoshida operator J_λ, which acts on the functions from X into $\overline{\mathbb{R}}$, according to the following definition:

$$J_\lambda f(x) = \inf_{\xi\in X} (f(\xi) + \lambda d(x,\xi)) , \ \forall\, x \in X .$$

The following result holds.

If (f_n) is a sequence of lower equibounded functions from X into $\overline{\mathbb{R}}$, then

$$\Gamma(X^-) \liminf_{n\to\infty} f_n = \lim_{\lambda\to\infty} \ \liminf_{n\to\infty} J_\lambda f_n,$$

$$\Gamma(X^-) \limsup_{n\to\infty} f_n = \lim_{\lambda\to\infty} \ \limsup_{n\to\infty} J_\lambda f_n.$$

The first generalization of the Γ convergence is due to R.Peirone, who has considered a complete lattice instead of $\overline{\mathbb{R}}$. More particularly, he has looked for a notion of convergence, for functions with values into $\overline{\mathbb{R}}^n$, in such a way that the analoguous of the theorem of the convergence of minima still holds for Pareto's minima, with respect to the order of $\bar{\mathbb{R}}^n$ induced by a convex sharp cone.

The Kuratowski convergence of the graphs of functions f_n (with some boundness condition) insures that

$$K \liminf_{n\to\infty} \text{MIN } f_h \subseteq \text{MIN } f \quad \text{(if the cone is closed)}$$

$$K \limsup_{n\to\infty} \text{MIN } f_h \supseteq \text{MIN } f \quad \text{(if the cone is open)}$$

(where K means that the Kuratowski limits are considered,and MIN denotes the set of Pareto's minima).

For this convergence, the limit of a sequence of functions from X into $\bar{\mathbb{R}}$ can be a multivalued function from X into $\bar{\mathbb{R}}$.

So, it seems to be convenient to consider the following notion of convergence, for which the preceding result on Pareto's minima still holds:

(f_n) tends to f if, and only if:

i) $f(x) \in K \liminf_{n\to\infty} f_n(x)$,

ii) $\forall\, t \in K \limsup_{n\to\infty} f_n(x)$ we have $t \geq f(x)$, for all x in X.

If the cone is closed, this convergence is equivalent to the $\Gamma(X^-)$ convergence of $(\phi \circ f_n)$ to $(\phi \circ f)$, for all linear functional ϕ, which is positive on the cone.

Unfortunately, at the moment we have no compactness theorem for this convergence.

The following step in the generalization of Γ convergence is the definition of G operator given by E.De Giorgi.

Let A be a set, and let L be a family of complete lattices.

For all $L \in L$, let adm(A,L) be the set of function f such that

$dom(f) \subseteq A$ and $range(f) \subseteq L$. g is a $G(A,\mathcal{L})$ operator if:

1) $$dom(g) = \{(f,L) : L \in \mathcal{L}, \quad f \in adm(A,L)\} ;$$

2) $\forall\, L_1, L_2 \in \mathcal{L}, \quad \forall\, f_1 \in adm(A,L_1), \quad \forall\, f_2 \in adm(A,L_2)$ such that

$$dom(f_1) \subseteq dom(f_2) ,$$

then

$$dom\ g(f_1,L_1) \subseteq dom\ g(f_2,L_2);$$

3) $\forall\, L \in \mathcal{L}, \quad \forall\, f_1, \ f_2 \in adm(A,L)$ such that

$$dom(f_1) = dom(f_2) \quad , \quad f_1 \leq_L f_2$$

then

$$g(f_1,L) \leq g(f_2,L) ;$$

4) $\forall\, L_1, L_2 \in \mathcal{L}, \quad \forall\, f \in adm(A,L_1)$, if $\phi : L_1 \to L_2$ is a complete morphism of lattices, then

$$\phi \circ g(f,L_1) = g(\phi \circ f, L_2);$$

3) and 4) implies the following property:

4') $\forall\, L \in \mathcal{L}$, if $f \in adm(A,L)$ and $f(x) = c$, $\forall\, x \in dom(f)$, then

$$g(f,L)\ (y) = c \quad , \quad \forall\, y \in dom\ g(f,L).$$

If g satisfies 1), 2), 3), 4'), then g is said to be a $G'(A,\mathcal{L})$ operator.

An example of a $G'(A,\mathcal{L})$ operator which is not a $G(A,\mathcal{L})$ is the Yosida operator previously considered.

The first results on the G operators were obtained by G.H.Greco.He has proved that if two G operators coincide on the function with values into {0,1}, then they coincide on all functions with values in any completely distributive lattice (viceversa, if a lattice L is not completely distributive, then there exist two G operators which coincide on the functions with values into {0,1} but which do not coincide on all L-valued functions).

He has also proved a representation theorem for G operators on completely distributive lattices.

APPENDIX: Γ-LIMITS OF FUNCTIONS OF TWO VARIABLES

T. Franzoni - S. Francaviglia

We give here some recent results on Γ-limits of functions of two variables. These results have been obtained with a study "internal" to Γ-convergence; so we give only some indication of the possible applications.

We plan the work with notations similar to that of Rockafellar and apply some results, preceding or contemporany, of Attouch-Wets, Buttazzo, Cavazzuti and G.H.Greco.

We denote by $\mathcal{B}$ the lattice of functions from $X \times Y$ (X,Y topological spaces) into $\overline{\mathbb{R}}$. We study on $\mathcal{B}$ Γ operators of type "lim inf" with respect to X and of type "lim sup" with respect to Y.

More precisely, we consider on $\mathcal{B}$ the operators $\Gamma(X^-)$, $\Gamma(Y^+,X^-)$, $\Gamma(X^-,Y^+)$, $\Gamma(X^+)$, which we call elementary and which are defined in the following way:

$$\Gamma(X^-)f(x,y) = \sup_{U \in \mathcal{I}(x)} \inf_{\xi \in U} f(\xi,y) ,$$

$$\Gamma(Y^+,X^-)f(x,y) = \sup_{U \in \mathcal{I}(x)} \inf_{V \in \mathcal{I}(y)} \sup_{\eta \in V} \inf_{\xi \in U} f(\xi,\eta),$$

$$\Gamma(X^-,Y^+)f(x,y) = \inf_{V\in I(y)} \sup_{U\in I(x)} \inf_{\xi\in U} \sup_{\eta\in V} f(\xi,\eta),$$

$$\Gamma(Y^+)f(x,y) = \inf_{V\in I(y)} \sup_{\eta\in V} f(x,\eta)$$

for all $(x,y) \in X \times Y$ and $f \in \mathcal{B}$ ($I(x)$ and $I(y)$ denote the set of neighbourhoods of x in X and y in Y).

We set the following results:

a) for all $f \in \mathcal{B}$, fixed a point y in Y, the functions $\Gamma(X^-)f$ and $\Gamma(Y^+,X^-)f$ are lower semicontinuous from X into $\overline{\mathbb{R}}$ and, analogously, fixed a point x in X, $\Gamma(X^-,Y^+)f$ and $\Gamma(Y^+)$ are upper semicontinuous from Y into $\overline{\mathbb{R}}$.

b) Given $f \in \mathcal{B}$ and any ordinal number $\alpha > 1$, we define

$$\Gamma(Y^+,X^-)^\alpha f = \begin{cases} \Gamma(Y^+,X^-)(\Gamma(Y^+,X^-)^{\alpha-1}f) & \text{if } \alpha - 1 \text{ exists} \\ \sup_{\beta<\alpha} \Gamma(Y^+,X^-)^\beta & \text{if } \alpha - 1 \text{ does not exist} \end{cases}$$

Analogously we can define $\Gamma(X^-,Y^+)^\alpha f$.

Then we have the following inequalities:

$$\Gamma(X^-) \le \Gamma(Y^+,X^-) \le \Gamma(Y^+,X^-)^\beta \le \Gamma(Y^+,X^-)^\alpha \le$$

$$\le \Gamma(X^-)\Gamma(Y^+)\Gamma(X^-) \le \begin{matrix} \Gamma(Y^+)\Gamma(X^-) \\ \\ \Gamma(X^-)\Gamma(Y^+) \end{matrix} \le \Gamma(Y^+)\Gamma(X^-)\Gamma(Y^+) \le \Gamma(X^-,Y^+)^\alpha \le$$

$$\le \Gamma(X^-,Y^+)^\beta \le \Gamma(X^-,Y^+) \le \Gamma(Y^+)$$

for all pairs of ordinal numbers α, β such that $1 < \beta \le \alpha$; we have also $\Gamma(X^-) \le Id \le \Gamma(Y^+)$.

c) $\Gamma(X^-)^2 = \Gamma(X^-)$; $\Gamma(Y^+)^2 = \Gamma(Y^+)$;

$(\Gamma(X^-)\Gamma(Y^+))^2 = \Gamma(X^-)\Gamma(Y^+)$; $(\Gamma(Y^+)\Gamma(X^-))^2 = \Gamma(Y^+)\Gamma(X^-)$;

$\Gamma(X^-)\Gamma(Y^+,X^-) = \Gamma(Y^+,X^-)\Gamma(X^-) = \Gamma(Y^+,X^-)$;

$\Gamma(Y^+)\Gamma(X^-,Y^+) = \Gamma(X^-,Y^+)\Gamma(Y^+) = \Gamma(X^-,Y^+)$;

$\Gamma(X^-)\Gamma(X^-,Y^+) = \Gamma(Y^+,X^-)\Gamma(X^-,Y^+) = \Gamma(Y^+,X^-)\Gamma(Y^+) = \Gamma(X^-)\Gamma(Y^+)$;

$\Gamma(Y^+)\Gamma(Y^+,X^-) = \Gamma(X^-,Y^+)\Gamma(Y^+,X^-) = \Gamma(X^-,Y^+)\Gamma(X^-) = \Gamma(Y^+)\Gamma(X^-)$.

If X and Y satisfy the first countability axiom, then for all f there exists an ordinal number $\gamma < \omega_1$ (the first uncountable ordinal number) such that

$$\Gamma(X^-,Y^+)^{\gamma+1}f = \Gamma(X^-,Y^+)^{\gamma}f \quad , \quad \Gamma(Y^+,X^-)^{\gamma+1}f = \Gamma(Y^+,X^-)^{\gamma}f.$$

R.Peirone has shown that if we pose $X = Y = [0,1]$, then for all ordinal number γ less then ω_1 there exists $f \in \mathcal{B}$ such that $\Gamma(X^-,Y^+)^{\gamma+1}f \neq$ $\neq \Gamma(X^-,Y^+)^{\gamma}f$.

According to these results, if X and Y satisfy the first countability axiom, then $\Gamma(X^-,Y^+)^{\gamma} = \Gamma(X^-,Y^+)^{\omega_1}$ and $\Gamma(Y^+,X^-)^{\gamma} = \Gamma(Y^+,X^-)^{\omega_1}$ for all $\gamma > \omega_1$.

If $f \in \mathcal{B}$, we say that $(x,y) \in X \times Y$ is a saddle point of f if

$$f(x,\eta) \le f(x,y) \le f(\xi,y)$$

for all $\xi \in X$ and $\eta \in Y$.

The following result hold:

1) if (x,y) is a saddle point of f, then

$$\Gamma(X^-)\, f(x,y) = f(x,y) = \Gamma(Y^+)\, f(x,y);$$

2) if f, $g \in \mathcal{B}$, $\Gamma(X^-)\, f \le g \le \Gamma(Y^+)\, f$ and (x,y) is a saddle point of f , then (x,y) is a saddle point of g, and $g(x,y) = f(x,y)$.

If $g, h \in \mathcal{B}$, we call (g,h) a stable pair if

$$g = \Gamma(X^-)\,h \quad \text{and} \quad h = \Gamma(Y^+)\,g .$$

As G.H.Greco pointed out, the following result holds.

Let X, Y be two open convex subsets of topological vectorial spaces, let A, B be convex subsets of X, Y respectively, and suppose A or B compact. If f is a quasi convex-concave function from X into $\overline{\mathbb{R}}$, then we have

$$\inf_{\xi\in A}\ \sup_{\eta\in B}\ \Gamma(X^-)\Gamma(Y^+)f(\xi,\eta) = \sup_{\eta\in B}\ \inf_{\xi\in A}\ \Gamma(Y^+)f(\xi,\eta).$$

So, if (g,h) is a stable pair, and $f \in \mathcal{B}$ satisfies $g \le f \le h$, in the previous hypothesis, then we have

$$\inf_{\xi\in A}\ \sup_{\eta\in B}\ f(\xi,\eta) = \inf_{\eta\in B}\ \sup_{\xi\in A}\ f(\xi,\eta).$$

If (f_n) is a sequence of elements of $\mathcal{B}$, it is possible to define the following functions from $X \times Y$ into $\overline{\mathbb{R}}$, defined by

$$\Gamma(Y^+,X^-)\liminf_{n\to\infty} f_n(x,y) = \sup_{U\in\mathcal{I}(x)}\ \inf_{V\in\mathcal{I}(x)}\ \liminf_{n\to\infty}\ \sup_{\eta\in V}\ \inf_{\xi\in U} f_n(\xi,\eta),$$

$$\Gamma(X^+,Y^-)\limsup_{n\to\infty} f_n(x,y) = \sup_{U\in\mathcal{I}(x)}\ \inf_{V\in\mathcal{I}(x)}\ \limsup_{n\to\infty}\ \sup_{\eta\in V}\ \inf_{\xi\in U} f_n(\xi,\eta),$$

$$\Gamma(X^-,Y^+)\limsup_{n\to\infty} f_n(x,y) = \inf_{V\in\mathcal{I}(x)}\ \sup_{U\in\mathcal{I}(x)}\ \liminf_{n\to\infty}\ \inf_{\xi\in U}\ \sup_{\eta\in V} f_n(\xi,\eta),$$

$$\Gamma(X^-,Y^+)\limsup_{n\to\infty} f_n(x,y) = \inf_{V\in\mathcal{I}(x)}\ \sup_{U\in\mathcal{I}(x)}\ \limsup_{n\to\infty}\ \inf_{\xi\in U}\ \sup_{\eta\in V} f_n(\xi,\eta).$$

We say that a sequence (g_n,h_n) of a stable pair converges to the stable pair (g_∞,h_∞) if

$$g_\infty = \Gamma(Y^+,X^-) \limsup_{n\to\infty} f_n \quad , \quad h_\infty = \Gamma(X^-,Y^+) \limsup_{n\to\infty} f_n \ ,$$

where (f_n) is any sequence of elements of $\mathcal{B}$ such that $g_n \le f_n \le h_n$.

We have for this convergence the following compactness result.

Let (g_n,h_n) be a sequence of stable pairs of quasi convex-concave functions of $X\times Y$ in $\overline{\mathbb{R}}$, and let us suppose X and Y to be open subsets of two locally compact topological vector spaces; then there exists a sequence $(\tilde{g}_n,\tilde{h}_n)$, extracted from (g_n,h_n), that converges in the sense of preceding definitions. Moreover, the limit so obtained is a stable pair of quasi convex-concave functions.

By a theorem of Attouch and Wets, if (x_n,y_n) is a sequence of saddle points of f_n, and (x_n), (y_n) converge to x_∞ and y_∞, then (x_∞,y_∞) is a saddle point of any function f_∞ satisfying $g_\infty \le f_\infty \le h_\infty$, and we have also $f(x_\infty,y_\infty) = \lim_{n\to\infty} f_n(x_n,y_n)$.

REFERENCES

[1] H.Attouch, R.J.B.Wets. 'A convergence theorem for saddle functions'. *Transactions of the A.M.S.*, vol.280, n.1 (1983).

[2] G.Buttazzo. 'Su una definizione generale dei Γ-limiti'. *Boll.Un.Mat. Ital.*, 5(14b), (1977), 722-744.

[3] G.Choquet. 'Convergence'. *Ann. Univ. Grenoble*, 23, 59,111 (1974).

[4] E.Cavazzuti. 'Γ-limiti multipli e loro caratterizzazioni'. Atti Convegno "Studi di problemi-limite della Analisi Funzionale", Bressanone 7-9 settembre (1981).

[5] E.Cavazzuti. 'Γ-convergenze multiple, convergenza dei punti di maxmin'. To appear in *Boll. Un. Mat. Ital.*

[6] E.Cavazzuti. 'Alcune caratterizzazioni della Γ-convergenza multipla'. To appear.

[7] E.De Giorgi. 'Γ-convergenza e G-convergenza'. *Boll. Un. Mat. Ital.*, (5) 14-A (1977).

[8] E.De Giorgi. 'Convergence problems for functionals and operators'. Proceed. Int. Meeting on "Recent Methods in Nonlinear Analysis". Roma 8-12 maggio 1979, ed. by E.De Giorgi, E.Magenes, U.Mosco, Pitagora ed., Bologna (1979).

[9] E.De Giorgi. 'Generalized limits in Calculus of Variations'. Quaderni della Scuola Normale Superiore.

[10] D.Dolecki, G.H.Greco. 'Convergence and Sequential Convergence'. *U.T.M.* 106, Agosto (1982).
[11] G.H.Greco. 'Limitoidi e reticoli completi'. Rapporti Dipart. Mat. di Trento (1983).
[12] R.T.Rockafellar. 'Convex Analysis'. Princeton Univ. Press (1970).
[13] R.Peirone. 'Γ-limiti e limiti di Pareto'. *Rend. Acc. Naz.Lincei* , (1983).

Chapter 11

Constructive Aspects in Time Optimal Control

R. Hoppe

Abstract. Approximations of time optimal control problems are considered in the framework of discrete convergence in discrete approximations. The control systems are formulated in an abstract Banach space setting including both the case of distributed and boundary control. Controllability of the given and the approximating systems is studied in terms of the corresponding input maps and general convergence results are established for the reachable sets, optimal controls and minimum times.

1. INTRODUCTION

Given an initial state u^o in a reflexive, separable Banach space E, we consider a control system (C) evolving according to

(1.1) $$u(t) = S(t)u^o + L_t f, \quad t \geq 0$$

where $S(t) : E \to E, t \geq 0$, is a C_o-semigroup of type (M,ω) with infinitesimal generator $A : D(A) \subset E \to E$, the operator L_t is a bounded linear map from $L^\infty((0,t);V)$ in E, V being another reflexive, separable Banach space, and the input f is taken from the class of admissible controls

(1.2) $$F_t = \{f \in L^\infty([0,t];V) \mid \|f(\tau)\|_V \leq 1 \text{ a.e. in } [0,t]\},$$

A state $u^1 \in E$ is said to be approximately controllable if there exist

$t^\circ > 0$ and an admissible control $f \in F_{t^\circ}$ transferring the system from u° to $B(u^1,\varepsilon) = \{u \in E \mid \| u-u^1 \|_E \le \varepsilon\}$, $\varepsilon > 0$, in time t°, i.e.

$$(1.3) \qquad u(0) = u^\circ, \quad u(t^\circ) \in B(u^1,\varepsilon)$$

where $u(t)$, $t \in [0,t^\circ]$, is the corresponding admissible trajectory obtained from (1.1). The smallest t° for which (1.3) holds true is called the transition time of the admissible control f and the infimum t^* of the transition times of all admissible controls is called the minimum time with respect to u°, $B(u^1,\varepsilon)$ and F. Finally,if there exists an $f^* \in F_{t^*}$ such that (1.3) is satisfied with transition time t^*,then f^* will be denoted as optimal control.

The abstract control system (C) can serve as a model for both distributed and boundary control.In fact,if $V = E$ and L_t is given by

$$(1.4) \qquad L_t^d f = \int_0^t S(t-\tau)f(\tau)d\tau,$$

then $u(t)$,$t \ge 0$,represents the mild solution of the evolution equation

$$(1.5) \qquad \frac{d}{dt}u(t) = Au(t) + f(t), \quad t \ge 0,$$

and we may interpret (C) as a distributed control problem.On the other hand,if L_t is given by

$$(1.6) \qquad L_t^b f = - \int_0^t AS(t-\tau)Df(\tau)d\tau,$$

where $S(t)$, $t \ge 0$, is additionally supposed to be analytic and D is a bounded linear map from V in E such that

$$(1.7) \qquad \| AS(t)D\| = O(t^{\theta - 1})$$

for some $0 < \theta < 1$,then (C) may be viewed as the Banach space formulation of a boundary control problem,the operator D denoting for example the Dirichlet map (cf.[19]).

Remark. Note that in view of [19;Thm.3] condition (1.7) ensures that the input map L_t^b given by (1.6) is indeed a bounded linear operator from $L^\infty([0,t];V)$ in E. In both cases it is well known (cf. e.g. [1], [5], [10]) that if u^1 is approximately controllable, then there exists an optimal control f^* which, under some additional assumptions, is uniquely determined and satisfies the bang-bang principle.

In studying the above control problems a decisive role will be played by the adjoint operators L_t^* which can be interpreted as observability maps for the corresponding dual observed systems (see e.g. [4]). The maps L_t^* can be shown to be bounded linear operators from E^* in $L^1([0,t];V^*)\subset(L^\infty([0,t];V)^*$ given by

$$(1.8)\qquad (L_t^d)^* = S^*(t-\cdot)$$

for distributed control and by

$$(1.9)\qquad (L_t^b)^* = D^*S^*(t-\cdot)A^*$$

in case of boundary control.

For notational convenience the spaces $L^\infty([0,t];V)$ respectively $L^1([0,t];V^*)$ will henceforth be denoted by W^∞ respectively W^1.

The approximate solution of time optimal control problems both in case of distributed and boundary control has been studied by various authors (cf. e.g. [3],[8],[11],[12],[13],[14],[15]). In the sequel, following the approach in [8], we will develop a unified theory based on the concept of discrete convergence in discrete approximations. For this purposé, let us assume that $(E_n)_{\mathbb{N}}$ and $(V_n)_{\mathbb{N}}$ are sequences of reflexive Banach spaces approximating E and V in a sense which will be made precise in the next section. Furthermore, let $(S_n(t))_{\mathbb{N}}$ be a sequence of C_0-semigroups $S_n(t) : E_n \to E_n$, $t \geq 0$, $n\in\mathbb{N}$, of type (M_n,ω_n) with infinitesimal generators $A_n : D(A_n)\subseteq E_n \to E_n$ and let $(L_{t,n})_{\mathbb{N}}$ be a sequense of input maps $L_{t,n} : W_n^\infty \to E_n$, $n\in\mathbb{N}$. Given initial states u_n° and final states u_n^1

both in E_n, $n \in \mathbb{N}$, we consider control systems (C_n) given by

$$(1.10) \qquad u_n(t) = S_n(t)u_n^o + L_{t,n}f_n, \quad t \geq 0,$$

and we are looking for admissible controls $f_n \in F_{t,n}$ within the class

$$(1.11) \qquad F_{t,n} = \{f_n \in W_n^\infty = L^\infty([0,t];V_n) \mid \| f_n(\tau) \| \leq 1 \text{ a.e. in } [0,t]\}$$

steering the system from u_n^o to $B_n(u_n^1, \varepsilon)$ in some finite time t_n^o, i.e.

$$(1.12) \qquad u_n(0) = u_n^o, \qquad u_n(t_n^o) \in B_n(u_n^1, \varepsilon).$$

For distributed control the input maps $L_{t,n}$ are specified by

$$(1.13) \qquad L_{t,n}^d f_n = \int_0^t S_n(t - \tau) f_n(\tau) d\tau,$$

while for boundary control

$$(1.14) \qquad L_{t,n}^b f_n = - \int_0^t A_n S_n(t - \tau) D_n f_n(\tau) d\tau$$

assuming $S_n(t)$, $t \geq 0$, $n \in \mathbb{N}$, analytic and $D_n : V_n \to E_n$, $n \in \mathbb{N}$, bounded with

$$(1.15) \qquad \| A_n S_n(t) D_n \| = O(t^{\theta_n - 1}), \quad 0 < \theta_n < 1.$$

2. DISCRETE CONVERGENCE IN DISCRETE APPROXIMATIONS

We will shortly review some highlights in the theory of discrete convergence in discrete approximations which will serve as a basic tool in the subsequent sections. For details we refer to [6],[7],[16] and [17]. Given real Banach spaces E, E_n, $n \in \mathbb{N}$, and a sequence $R = (R_n)_{\mathbb{N}}$ of restriction operators $R_n : E \to E_n$, $n \in \mathbb{N}$, the triple $(E, \Pi E_n, R)$ is called a discrete approximation with convergent norms (cn-approximation) iff

(i) $$\| R_n(\alpha u + \beta v) - \alpha R_n u - \beta R_n v \|_{E_n} \to 0 \quad (n \in \mathbb{N}),$$
$$\alpha, \beta \in \mathbb{R}; \quad u, v \in E,$$

(ii) $$\| R_n u \|_{E_n} \to \| u \|_E \quad (n \in \mathbb{N}), \quad u \in E,$$

(iii) $$\sup_{n \in \mathbb{N}} \| R_n u \|_{E_n} < \infty, \quad u \in E.$$

A sequence $(u_n)_{\mathbb{N}'}$, of elements $u_n \in E_n$, $n \in \mathbb{N}' \subset \mathbb{N}$, is said to converge discretely strongly to an element $u \in E$ ($s\text{-}\lim_E u_n = u (n \in \mathbb{N}')$) iff $\| u_n - R_n u \|_{E_n} \to 0 \ (n \in \mathbb{N}')$.

Obviously, if $s\text{-}\lim_E u_n = u \ (n \in \mathbb{N}')$, we also have norm-convergence, i.e. $\| u_n \|_{E_n} \to \| u \|_E \ (n \in \mathbb{N}')$, which justifies to call $(E, \Pi E_n, R)$ a cn-approximation.

A dual concept is the discrete weak convergence of functionals: A sequence $(u_n^*)_{\mathbb{N}'}$, of elements $u_n^* \in E_n^*$, $n \in \mathbb{N}' \subset \mathbb{N}$, converges discretely weakly to an element $u^* \in E^*$ ($w\text{-}\lim_{E^*} u_n = u \ (n \in \mathbb{N}')$) iff for each $u \in E$ and any sequence $(u_n)_{\mathbb{N}'}$, $u_n \in E_n$, $n \in \mathbb{N}'$, we have

$$s\text{-}\lim_E u_n = u \ (n \in \mathbb{N}') \Leftrightarrow \langle u_n^*, u_n \rangle_{E_n^*, E_n} \to \langle u^*, u \rangle_{E^*, E} \quad (n \in \mathbb{N}'),$$

where $\langle \cdot, \cdot \rangle_{E_n^*, E_n}$ respectively $\langle \cdot, \cdot \rangle_{E^*, E}$ refers to the dual pairing between E_n and E_n^* resp. E and E^*.

It is easy to see that if $w\text{-}\lim_{E^*} u_n^* = u^* \ (n \in \mathbb{N}')$ then $\| u^* \|_{E^*} \le \liminf \| u_n^* \|_{E_n^*}$.

If $(E^*, \Pi E_n^*, Q)$ is a cn-approximation of the dual space E^*, and the Banach spaces E, E_n, $n \in \mathbb{N}$, are reflexive ones, we may likewise define a discrete weak convergence of sequences of elements in E_n (cf. [9]) : A sequence $(u_n)_{\mathbb{N}'}$, $u_n \in E_n$, $n \in \mathbb{N}' \subset \mathbb{N}$, converges discretely weakly to $u \in E$ ($\text{w-lim}_E\, u_n = u$ $(n \in \mathbb{N}')$) iff for each $u^* \in E^*$ and any sequence $(u_n^*)_{\mathbb{N}'}$, $u_n^* \in E_n^*$, $n \in \mathbb{N}'$, there holds

$$\text{s-lim}_{E^*}\, u_n^* = u^* \ (n \in \mathbb{N}') \Rightarrow \langle u_n^*, u_n \rangle_{E_n^*, E_n} \to \langle u^*, u \rangle_{E^*, E} \ (n \in \mathbb{N}')$$

If E is separable, we have the following equivalent characterizations of discrete strong resp. discrete weak convergence (cf. [9], [16]):

Lemma 2.1. Let $(E, \Pi E_n, R)$ and $(E^*, \Pi E_n^*, Q)$ be cn-approximations where E, E_n, $n \in \mathbb{N}$, are reflexive Banach spaces and E is separable. Then for each $u \in E$ and any sequence $(u_n)_{\mathbb{N}'}$, $u_n \in E_n$, $n \in \mathbb{N}' \subset \mathbb{N}$; we have $\text{s-lim}_E\, u_n = u$ $(n \in \mathbb{N}')$ [resp. $\text{w-lim}_E\, u_n = u$ $(n \in \mathbb{N}')$] iff each $u^* \in E^*$ and any sequence $(u_n^*)_{\mathbb{N}'}$, $u_n^* \in E_n^*$, $n \in \mathbb{N}'$, such that $\text{w-lim}_{E^*} u_n^* = u^*$ $(n \in \mathbb{N}')$ [resp. $\text{s-lim}_{E^*} u_n^* = u^*$ $(n \in \mathbb{N}')$] there holds $\langle u_n^*, u_n \rangle_{E_n^*, E_n} \to \langle u^*, u \rangle_{E^*, E}$ $(n \in \mathbb{N}')$.

Moreover, we have the following discrete weak compactness of bounded sequences in E_n resp. E_n^* (cf. [6]):

Lemma 2.2. Under the same hypotheses as in Lemma 2.1, for any bounded sequence $(u_n)_{\mathbb{N}'}$, $u_n \in E_n$, $n \in \mathbb{N}' \subset \mathbb{N}$ [respectively $(u_n^*)_{\mathbb{N}'}$, $u_n \in E_n^*$, $n \in \mathbb{N}' \subset \mathbb{N}$] there exist a subsequence $\mathbb{N}'' \subset \mathbb{N}'$ and an element $u \in E$ [resp. $u^* \in E^*$] such that $\text{w-lim}_E\, u_n = u$ $(n \in \mathbb{N}'')$ and $\| u_n \|_{E_n} \to \| u \|_E$ $(n \in \mathbb{N}'')$ [resp. $\text{w-lim}_{E^*}\, u_n^* = u^*$ $(n \in \mathbb{N}'')$ and $\| u_n^* \|_{E_n^*} \to \| u^* \|_{E^*}$ $(n \in \mathbb{N}'')$].

We also need the notions of strong resp. weak limits of subsets, $\Omega_n \subset E_n$, $n \in \mathbb{N}' \subset \mathbb{N}$. We define

$$\text{s-Lim sup}_E\, \Omega_n = \{ u \in E \mid \exists\, (u_n)_{\mathbb{N}''}, u_n \in \Omega_n, n \in \mathbb{N}'' \subset \mathbb{N}' : \text{s-lim}_E\, u_n = u \ (n \in \mathbb{N}'') \},$$

$$\text{s-Lim inf}_E\, \Omega_n = \{ u \in E \mid \exists\, (u_n)_{\mathbb{N}'}, u_n \in \Omega_n, n \in \mathbb{N}' : \text{s-lim}_E\, u_n = u \ (n \in \mathbb{N}') \}.$$

Similarly, we introduce the sets w-Lim $\sup_E \Omega_n$ and w-Lim $\inf_E \Omega_n$ replacing discrete strong by discrete weak convergence in the above definitions.

Clearly, we have the following inclusions

(i) w-Lim $\inf_E \Omega_n \subseteq$ w-Lim $\sup_E \Omega_n$, s-Lim $\inf_E \Omega_n \subseteq$ s-Lim $\sup_E \Omega_n$,

(ii) s-Lim $\inf_E \Omega_n \subseteq$ w-Lim $\inf_E \Omega_n$, s-Lim $\sup_E \Omega_n \subseteq$ w-Lim $\sup_E \Omega_n$.

If in (i) equality holds, we simply write w-$\mathrm{Lim}_E\, \Omega_n$ respectively s-$\mathrm{Lim}_E\, \Omega_n$.

Moreover, if w-Lim $\sup_E \Omega_n \subseteq$ s-Lim $\inf_E \Omega_n$, all the limit sets coincide and will be denoted by $\mathrm{Lim}_E\, \Omega_n$.

We denote by $B(E,F)$ the Banach algebra of bounded linear operators $B : E \to F$ and by $C(E,F)$ the class of densely defined linear operators A with domain $D(A)$ in E range $R(A)$ in F. Let us assume that $(E,\Pi E_n,R^E)$, $(E^*,\Pi E_n^*,Q^{E^*})$ and $(F,\Pi F_n,R^F)$, $(F^*,\Pi F_n^*,Q^{F^*})$ are cn-approximations satisfying the assumptions of Lemma 2.1. A sequence $(A_n)_{\mathbb{N}}$ of operators $A_n \in C(E_n,F_n)$, $n\in\mathbb{N}$, is said to converge discretely strongly [discretely weakly] to an operator $A\in C(E,F)$ ($A_n \to A$ ($n\in\mathbb{N}$))[respectively $A_n \rightharpoonup A$ ($n\in\mathbb{N}$)] iff for each $u\in D(A)$ there is a sequence $(u_n)_{\mathbb{N}}$, $u_n\in D(A_n)$, $n\in\mathbb{N}$, such that s-$\lim_E u_n = u$ ($n\in\mathbb{N}$) and s-$\lim_F A_n u_n = Au$ ($n\in\mathbb{N}$) [resp. w-$\lim_E u_n = u$ ($n\in\mathbb{N}$) and w-$\lim_F A_n u_n = Au$ ($n\in\mathbb{N}$)].

The sequence $(A_n)_{\mathbb{N}}$ is consistent with A iff for each $u\in D(A)$ there exists a sequence $(u_n)_{\mathbb{N}}$, $u_n\in D(A_n)$, $n\in\mathbb{N}$, such that s-$\lim_E u_n = u(n\in\mathbb{N})$ and s-$\lim_F A_n u_n = Au$ ($n\in\mathbb{N}$). If $B_n\in B(E_n,F_n)$, $n\in\mathbb{N}$, the sequence $(B_n)_{\mathbb{N}}$ is called stable iff $\limsup_{n\in\mathbb{N}} \|B_n\| < \infty$ and inversely stable iff there exist a positive constant γ and a final piece $\mathbb{N}_1 \subset \mathbb{N}$ such that $\|B_n u_n\|_{F_n} \geq \gamma \|u_n\|_{E_n}$, $u_n\in E_n$, $n\in\mathbb{N}_1$.

Lemma 2.3. (<u>cf</u>. [16]). Let $B_n \in B(E_n,F_n)$, $n\in\mathbb{N}$, <u>and</u> $B\in B(E,F)$. <u>Then there holds:</u>

(i) $B_n \to B$ $(n\in\mathbb{N})$ <u>if and only if the sequence</u> $(B_n)_{\mathbb{N}}$ <u>is stable and consistent with</u> B.

(ii)
$$B_n \to B\ (n\in\mathbb{N}) \Longleftrightarrow B_n^* \to B^*\ (n\in\mathbb{N})$$
$$B_n \to B\ (n\in\mathbb{N}) \Longleftrightarrow B_n^* \to B^*\ (n\in\mathbb{N}).$$

(iii) If $(B_n)_{\mathbb{N}}$ <u>is inversely stable and consistent with</u> B, <u>then</u> B <u>is injective</u>.

(iv) <u>The sequence</u> $(B_n)_{\mathbb{N}}$ <u>is inversely stable, consistent with</u> B <u>and</u> s-lim $\sup_F R(B_n) \subset R(B)$ <u>if and only if</u> s-$\lim_F R(B_n) = R(B)$ <u>and</u> s-$\lim_F B_n u_n = Bu$ $(n\in\mathbb{N}) \Rightarrow$ s-$\lim_E u_n = u$ $(n\in\mathbb{N})$.

Another important concept is that of a-regularity (cf.[6],[7]) : A pair $A,(A_n)_{\mathbb{N}}$ of operators $A\in C(E,F)$, $A_n\in C(E_n,F_n)$, $n\in\mathbb{N}$, is said to be a-regular iff for any bounded sequence $(u_n)_{\mathbb{N}'}$, $u_n\in D(A_n)$, $n\in\mathbb{N}'\subset\mathbb{N}$, such that s-$\lim_F A_n u_n = w$ $(n\in\mathbb{N}')$ for some $w\in F$ there exist a subsequence $\mathbb{N}''\subset\mathbb{N}'$ and an $u\in D(A)$ such that s-$\lim_E u_n = u$ $(n\in\mathbb{N}'')$ and $w = Au$.

Obviously, a-regular operators satisfy s-Lim $\sup_F R(A_n) \subseteq R(A)$.

Moreover, we have (cf. [6],[7]):

Lemma 2.4. <u>Suppose</u> $B\in B(E,F)$ <u>and</u> $B_n\in B(E_n,F_n)$, $n\in\mathbb{N}$, <u>to be a-regular. Then there holds:</u>

(i) <u>If</u> B <u>is injective, then</u> $(B_n)_{\mathbb{N}}$ <u>is inversely stable</u>.

(ii) <u>Suppose</u> $(B_n)_{\mathbb{N}}$ <u>to be stable and let</u> $A\in C(E,F)$, $A_n\in C(E_n,F_n)$, $n\in\mathbb{N}$. <u>Then, if the pair</u> $A,(A_n)_{\mathbb{N}}$ <u>is a-regular, so is the pair</u> $BA,(B_nA_n)_{\mathbb{N}}$.

(iii) <u>If</u> B <u>is injective and the pair</u> $B,(B_n)_{\mathbb{N}}$ <u>is a-regular and consistent, then there exists a constant</u> $\gamma>0$ <u>such that</u> $\|Bu\|_F \ge \gamma\|u\|_E$, $u\in E$.

We close this section with the notion of discrete compactness of operators : A sequence $(B_n)_{\mathbb{N}}$ of operators $B_n \in \mathcal{B}(E_n, F_n)$, $n \in \mathbb{N}$ is called discretely compact iff given a bounded sequence $(u_n)_{\mathbb{N}}$, $u_n \in E_n$, $n \in \mathbb{N}$; for any subsequence $\mathbb{N}' \subset \mathbb{N}$ the sequence $(B_n u_n)_{\mathbb{N}'}$ contains a discretely strongly convergent subsequence.

3. CONVERGENCE OF INPUT MAPS AND REACHABLE SETS

Throughout the following we will assume that $(E, \Pi E_n, R^E), (E^*, \Pi E_n^*, Q^{E^*})$, $(V, \Pi V_n, R^V)$ and $(V^*, \Pi V_n^*, Q^{V^*})$ are cn-approximations of reflexive, separable Banach spaces E, V respectively their duals by sequences of reflexive Banach spaces E_n, V_n $n \in \mathbb{N}$, respectively their duals. Then, we canonically get cn-approximations $(W^\infty, \Pi W_n^\infty, R^{W^\infty})$ $(W^1, \Pi W_n^1, Q^{W^1})$ by setting $(R_n^{W^\infty} f(\tau) = R_n^V f(\tau)$, $(Q_n^{W^1} f^*)(\tau) = Q_n^{V^*} f^*(\tau)$, $\tau \in [0,t]$, $n \in \mathbb{N}$.

We will begin with some basic controllability results. The control systems (C) abd (C_n) are exactly controllable iff the input maps L_t and $L_{t,n}$, $n \in \mathbb{N}$, $t > 0$, are surjective, i.e. $R(L_t) = E$ and $R(L_{t,n}) = E_n$, $n \in \mathbb{N}$, and approximately controllable iff $\mathrm{cl}\, R(L_t) = E$ and $\mathrm{cl}\, R(L_{t,n}) = E_n$, $n \in \mathbb{N}$. A necessary and sufficient condition for exact controllability is the existence of positive constants $\gamma(t)$ and $\gamma_n(t)$, $n \in \mathbb{N}$, such that

(3.1a) $$\| L_t^* u^* \|_{W^1} \geq \gamma(t) \| u^* \|_{E^*}, \quad u^* \in E^*$$

(3.1b) $$\| L_{t,n}^* u_n^* \|_{W_n^1} \geq \gamma_n(t) \| u_n^* \|_{E_n^*}, \quad u_n^* \in E_n^*,$$

while approximate controllability holds iff $N(L_t^*) = \{0\}$ and $N(L_{t,n}^*) = \{0\}$, $N(L_t^*)$ and $N(L_{t,n}^*)$ denoting the null spaces of L_t^* and $L_{t,n}^*$ respectively.

Clearly, (3.1a) resp. (3.1b) holds true if and only if $B_E(0,\gamma(t)) \subseteq L_t B_{W^\infty}(0,1)$ resp. $B_{E_n}(0,\gamma_n(t)) \subseteq L_{t,n} B_{W_n^\infty}(0,1)$. Due to this fact, the control systems (C_n) are said to be asymptotically uniformly exactly controllable if there exist $\gamma_0(t) > 0$ and a final piece $\mathbb{N}_1 \subset \mathbb{N}$ such that (3.1b) is satisfied for all $n \in \mathbb{N}_1$ with $\gamma_n(t)$ replaced by $\gamma_0(t)$. Consequently, we have the following obvious criterion for asymptotic uniform exact controllability:

Lemma 3.1. The control systems (C_n) are asymptotically uniformly exactly controllable if and only if the sequence $(L^*_{t,n})_{\mathbb{N}}$ is inversely stable.

Moreover, in view of Lemma 2.3(iii) and Lemma 2.4(i),(iii) we get the following relationship between approximate controllability of (C) and asymptotic uniform exact controllability of (C_n):

Theorem 3.2.

(i) If (C) is approximately controllable and the pair $L^*_t, (L^*_{t,n})_{\mathbb{N}}$ is a-regular, then (C_n) is asymptotically uniformly exactly controllable.

(ii) Conversely, if for (C_n) asymptotic uniform exact controllability holds true and $(L^*_{t,n})$ is consistent with L^*_t, then (C) is approximately controllable. If, additionally, the pair $L^*_t, (L^*_{t,n})$ is a-regular, then (C) is exactly controllable.

In finite dimensional spaces the notions of exact and approximate controllability coincide while in the infinite dimensional case it is well known that many control systems are not exactly but only approximately controllable (cf. [18]). So, with regard to applications, where (C_n) is usually obtained from (C) by finite difference or finite element techniques, the standard situation will be that (C) is approximately controllable while (C_n) is asymptotically uniformly exactly controllable.

The results of Theorem 3.2 require a detailed study of the input maps and their adjoints. As a first step in this direction we will establish convergence criteria for the semigroups $S_n(t)$, $n\in\mathbb{N}$, and their adjoints. For this purpose let us make the following assumptions: $S(t)$ and $S_n(t)$, $n\in\mathbb{N}$, $t \geq 0$, are C_o-semigroups of type (M,ω) and (M_n,ω_n) with infinitesimal generators $A\in C(E,E)$ and $A_n\in C(E_n,E_n)$ such that

(A_1) $\bar{M} = \limsup M_n < \infty$, $\bar{\omega} = \limsup \omega_n < \infty$,

(A_2) the pair $(\lambda I - A),(\lambda I_n - A_n)_{\mathbb{N}}$, $\lambda > \max(\omega,\bar{\omega})$, is a-regular and consistent,

(A_3) the pair $(\lambda I^* - A^*),(\lambda I_n^* - A_n^*)_{\mathbb{N}}$, $\lambda > \max(\omega,\bar{\omega})$ is a-regular and consistent.

Theorem 3.3.
Suppose that assumptions (A_1),(A_2) [respectively (A_1),(A_3)] hold true. Then $S_n(t) \to S(t)$ $(n\in\mathbb{N})$ [respectively $S_n^*(t)\to S^*(t)$ $(n\in\mathbb{N})$] uniformly on finite subintervals of $[0,\infty)$.

Proof. The a-regularity of $(\lambda I - A),(\lambda I_n - A_n))_{\mathbb{N}}$, $\lambda > \max(\omega,\bar{\omega})$, yields $\text{s-lim sup}_E R(\lambda I_n - A_n) \leq R(\lambda I - A)$ while (A_1) implies the inverse stability of the sequence $(\lambda I_n - A_n)_{\mathbb{N}}$. Together with the consistency of $(\lambda I - A),(\lambda I_n - A_n)_{\mathbb{N}}$, Lemma 2.3 (iv) gives discrete strong convergence of the resolvents, i.e. $(\lambda I_n - A_n)^{-1} \to (\lambda I - A)^{-1}$ $(n\in\mathbb{N})$. Then, by standard arguments, one can easily deduce uniform discrete strong convergence of the semigroups $S_n(t) \to S(t)$ $(n\in\mathbb{N})$. To establish uniform discrete strong convergence of the adjoint semigroups, exactly the same arguments as before do apply. #

An immediate consequence of the preceding result is :

Corollary 3.4. Under the hypotheses of Theorem 3.3 we have $L^d_{t,n} \to L^d_t$

$(n \in \mathbb{N})$ <u>[resp.</u> $(L^d_{t,n})^* \to (L^d_t)^*$ $(n \in \mathbb{N})]$ <u>uniformly on bounded subintervals of</u> $[0,\infty)$.

In order to get convergence results for the input maps $L^b_t, L^b_{t,n}$ of the abstract boundary control systems let us state another set of assumptions:

(B_1) The sequence $(D_n)_{\mathbb{N}}$ is stable and

$$0 < \liminf \theta_n < \limsup \theta_n < 1,$$

(B_2) The pair $D,(D_n)_{\mathbb{N}}$ is consistent,

(B_3) The pair $D^*,(D^*_n)_{\mathbb{N}}$ is consistent,

(B_4) $A_n \to A$ $(n \in \mathbb{N})$,

(B_5) The pair $A^*,(A^*_n)_{\mathbb{N}}$ is consistent

(B_6) The pairs $D^*,(D^*_n)_{\mathbb{N}}$, $(L^d_t)^*,((L^d_{t,n})^*)_{\mathbb{N}}$ and $A^*,(A_n)^*$ are a-regular,

(B_7) The sequence $(D_n)_{\mathbb{N}}$ is discretely compact,

(B_8) The pair $D,(D_n)_{\mathbb{N}}$ is a-regular.

Theorem 3.5.

<u>Under conditions</u> $(A_1),(B_1)$ <u>there holds:</u>

(i) <u>If assumptions</u> $(A_2),(B_2),(B_4)$ <u>[resp.</u> $(A_3),(B_3),(B_5)]$ <u>are satisfied then</u> $L^b_{t,n} \to L^b_t$ $(n \in \mathbb{N})$ <u>[resp.</u> $(L^b_{t,n})^* \to (L^b_t)^*$ $(n \in \mathbb{N})]$ <u>uniformly on bounded subintervals of</u> $[0,\infty)$.

(ii) <u>If assumption</u> (B_6) <u>holds true, then the pair</u> $(L^b_t)^*,((L^b_{t,n})^*)_{\mathbb{N}}$ <u>is a-regular.</u>

(iii) Under conditions (A_2), (B_4) and (B_7) the sequence $(L^b_{t,n}\cdot)_{\mathbb{N}}$ is discretely compact.

Proof. (i) In view of (B_1), (B_2), Lemma 2.2 (i) gives $D_n \to D(n\in\mathbb{N})$.

Since $S_n(t) \to S(t)$ $(n\in\mathbb{N})$ by means of Theorem 3.3 and $A_n \to A(n\in\mathbb{N})$ because of (B_4), we immediately get $A_nS_n(t - \cdot)D_n \to AS(t - \cdot)D(n\in\mathbb{N})$ which in turn yields $L^b_{t,n} \to L^b_t$ $(n\in\mathbb{N})$.

In order to prove convergence of the adjoints, we claim that $((L^b_{t,n})^*)_{\mathbb{N}}$ is consistent with $(L^b_t)^*$. In fact, by (B_5) for each $u^*\in E^*\subseteq$ $\subseteq D(A^*)$ there exists $(u^*_n)_{\mathbb{N}}$, $U^*_n\in D(A^*_n)$, $n\in\mathbb{N}$, such that $\text{s-lim}_{E^*} u^*_n = u^*(n\in\mathbb{N})$ and $\text{s-lim}_{E^*} A^*_n u^*_n = A^*u^*$ $(n\in\mathbb{N})$. But $S^*_n(t - \cdot) \to S^*(t - \cdot)$ $(n\in\mathbb{N})$ and $D^*_n \to D^*(n\in\mathbb{N})$ because of Theorem 3.3 and Lemma 2.2 (i) whence $\text{s-lim}_{V^*} D^*_n S^*_n(t - \cdot)A^*_n u^*_n = D^*S^*(t - \cdot)A^*u^*$ $(n\in\mathbb{N})$ and thus $\text{s-lim}_{W^1}(L^b_{t,n})^* u^*_n = (L^b_t)^* u^*$ $(n\in\mathbb{N})$. On the other hand, (B_1) implies the stability of $((L^b_{t,n})^*)_{\mathbb{N}}$ and consequentely, the assertion follows again from Lemma 2.2 (i).

Obviously, in both cases the discrete convergence is uniform on bounded subintervals of $[0,\infty)$.

(ii) The a-regularity of the adjoint input maps follows by applying Lemma (2.4) (ii) twice.

(iii) If $(f_n)_{\mathbb{N}}$, $f_n\in W^\infty_n$, $n\in\mathbb{N}$, is a bounded sequence, then $(f_n(\tau))_{\mathbb{N}}$ is bounded for almost all $\tau\in[0,t]$ and hence, by (B_7) for any $\mathbb{N}'\subset\mathbb{N}$ there exist $\mathbb{N}''\subset\mathbb{N}'$ and $v(\tau)\in E$ such that $\text{s-lim}_E D_nf_n(\tau) = v(\tau)$ $(n\in\mathbb{N}'')$. Since $S_n(t) \to S(t)$ $(n\in\mathbb{N})$ and because of (B_4), we arrive at $\text{s-lim}_E A_nS_n(t-\tau)D_nf_n(\tau) =$ $= AS(t - \tau)v(\tau)$ $(n\in\mathbb{N}'')$ whence $\text{s-lim}_E L^b_{t,n} f_n = w$ $(n\in\mathbb{N}'')$ where

$$w = \int_0^t AS(t - \tau)v(\tau)d\tau . \qquad \#$$

We now consider the reachable sets

(3.2a) $$R_t = \{u\in E \mid u = S(t)u^0 + L_t f,\ f\in F_t\},$$

(3.2b) $$R_{t,n} = \{u_n\in E_n \mid u_n = S_n(t)u_n^0 + L_{t,n}f_n,\ f_n\in F_{t,n}\}.$$

If $u^0 = 0$, we will write R_t^0, and if $L_t = L_t^d$ respectively $L_t = L_t^b$, the corresponding sets will be denoted by $R_t^d, R_t^{d,0}$ respectively $R_t^b, R_t^{b,0}$.

The following results establish convergence of the reachable sets both in case of distributed and boundary control (the terms in brackets always will refer to the boundary control systems, i.e. $R_t = R_t^b$ etc.):

Theorem 3.6.

Under assumption (A_1) we have for each $t > 0$:

(i) If condition (A_3) is satisfied [respectively conditions $(A_3),(B_1),(B_3),(B_5)$], then the sequence $(R_{t,n}^0)_{\mathbb{N}}$ is discretely weakly compact and there holds

$$\text{w-Lim sup}_E\, R_{t,n}^0 \subseteq R_t^0.$$

Moreover, if $\text{w-lim}_E\, u_n^0 = u^0$ $(n\in\mathbb{N})$, then we also have

$$\text{w-Lim sup}_E\, R_{t,n} \subseteq R_t\ .$$

(ii) If conditions $(A_2),(B_4),(B_7),(B_8)$ hold true, then the sequence $(R_{t,n}^0)_{\mathbb{N}}$ is discretely compact and

$$\text{s-Lim sup}_E\, R_{t,n}^{b,0} \subseteq R_t^{b,0}\ .$$

(iii) Suppose that condition (A_2) is fulfilled [respectively conditions $(A_2),(B_2),(B_4)$]. Then there holds

$$R_t^0 \subseteq \text{s-Lim inf}_E\, R_{t,n}^0\ .$$

If additionally $\text{s-lim}_E\, u_n^o = u^o$ $(n \in \mathbb{N})$, then also

$$R_t \subseteq \text{s-Lim}\inf_E R_{t,n} \quad .$$

(iv) If assumptions $(A_2),(A_3)$ [respectively $(A_2),(A_3),(B_1)$-(B_5)] are met, then

$$\text{Lim}_E R_{t,n}^o = R_t^o \quad .$$

Furthermore, if $\text{s-lim}_E u_n^o = u^o$ $(n \in \mathbb{N})$, then also

$$\text{Lim}_E R_{t,n} = R_t \quad .$$

Proof. Assertions (i),(iii) and (iv) will only be shown in case of boundary control, because the corresponding proofs for distributed control follow the same pattern:

(i) Let $(u_n)_{\mathbb{N}}$ be a bounded sequence of reachable states $u_n \in R_{t,n}^{b,o}$, $n \in \mathbb{N}$, and let $\mathbb{N}' \subset \mathbb{N}$. Then there exist $f_n \in F_{t,n}$, $n \in \mathbb{N}'$, such that $u_n = L_{t,n}^b f_n$. After a correction on sets of measure zero, we may assume that for each $\tau \in [0,t]$ the sequence $(f_n(\tau))_{\mathbb{N}}$ is bounded and hence, by Lemma 2.2 there exist a subsequence $\mathbb{N}'' \subset \mathbb{N}'$ and $f(\tau) \in V$ such that $\text{w-lim}_V f_n(\tau) = f(\tau)$ $(n \in \mathbb{N}'')$ and $\| f(\tau) \|_V \le \liminf \| f_n(\tau) \|_{V_n} \le 1$, i.e. $f \in F_t$. Now, let $u_n^* \in E_n^*$, $n \in \mathbb{N}''$, and $u^* \in E^*$ such that $\text{s-lim}_{E^*} u_n^* = u^*$ $(n \in \mathbb{N}'')$. Then we have

$$\langle u_n^*, L_{t,n}^b f_n \rangle_{E_n^*, E_n} = \langle (L_{t,n}^b)^* u_n^*, f_n \rangle_{W_n^1, W_n^\infty} =$$

$$= \int_0^t \langle D_n^* S_n^*(t-\tau) A_n^* u_n^*, f_n(\tau) \rangle_{V_n^*, V_n} \, d\tau .$$

Since $\text{s-lim}_{V^*} D_n^* S_n^*(t-\tau) A_n^* u_n^* = D^* S^*(t-\tau) A^* u^*$ $(n \in \mathbb{N}'')$ because of Theorem 3.5(i), Lemma 2.1 implies that the integrand converges pointwise to $< D^*S^*(t-\tau)A^*u^*, f(\tau)>_{V^*,V}$. Moreover, the integrand is uniformly bounded and consequently, the integral converges to

$$\int_0^t < D^*S^*(t-\tau)A^*u^*, f(\tau)>_{V^*,V} d\tau =$$

$$= <(L_t^b)^* u^*, f>_{W^1,W^\infty} = < u^*, L_t^b f>_{E^*,E} .$$

We have thus shown $w\text{-}\lim_E u_n = u$ $(n\in\mathbb{N}'')$ where $u = L_t^b f$, $f\in F_t$.

Let us now assume that $u\in w\text{-}\lim\sup_E R_{t,n}^{b,0}$, i.e. there exists a sequence $(u_n)_{\mathbb{N}'}$, $u_n\in R_{t,n}^{b,0}$, $n\in\mathbb{N}'\subset\mathbb{N}$, such that $w\text{-}\lim_E u_n = u$ $(n\in\mathbb{N}')$. Because discretely weakly convergent sequences are bounded, we may use the above result to conclude that there are a subsequence $\mathbb{N}''\subset\mathbb{N}'$ and an element $f\in F_t$ such that $w\text{-}\lim_E u_n = L_t^b f$ $(n\in\mathbb{N}'')$. Since limits of discretely weakly convergent sequences are unique, we get $u = L_t^b f$, i.e. $u\in R_t^{b,0}$.

Moreover, since $S_n^*(t)\to S^*(t)$ $(n\in\mathbb{N})$, $t\geq 0$, because of Theorem 3.3, Lemma 2.1(ii) tells us $S_n(t)\rightharpoonup S(t)$ $(n\in\mathbb{N})$. So, if $w\text{-}\lim_E u_n^0 = u^0$ $(n\in\mathbb{N})$ we get $w\text{-}\lim_E S_n(t)u_n^0 = S(t)u^0$ $(n\in\mathbb{N})$. Combined with that what we have shown before, we have $w\text{-}\mathrm{Lim}\sup_E R_{t,n}^b \subset R_t^b$.

(ii) Again, taking $(u_n)_{\mathbb{N}'}$, $u_n\in R_{t,n}^{b,0}$, $n\in\mathbb{N}'\subset\mathbb{N}$, as a bounded sequence of reachable states $u_n = L_{t,n}^b f_n$, $f_n\in F_{t,n}$, $n\in\mathbb{N}'$, we may assume $(f_n(\tau))_{\mathbb{N}'}$ to be bounded for each $\tau\in[0,t]$. Assumption (B_7) implies the existence of a subsequence $\mathbb{N}''\subset\mathbb{N}'$ and an element $v(\tau)\in E$ such that $s\text{-}\lim_E D_n f_n(\tau) = v(\tau)$ $(n\in\mathbb{N}'')$. Then, by (B_8) we deduce the existence of another subsequence $\mathbb{N}'''\subset\mathbb{N}''$ and an element $f(\tau)\in V$ such that $s\text{-}\lim_V f_n(\tau) = f(\tau)$ $(n\in\mathbb{N}''')$ and $v(\tau) = Df(\tau)$. Since $\| f_n(\tau)\|_{V_n} \leq 1$, we also have

$\| f(\tau)\|_V \leq 1$, i.e. $f \in F_t$. Moreover, $S_n(t) \to S(t)$ $(n \in \mathbb{N})$, $t \geq 0$, by Theorem 3.3 and $A_n \to A$ $(n \in \mathbb{N})$ because of (B_4) whence s-$\lim_E A_n S_n(t-\tau) D_n f_n(\tau) = AS(t-\tau)Df(\tau)$ $(n \in \mathbb{N}''')$ and thus also s-$\lim_E L^b_{t,n} f_n = L^b_t f$ $(n \in \mathbb{N}''')$. If $u \in$ s-Lim $\sup_E R^{b,0}_{t,n}$, the above arguments and the uniqueness of discrete strong limits imply the existence of an $f \in F_t$ such that $u = L^b_t f$, i.e. $u \in R^{b,0}_t$.

(iii) If $u \in R^{b,0}_t$ there exists $f \in F_t$ such that $u = L^b_t f$. Assuming $\| f(\tau)\|_V \leq 1$ for all $\tau \in [0,t]$, by (B_2) for each $\tau \in [0,t]$ there exists a sequence $(f_n(\tau))_{\mathbb{N}}$, $f_n(\tau) \in V_n$, $n \in \mathbb{N}$, such that s-$\lim_V f_n(\tau) = f(\tau)$ $(n \in \mathbb{N})$ and s-$\lim_E D_n f_n(\tau) = Df(\tau)$ $(n \in \mathbb{N})$. Note that we also have norm convergence, i.e. $\| f_n(\tau)\|_{V_n} \to \| f(\tau)\|_V$ $(n \in \mathbb{N})$. So, if $\| f(\tau)\|_V < 1$, for a final piece $\mathbb{N}_1 \subset \mathbb{N}$ we also have $\| f_n(\tau)\|_{V_n} < 1$, $n \in \mathbb{N}_1$. We set $g_n(\tau) = \| f_n(\tau)\|^{-1}_{V_n} f_n(\tau)$, $n \in \mathbb{N} \setminus \mathbb{N}_1$, and $g_n(\tau) = f_n(\tau)$, $n \in \mathbb{N}_1$. If $\| f(\tau)\|_V = 1$, we define $g_n(\tau) = f_n(\tau)$, if $\| f_n(\tau)\|_{V_n} \leq 1$, and $g_n(\tau) = \| f_n(\tau)\|^{-1}_{V_n} f_n(\tau)$ otherwise. Consequently, in any case $\| g_n(\tau)\|_{V_n} \leq 1$ and s-$\lim_V g_n(\tau) = f(\tau)$ $(n \in \mathbb{N})$ as well as s-$\lim_E D_n g_n(\tau) = Df(\tau)$ $(n \in \mathbb{N})$. But $S_n(t) \to S(t)$ $(n \in \mathbb{N})$, $t \geq 0$, by Theorem 3.3 and $A_n \to A$ $(n \to \mathbb{N})$ in view of (B_4) which yields s-$\lim_E A_n S_n(t-\tau) D_n g_n(\tau) = AS(t-\tau)Df(\tau)$ $(n \in \mathbb{N})$ and also s-$\lim_E L^b_{t,n} g_n = L^b_t f$ $(n \in \mathbb{N})$. Since $g_n \in F_{t,n}$, we have $L^b_{t,n} g_n \in R^{b,0}_{t,n}$ which gives the assertion.

If s-$\lim_E u^o_n = u^o$ $(n \in \mathbb{N})$, we get s-$\lim_E S_n(t) u^o_n = S(t) u^o$ $(n \in \mathbb{N})$ and consequently, taking the above result into account, we arrive at $R^b_t \subseteq$ s-Lim $\inf_E R^b_{t,n}$.

(iv) The assertion is an immediate consequence of (i) and (iii). #

4. CONVERGENCE OF MINIMUM TIMES AND OPTIMAL CONTROLS

Based upon the convergence results for the input maps and reachable sets we are now able to establish convergence of minimum times and optimal controls. The first result concerns the approximability of (C) by (C_n):

Theorem 4.1.

Let $u^1 \in E$ be an approximately controllable state in the sense that (1.3) holds true for some $\varepsilon > 0$, let conditions $(A_1),(A_2)$ [respectively $(A_1),(A_2)$ $(B_2),(B_4)$] be satisfied and assume that (C_n) is asymptotically uniformly exactly controllable. Moreover, let $u_n^o \in E_n$, $u_n^1 \in E_n$, $n \in \mathbb{N}$, such that $\text{s-lim}_E u_n^o = u^o$ $(n \in \mathbb{N})$ and $\text{s-lim}_E u_n^1 = u^1$ $(n \in \mathbb{N})$. Then there holds:

(i) For any $\delta > 0$ there exist $\eta(\delta) > 0$ with $\eta(\delta) \to 0$ as $\delta \to 0$, a sequence $(\varepsilon_n)_{\mathbb{N}}$ of positive real numbers with $\lim_{n\to\infty} \varepsilon_n \le \varepsilon$ and a final piece $\mathbb{N}_1 \subset \mathbb{N}$ such that for each $n \in \mathbb{N}_1$ there is an $u_n \in R_{t_\varepsilon^* + \delta, n}$ satisfying

$$(4.1) \qquad \| u_n^1 - u_n \|_{E_n} \le \eta_n(\delta) = \varepsilon_n + \eta(\delta).$$

(ii) Let additionally condition (A_3) [respectively conditions (A_3), $(B_1),(B_3),(B_5)$] be fulfilled. Then, for any null sequence $(\delta_n)_{\mathbb{N}}$ of positive real numbers, denoting by t_n^* and f_n^*, $n \in \mathbb{N}_1$, the minimum time and an optimal control with respect to $u_n^o, u_n^1, B_{E_n}(u_n^1, \eta_n(\delta_n))$, we have $t_n^* \to t_\varepsilon^*$ $(n \in \mathbb{N}_1)$ and there exist a subsequence $\mathbb{N}' \subset \mathbb{N}_1$ and an optimal control $f^* \in F_{t_\varepsilon^*}$ such that $\text{w-lim}_V\, g_n^*(\tau) = f^*(\tau)$ $(n \in \mathbb{N}')$ for almost all $\tau \in [0, t^*]$ where $g_n^*(\tau) = f_n^*(\tau)$, $0 \le \tau \le t_\varepsilon^*$, and $g_n^*(\tau) = 0$ elsewhere.

(iii) In the case of boundary control, if in addition to the assumptions in (i),(ii) conditions $(B_7),(B_8)$ are met, then the same statements as in (ii) hold true with discrete weak replaced by discrete strong convergence.

Proof. Let $u^*(t)$, $0 \le t \le t_\varepsilon^*$, be the optimal trajectory with respect to u^o, u^1 and $B_E(u^1,\varepsilon)$. Since $S(t)u \to u$ as $t \to 0^+$, for given $\delta > 0$ there exists $\eta(\delta) > 0$ such that

(4.2) $$\| S(\delta)u^*(t_\varepsilon^*) - u^*(t_\varepsilon^*) \|_E < \eta(\delta).$$

Setting $v_n = R_n u^*(t_\varepsilon^*)$, $n \in \mathbb{N}$, we obviously have $\text{s-lim}_E v_n = u^*(t_\varepsilon^*)$ $(n \in \mathbb{N})$. But $S_n(\delta) \to S(\delta)$ $(n \in \mathbb{N})$ and hence, there is an $n_1(\delta) \in \mathbb{N}$ such that for all $n \ge n_1(\delta)$

(4.3) $$\| S_n(\delta)v_n - v_n \|_{E_n} \le \eta(\delta).$$

On the other hand, by Theorem 3.6(iii) there exists $(f_n)_{\mathbb{N}}$, $f_n \in F_{t_\varepsilon^*,n}$, $n \in \mathbb{N}$, such that $\text{s-lim}_E w_n = u^*(t_\varepsilon^*)$ $(n \in \mathbb{N})$ where $w_n = S_n(t_\varepsilon^*)u_n^o + L_{t_\varepsilon^*,n} f_n$. Consequently, there exists $n_2(\delta) \in \mathbb{N}$ such that for all $n \ge n_2(\delta)$

(4.4) $$\| S_n(\delta)(w_n - v_n) \|_{E_n} \le \gamma_o(\delta)$$

and thus, for $n \ge n_2(\delta)$ we find $g_n \in F_{\delta,n}$ satisfying $S_n(\delta)\,(w_n - v_n) = L_{\delta,n} g_n$. If we take $u_n = S_n(\delta)v_n$ and $h_n(\tau) = f_n(\tau)$, $0 \le \tau \le t_\varepsilon^*$, $h_n(\tau) = g_n(\tau - t_\varepsilon^*)$, $t_\varepsilon^* < \tau \le t_\varepsilon^* + \delta$, then $h_n \in F_{t_\varepsilon^*+\delta,n}$ and $u_n = S_n(t_\varepsilon^* + \delta)\, u_n^o + L_{t_\varepsilon^*+\delta,n} h_n$, i.e. $u_n \in R_{t_\varepsilon^*+\delta,n}$. Moreover,

$$\| u_n^1 - u_n \|_{E_n} \le \| u_n^1 - v_n \|_{E_n} + \| v_n - u_n \|_{E_n}.$$

But $\| u_n^1 - v_n \|_{E_n} \to \| u^1 - u^*(t^*) \|_E \le \varepsilon$ as $n \to \infty$ and thus, taking (4.3) into account, (4.1) holds true for $n \ge \max(n_1(\delta), n_2(\delta))$ with

$\varepsilon_n = \|u_n^1 - v_n\|_{E_n}$.

(ii) It follows directly from part (i) of the proof that $t_n^* \le t_\varepsilon^* + \delta_n$ for sufficiently large $n \in \mathbb{N}$, let's say $n \in \mathbb{N}_1 \subset \mathbb{N}$, and thus $\limsup t_n^* \le t_\varepsilon^*$. On the other hand, if $u_n^*(t)$, $0 \le t \le t_n^*$, is the trajectory corresponding to $f_n^* \in F_{t_n^*,n}$ and $\hat{t}_\varepsilon = \liminf t_n^*$, then $g_n^*|_{[0,\hat{t}_\varepsilon]} \in F_{\hat{t}_\varepsilon,n}$ and $u_n^*(\hat{t}_\varepsilon) \in R_{\hat{t}_\varepsilon,n}$. By Theorem 3.6(i) there exist a subsequence $\mathbb{N}' \subset \mathbb{N}_1$ and an $\hat{f} \in F_{\hat{t}_\varepsilon}$ such that $\text{w-lim}_V\, g_n^*(\tau) = \hat{f}(\tau)$ $(n \in \mathbb{N}')$ a.e. and $\text{w-lim}_E\, u_n^*(\hat{t}_\varepsilon) = u(\hat{t}_\varepsilon)$ $(n \in \mathbb{N}')$ where $u(t)$, $0 \le t \le \hat{t}_\varepsilon$, is the trajectory corresponding to $\hat{f}$. Since discrete strong implies discrete weak convergence, we also have $\text{w-lim}_E\, u_n^1 = u^1$ $(n \in \mathbb{N})$ and hence, $\text{w-lim}_E\, (u_n^1 - u_n^*(\hat{t}_\varepsilon)) = u^1 - u(\hat{t}_\varepsilon)$ $(n \in \mathbb{N}')$ whence

$$\|u^1 - u(\hat{t}_\varepsilon)\|_E \le \liminf \|u_n^1 - u_n^*(\hat{t}_\varepsilon)\|_{E_n} \le \varepsilon .$$

It follows that $t_\varepsilon^* \le \hat{t}_\varepsilon = \liminf t_n^*$ and consequently, combined with that what we have shown before, we have $t_n^* \to t_\varepsilon^*$ $(n \in \mathbb{N}_1)$.

(iii) The assertion follows along the same lines by applying Theorem 3.6(ii). #

The next result provides a partial characterization of the sets of approximately controllable states in E:

Theorem 4.2.

Under conditions $(A_1),(A_3)$ [respectively $(A_1),(A_3),(B_1),(B_3),(B_5)$] let $(u_n^o)_{\mathbb{N}}$ be a sequence of initial states $u_n^o \in E_n$, $n \in \mathbb{N}$, such that $\text{w-lim}_E\, u_n^o = u^o$ $(n \in \mathbb{N})$, and let $(u_n^1)_{\mathbb{N}}$ be a sequence of approximately controllable states $u_n^1 \in E_n$, $n \in \mathbb{N}$, such that (1.12) holds true for any sequence $(\varepsilon_n)_{\mathbb{N}}$ of positive real numbers with $\lim_{n\to\infty} \varepsilon_n = \varepsilon$ for some $\varepsilon > 0$ and assume that the corresponding sequence $(t_{\varepsilon_n}^*)_{\mathbb{N}}$ of minimum times is bounded.

Then, each $u^1 \in$ w-lim $\sup_E \{(u_n^1)_{\mathbb{N}}\}$ is an approximately controllable state in E in the sense that (1.3) holds true. In particular, there exists an $u^1 \in$ w-Lim $\sup_E \{(u_n^1)_{\mathbb{N}}\}$ with minimum time $t_\varepsilon^* \leq \liminf t_{\varepsilon_n}^*$.

If additionally condition (A_2) [respectively conditions (A_2), (B_2), (B_4)] is satisfied, (C_n) is asymptotically uniformly exactly controllable, s-$\lim_E u_n^0 = u^0$ $(n \in \mathbb{N})$ and $u^1 \in$ s-Lim $\sup_E \{(u_n^1)_{\mathbb{N}}\}$, then for a sequence $\mathbb{N}' \subset \mathbb{N}$ we have $t_{\varepsilon_n}^* \to t_\varepsilon^*$ $(n \in \mathbb{N}')$ and, denoting by $f_n^* \in F_{t_{\varepsilon_n}^*, n}$ corresponding optimal controls, there exist another subsequence $\mathbb{N}'' \subset \mathbb{N}'$ and an optimal control $f^* \in F_{t_\varepsilon^*}$ such that w-$\lim_V g_n^*(\tau) = f^*(\tau)$ $(n \in \mathbb{N}'')$ a.e. where g_n^* is defined as in Theorem 4.1(ii).
Under assumptions (B_7), (B_8) the same holds true with discrete weak replaced by discrete strong convergence.

Proof. If $u^1 \in$ w-Lim $\sup_E \{(u_n^1)_{\mathbb{N}}\}$ then there exists a subsequence $\mathbb{N}' \subset \mathbb{N}$ such that $u^1 =$ w-$\lim_E u_n^1$ $(n \in \mathbb{N}')$. Since $(t_{\varepsilon_n}^*)_{\mathbb{N}'}$ is bounded, we find $\mathbb{N}'' \subset \mathbb{N}'$ with $t_{\varepsilon_n}^* \to \hat{t}_\varepsilon$ $(n \in \mathbb{N}'')$. By Theorem 3.6(iii) there are another subsequence $\mathbb{N}''' \subset \mathbb{N}''$ and an $\hat{f} \in F_{\hat{t}_\varepsilon}$ such that w-$\lim_E u_n^*(\hat{t}_\varepsilon) = \hat{u}(\hat{t}_\varepsilon)$ $(n \in \mathbb{N}''')$ where $u_n^*(t)$ and $\hat{u}(t)$ are the trajectories corresponding to f_n^* and $\hat{f}$ respectively, and we have

$$\| u^1 - \hat{u}(\hat{t}_\varepsilon) \|_E \leq \liminf \| u_n^1 - u_n^*(\hat{t}_\varepsilon) \|_{E_n} \leq \varepsilon .$$

Now, let $\mathbb{N}' \subset \mathbb{N}$ be such that $t_{\varepsilon_n}^* \to \hat{t}_\varepsilon = \liminf t_{\varepsilon_n}^*$ $(n \in \mathbb{N}')$. But $(u_n^1)_{\mathbb{N}'}$ is bounded and hence, due to the discrete weak compactness of bounded sequences of elements in E_n, there exist $\mathbb{N}'' \subset \mathbb{N}'$ and $u^1 \in E$ with w-$\lim_E u_n^1 =$ $= u^1$ $(n \in \mathbb{N}'')$. By the same reasoning as above we conclude that u^1 is an approximately controllable state with minimum time $t_\varepsilon^* \leq \hat{t}_\varepsilon$.

Under the additional assumptions there exist $\mathbb{N}' \subset \mathbb{N}$ and a sequence $(\tilde{\varepsilon}_n)_{\mathbb{N}'}$ of positive real numbers with $\lim \tilde{\varepsilon}_n = \varepsilon$ $(n \in \mathbb{N}')$ such that

$$\| u_n^1 - u_n^*(t_\varepsilon^*) \|_{E_n} \leq \tilde{\varepsilon}_n .$$

Let us assume $t_\varepsilon^* < \liminf t_{\varepsilon_n}^*$ $(n \in \mathbb{N}')$. If $\tilde{\varepsilon}_n \leq \varepsilon_n$ this contradicts the time minimality of $t_{\varepsilon_n}^*$. On the other hand, if $\tilde{\varepsilon}_n > \varepsilon_n$, we find $z_n \in \partial B_{E_n}(u_n^1, \varepsilon_n)$ such that $\| z_n - u_n^*(t_\varepsilon^*) \|_{E_n} \to 0$ $(n \in \mathbb{N}')$. Then, arguing as in the proof of Theorem 4.1(i), for any $\delta > 0$ and sufficiently large $n \in \mathbb{N}'$ there exist controls $g_n \in F_{t_\varepsilon^* + \delta, n}$ and corresponding trajectories $u_n(t)$, $0 \leq t \leq t_\varepsilon^* + \delta$, such that

$$\| u_n^1 - u_n(t_\varepsilon^* + \delta) \|_{E_n} \leq \| u_n^1 - z_n \|_{E_n} + \eta(\delta)$$

which, letting $\delta \to 0$, also contradicts the time minimality of $t_{\varepsilon_n}^*$. #

5. APPLICATIONS

As an example let us consider the parabolic equation

(5.1a) $$\frac{\partial}{\partial t} u(x,t) + Au(x,t) = 0 , \quad x \in \Omega , \quad t > 0$$

(5.1b) $$u(x,0) = u^0(x) , \quad x \in \Omega$$

where Ω is a smooth bounded domain in Euclidean space R^d, $d \in \mathbb{N}$, and A is a second order uniformly elliptic operator given by

$$A = -\sum_{i,j=1}^{d} \frac{\partial}{\partial x_i} \left(a_{ij}(x) \frac{\partial}{\partial x_j} \right) + \sum_{i=1}^{d} b_i(x) \frac{\partial}{\partial x_i} + c(x)$$

with sufficiently smooth real-valued coefficients $a_{ij} = a_{ji}$, b_i and c. We consider boundary control either in the Dirichlet data

$$(5.1c)_D \qquad u(x,t) = g(x,t) \quad , x \in \Gamma = \partial\Omega \quad , \quad t > 0$$

or in the Neumann data

$$(5.1c)_N \qquad \frac{\partial}{\partial \nu} u(x,t) = g(x,t) \quad , x \in \Gamma = \partial\Omega \ , \quad t > 0$$

where $\frac{\partial}{\partial \nu} = \sum_{i,j=1}^{d} n_i a_{ij} \frac{\partial}{\partial x_j}$.

Denoting by $a(.,.): H^1(\Omega) \times H^1(\Omega) \to \mathbb{R}$ the bilinear form

$$a(u,v) = \sum_{i,j=1}^{d} \int_\Omega a_{ij} \frac{\partial u}{\partial x_i} \frac{\partial v}{\partial x_j} dx + \sum_{i=1}^{d} \int_\Omega b_i \frac{\partial u}{\partial x_i} v \, dx + \int_\Omega cuv \, dx,$$

clearly, a(.,.) is bounded, i.e.

$$(5.2) \qquad |a(u,v)| \le C \, |u|_{1,\Omega} \, |v|_{1,\Omega} \, , \qquad u,v \in H^1(\Omega) \ .$$

Moreover, let us assume for simplicity that a(.,.) is strictly $H^1(\Omega)$ - coercive, i.e.

$$(5.3) \qquad a(u,u) \ge \gamma \, |u|^2_{1,\Omega} \quad , \quad u \in H^1(\Omega) \quad , \quad \gamma > 0 \ .$$

It follows that the spectrum of the operators $A_D : D(A_D) \subset L^2(\Omega) \to L^2(\Omega)$, $D(A_D) = \{u \in H^2(\Omega) \mid u|_\Gamma = 0\}$ respectively $A_N : D(A_N) \subset L^2(\Omega) \to L^2(\Omega)$, $D(A_N) = \{u \in H^2(\Omega) \mid \frac{\partial}{\partial \nu} u|_\Gamma = 0\}$ is confined to the positive half-axis and $-A_D$ respectively $-A_N$ generate strongly continuous semigroups $S_D(t): L^2(\Omega) \to L^2(\Omega)$, $t \ge 0$, respectively $S_N(t): L^2(\Omega) \to L^2(\Omega)$, $t \ge 0$, of type (M,ω)

for some $\omega < 0$.
Finally, let us denote by D the Dirichlet map $u = Dg$ where

$$(5.4)_D \qquad Au(x) = 0 \;,\quad x \in \Omega \;,\quad u(x) = g(x) \;,\quad x \in \Gamma$$

and by N the Neumann map $u = Ng$ where

$$(5.4)_N \qquad Au(x) = 0 \;,\quad x \in \Omega \;,\quad \frac{\partial}{\partial \nu} u(x) = g(x) \;,\quad x \in \Gamma \;.$$

It is well known that under the foregoing assumptions $D: L^2(\Gamma) \to L^2(\Omega)$ and $N : L^2(\Gamma) \to L^2(\Omega)$ are compact linear operators satisfying (1.7) for some $0 < \theta < 1$. Consequently, the parabolic boundary control problems (5.1a), (5.1b), $(5.1c)_D$ respectively $(5.1c)_N$ can be cast in the abstract setting of Section 1 with $E = (L^2(\Omega), |.|_{0,\Omega})$, $V = (L^2(\Gamma), |.|_{0,\Gamma})$, $A = -A_D$ respectively $A = -A_N$ and D denoting the Dirichlet and Neumann map respectively.
We remark that the adjoint operators $-A_D^*$ and $-A_N^*$ with $D(A_D^*) = \{u \in H^2(\Omega) \mid u|_\Gamma = 0\}$ respectively $D(A_N^*) = \{u \in H^2(\Omega) \mid (\frac{\partial}{\partial \nu} + n.b)u|_\Gamma = 0\}$ likewise generate strongly continuous semigroups $S_D^*(t)$ and $S_N^*(t)$, $t \geq 0$.
Furthermore, it folloxs easily from Green's formula

$$(5.5) \qquad (Au,v)_{0,\Omega} - (u,A^*v)_{0,\Omega} = \int_\Gamma u(\frac{\partial}{\partial \nu} v + n.bv)\, d\sigma - \int_\Gamma \frac{\partial}{\partial \nu} u\, v\, d\sigma$$

by choosing $u = Dg$, $v = (A_D^*)^{-1} w$ respectively $u = Ng$, $v = (A_N^*)^{-1} w$ that the adjoints D^* and N^* are given by

$$(5.6)_D \qquad D^* w = -(\frac{\partial}{\partial \nu} + n.b)\, (A_D^*)^{-1} w|_\Gamma \;,$$

$$(5.6)_N \qquad N^* w = (A_N^*)^{-1} w|_\Gamma \;.$$

We first consider the approximate solution of Neumann boundary control. Given a null sequence $(h_n)_{\mathbb{N}}$ of positive real numbers $h_n < 1$, $n \in N$, we choose $E_n = S^{2,1}_{h_n}(\Omega)$, where $S^{r,k}_h(\Omega)$, $r>k\geq 0$, denotes a regular (r,k)-system, i.e. for every $u \in H^l(\Omega)$, $l\geq 0$, and every s with $0\leq s\leq \min(l,k)$ there exists $v \in S^{r,k}_h(\Omega)$ such that

(5.7) $\quad |v - u|_{s,\Omega} \leq C\, h^{\mu}\, |u|_{l,\Omega}$, $\mu = \min(r-s,l-s)$.

Moreover, we assume E_n to satisfy the inverse assumption

(5.8) $\quad |u_n|_{1,\Omega} \leq C\, h_n^{-1}\, |u_n|_{0,\Omega}$, $u_n \in E_n$.

Then, denoting by R_n^E the orthogonal projection of $L^2(\Omega)$ onto E_n with respect to the inner product in $L^2(\Omega)$, we have

(5.9) $\quad |R_n^E u - u|_{0,\Omega} \leq Ch^r\, |u|_{r,\Omega}$, $0\leq r\leq 2$, $u \in H^r(\Omega)$

and consequently, $(H^r(\Omega), \Pi E_n, R^E)$, $0<r\leq 2$, define discrete cn- approximations with discrete convergence being equivalent to norm convergence in $L^2(\Omega)$. But $H^r(\Omega)$ is dense in $L^2(\Omega)$ and thus,these cn-approximations uniquely induce a cn-approximation $(L^2(\Omega), \Pi E_n, R^E)$ (cf. [6]).
We choose V_n as the space generated by the traces of functions in $S^{2,1}_{h_n}(\Omega)$. V_n is known to define an $S^{3/2,1/2}_{h_n}(\Gamma)$ system and hence, denoting by R_n^V the orthogonal projection of $L^2(\Gamma)$ onto V_n with respect to $(.,.)_{0,\Gamma}$ we have

(5.10) $\quad |R_n^V u - u|_{0,\Gamma} \leq C\, h^{\mu}\, |u|_{r,\Gamma}$, $\mu=\min(3/2-r,r)$.

By the same argument as above, we thus obtain a cn-approximation

$(L^2(\Gamma), \Pi V_n, R^V)$.

We define $A_{N,n}: E_n \to E_n$ by

(5.11) $$(A_{N,n} u_n, v_n)_{0,\Omega} = a(u_n, v_n) \quad , \quad u_n, v_n \in E_n$$

and discrete Neumann maps $N_n : V_n \to E_n$ by

(5.12) $$a(N_n g_n, v_n) = (g_n, v_n)_{0,\Gamma} \quad , \quad v_n \in E_n \quad .$$

By (5.2) and (5.3) it follows that for every $\lambda \geq 0$ the sequence $(\lambda I_n + A_{N,n})_{\mathbb{N}}$ is both stable and inversely stable. In particular, $-A_{N,n}$, $n \in N$, generates a semigroup $S_{N,n}(t): E_n \to E_n$, $t \geq 0$, of the same type as $S_N(t)$. Moreover, denoting by R_n^H the elliptic projection of $H = H^1(\Omega)$ onto E_n, i.e.

$$a(R_n^H u, v_n) = a(u, v_n) \quad , \quad v_n \in E_n \quad ,$$

it turns out immediately that $A_{N,n} R_n^H = R_n^E A$. Since

(5.13) $$|R_n^H u - u|_{0,\Omega} \leq Ch^r \; |u|_{r,\Omega} \quad , \quad 1 \leq r \leq 2 \quad , \quad u \in H^r(\Omega) \quad ,$$

the pair $(\lambda I + A_N), (\lambda I_n + A_{N,n})_{\mathbb{N}}$ is consistent for every $\lambda \geq 0$, and we thus obtain the biconvergence $(\lambda I_n + A_{N,n}) \to (\lambda I + A_N)$ $(n \in \mathbb{N})$ and $(\lambda I_n + A_{N,n})^{-1} \to (\lambda I + A_N)^{-1}$ $(n \in \mathbb{N})$. It is easy to check that the same statements hold true for the adjoint operators A_N^*, $A_{N,n}^*$, $n \in \mathbb{N}$. Consequently, assumptions $(A_1), (A_2), (A_3)$ and $(B_4), (B_5)$ of Section 3 are satisfied in this case.

As far as the Neumann maps are concerned, it follows from (5.12) that

$N_n = R_n^H N|_{V_n}$ and $N_n^* = R_n^V N^*|_{E_n}$ implying that $(N_n)_{\mathbb{N}}$ is stable and that the pairs $N,(N_n)_{\mathbb{N}}$ and $N^*,(N_n^*)_{\mathbb{N}}$ are consistent thus establishing conditions (B_1) (with $\theta_n=\theta$), (B_2) and (B_3).

The discrete compactness of $(N_n)_{\mathbb{N}}$ (condition (B_7)) follows readily from the compactness of N. To establish a-regularity of $N,(N_n)_{\mathbb{N}}$ let $(g_n)_{\mathbb{N}}, g_n \in V_n$, $n \in \mathbb{N}$, be bounded and assume $N_n g_n \to u$ $(n \in \mathbb{N})$ for some $u \in L^2(\Omega)$. Then there exist a subsequence $\mathbb{N}' \subset \mathbb{N}$ and a $g \in L^2(\Gamma)$ such that $g_n \rightharpoonup g$ $(n \in \mathbb{N}')$ and $N_n g_n \to Ng$ $(n \in \mathbb{N}')$ whence $u = Ng$. Since $(N_n)_{\mathbb{N}}$ is inversely stable, we further have $g_n \to g$ $(n \in \mathbb{N}')$.

For the approximate solution of Dirichlet boundary control we choose $E_n \subset H_0^1(\Omega)$ as a (2,1)-regular system $S^{2,1}_{h_n,0}(\Omega)$ and $V_n \subset H^{1/2}(\Gamma)$ as a (3/2, 1/2)-regular system $S^{3/2,1/2}_{h_n}(\Gamma)$ generated by the traces of a (2,1)-regular system $F_n = S^{2,1}_{h_n}(\Omega) \subset H^1(\Omega)$. Again, we denote by R_n^E respectively R_n^V the orthogonal projections of $L^2(\Omega)$ respectively $L^2(\Gamma)$ onto E_n respectively V_n and by R_n^H respectively $R_n^{H}O$ the elliptic projections of $H=H^1(\Omega)$ respectively $H_O=H_0^1(\Omega)$ onto F_n respectively E_n. We define

$A_{D,n}: E_n \to E_n$ by

$$(5.14) \qquad (A_{D,n} u_n, v_n)_{0,\Omega} = a(u_n, v_n) \quad , \quad u_n, v_n \in E_n$$

and discrete Dirichlet maps $D_n : V_n \to E_n$ by $D_n = R_n^E D_n^1$ where $D_n^1 : V_n \to F_n$ is given by

$$(5.15) \qquad a(D_n^1 g_n, v_n) = 0 \ , \ v_n \in E_n \ , \quad D_n^1 g_n|_\Gamma = g_n \ .$$

In view of $A_{D,n} R_n^E = R_n^H O A_D$, $D_n^1 = R_n^H D|_{V_n}$ and $D_n^* = R_n^V D^*|_{E_n}$, conditions

$(A_1)-(A_3),(B_1)-(B_5)$ can be established in a similar way as in the Neumann case. The discrete compactness of $(D_n)_{\mathbb{N}}$ (condition (B_7)) follows from the compactness of D which can also be used to deduce the a-regularity of the pair $D,(D_n)_{\mathbb{N}}$ (condition (B_8)).

We finally consider the use of nonconforming methods in Dirichlet boundary control (cf. [14]). We choose E_n as a regular (r,2)-system $S^{r,2}_{h_n}(\Omega)$, $r\geq 4$, satisfying additionally the following inverse assumption

$$(5.16)\qquad |u_n|_{k,\Omega} + |u_n|_{k,\Gamma} \leq C\, h_n^{s-k}\,(\,|u_n|_{s,\Omega} + |u_n|_{s,\Gamma})$$

where $0\leq k\leq 1$ and $0\leq s\leq \min(k,1)$.

We define $\tilde{A}_{D,n}:E_n \to E_n$ by the nonconforming Rayleigh-Ritz Galerkin technique as used by Bramble and Schatz in [2]:

$$(5.17)\qquad (\tilde{A}_{D,n}u_n,v_n)_{0,\Omega} = \tilde{a}(u_n,v_n) + h_n^{-3}(u_n,v_n)_{0,\Gamma}$$

where $\tilde{a}(u_n,v_n) = (Au_n,Av_n)_{0,\Omega}$, $u_n,v_n\in E_n$.

Using the approximation properties of $S^{r,2}_{h_n}(\Omega)$ and the inverse assumptions (5.16), conditions $(A_1)-(A_3)$ and $(B_4),(B_5)$ can be readily verified in view of [2;Thm.4.1].

Instead of choosing $V_n = E_n|_\Gamma$ and $D_n = R^E_n D^1_n$ where D^1_n is defined as in the conforming case, we may take a larger class of approximating boundary controls, namely we may choose $V_n\subset L^2(\Gamma)$ as the space of piecewise constant functions on Γ, and we may define discrete Dirichlet maps $\tilde{D}_n: V_n\to E_n$ by using the same nonconforming technique as in the definition of $\tilde{A}_{D,n}$, i.e.

(5.18) $$\tilde{a}(\tilde{D}_n g_n, v_n) + h_n^{-3}(g_n - \tilde{D}_n g_n, v_n)_{0,\Gamma} = 0 \ , \ v_n \in E_n$$

If follows from [2;Corollary 4.1] that for $4-r \le k \le 1/2$

(5.19) $$|\tilde{D}_n g_n - Dg_n|_{k,\Omega} \le C\, h_n^{1/2-k}\, |g_n|_{0,\Gamma}$$

which immediately gives stability of $(\tilde{D}_n)$ and consistency of $D,(\tilde{D}_n)$ (conditions $(B_1),(B_2)$). In particular, $\tilde{D}_n \to D(n\in\mathbb{N})$ and discrete compactness of $(\tilde{D}_n)_{\mathbb{N}}$ as well as a-regularity of $D,(\tilde{D}_n)_{\mathbb{N}}$ (conditions (B_7), (B_8)) can be established as in the conforming case using the compactness of D.

Finally, if we choose v_n by $A^*_{D,n}A_{D,n}v_n = w_n$ in (5.18), where $A_{D,n}$ is defined as in the conforming case with respect to an (r,)-regular system $S^{r,2}_{h_n,0}(\Omega)$, we find

$$(\tilde{D}^*_n w_n, g_n)_{0,\Gamma} = -(\tilde{D}_n g_n (\frac{\partial}{\partial\nu} + n.b) \quad (A^*_{D,n})^{-1} w_n)_{0,\Gamma}$$

from which we can deduce $\tilde{D}^*_n \to D^*(n\in\mathbb{N})$ and thus condition (B_3).

REFERENCES

[1] Balakrishnan, A.V.'Optimal control problems in Banach spaces'.*SIAM J.Control Optimizution* 3, 152-180 (1965)

[2] Bramble,J. and Schatz,A. 'Rayleigh-Ritz-Galerkin methods for Dirichlet's problem using subspaces without boundary conditions'.*Comm.Pure Appl.Math.* 23 ,653-675 (1970)

[3] Carja,O.'On variational perturbations of control problems:minimum time problem and minimum effort problem'.To appear in *Journal of Optimization Theory and Applications*

[4] Dolecki,S. an Russell,D.L.'A general theory of observation and control'.*SIAM J.Control Optimization* 15 ,185-220 (1977)

[5] Fattorini,H.O.'The time-optimal control problem in Banach spaces'. *Appl.Math.Optim.* 1 , 163-188 (1974)

[6] Grigorieff,R.D.'Zur Theorie linearer approximationsregulärer Operatoren I'.*Math.Nachr.* 55 ,223-249 (1972)

[7] Grigorieff,R.D.'Zur Theorie linearer approximationsregulärer Operatoren II'.*Math.Nachr.* 55 , 251-263 (1972)
[8] Hoppe,R.H.W.'On the approximate solution of time-optimal control problems.'*Appl.Math. Optim.* 9 ,263-290 (1983)
[9] Jeggle,H.'Zur Störungstheorie nichtlinearer Variationsungleichungen! in: Optimization and optimal control,Proc.Conf.Oberwolfach November 17-23,1974 (eds.:Bulirsch,R.Oettli,W. and Stoer,J.),*Lec. Notes in Mathematics* 477,Springer:Berlin-Heidelberg-New York,1975
[10]Knowles,G.'Time optimal control in infinite dimensional spaces. *SIAM J.Control Optimization* 14 , 919-933 (1976)
[11]Knowles,G.'Some problems in the control of distributed systems,and their numerical solution".*SIAM J.Control Optimization* 17,5-22 (1979)
[12]Knowles,G.'Finite element approximation of parabolic time optimal control problems'.*SIAM J. Control Optimization* 20 , 414-427 (1982)
[13]Lasiecka,I.'Finite element approximation of time optimal control problems for parabolic equations.in:System Modeling and Optimization,Proc. 10th IFIP conf. New York,August 31-September 4,1981(eds. Derrick,R.F. and Kozin,F.),*Lect. Notes in Control and Information* 38,Springer:Berlin-Heidelberg-New York,1982
[14]Lasiecka,I.'Ritz Galerkin approximation of time-optimal control problems for parabolic systems with Dirichlet boundary conditions. *SIAM J. Control Optimization* 22 ,477-500 (1984)
[15]Sasai,H. and Shimemura,E.'On the convergence of approximating solutions for linear distributed parameter optimal control problems'. *SIAM J. Control Optimization* 9, 263-273 (1971)
[16]Stummel,F.'Diskrete Konvergenz linearer Operatoren I'.*Math.Ann.* 190, 45-92 (1970)
[17]Stummel,F.'Diskrete Konvergenz linearer Operatoren II'.*Math. Z.* 120, 231-264 (1971)
[18]Triggiani,R.'On the lack of exact controllability for mild solutions in Banach spaces!*J.Math.Anal.Appl.* 50 ,438-446 (1975)
[19]Washburn,D.'A bound on the boundary input for parabolic equations with applications to time optimal control'.*SIAM J. Control Optimization* 17 ,652-671 (1979)

Chapter 12

Distributed Systems with Uncomplete Data and Lagrange Multipliers

J-L. Lions

1. INTRODUCTION

Some recent work in the theory of Optimal control for *Distributed Systems*[1] has been concerned with situations where the existence of a *Lagrange Multiplier*, or of *an adjoint state*, is not straightforward.

Let us mention first the optimal control of *multi-state systems*; in such systems, the "state equation"

$$Az = Bv \tag{1.1}$$

where z (resp. v) stands for the *state* (resp. the *control*) and where A is a *non linear partial differential operator*, may admit *several solutions*, and even an *infinite* number of solutions. For instance, in J.L. LIONS [1], Chapter 3, we have considered the "state equation"

$$\begin{cases} -\Delta z - z^3 = v \quad \text{in} \quad \Omega \subset \mathbb{R}^3, \\ z = 0 \quad \text{on} \quad \partial\Omega \end{cases} \tag{1.2}$$

[1] i.e. Systems Governed by Partial Differential Equations.

and for a given v, equation (1.2) will admit in general[2] an infinite number of solutions.

In such a situation, it is clearly difficult to think of the mapping

$$(1.3)\qquad \begin{cases} v \to z = z(v) \\ \\ \text{control} \to \text{state} \end{cases}$$

which is multi-valued and where the set of z's solution of (1.2) has a non trivial nature. It is then preferable to consider a priori *the set of couples* {v,z} which satisfy (1.1), in suitable function spaces and to consider a *cost function*

$$(1.4)\qquad v,z \to J(v,z)$$

to be minimized. More precisely, the problem is to find

$$(1.5)\qquad \begin{cases} \operatorname{Inf} J(v,z) \\ \\ v \in \mathcal{U}_{ad}, \quad \{v,z\} \text{ satisfy } (1.1), \end{cases}$$

where

$$(1.6)\qquad \mathcal{U}_{ad} = \text{set of } \textit{admissible controls}.$$

(2) We refer to J-L. LIONS [1] for a precise statement and for the bibliography concerning this result (A. BAHRI, H. BERESTICKY and other authors).

For example, in case (1.2) we can consider the set of couples $\{v,z\} \in L^2(\Omega) \times L^6(\Omega)$ (3) which satisfy (1.2), and we consider

$$(1.7) \qquad J(v,z) = \frac{1}{6}\| z-z_d \|^6_{L^6(\Omega)} + \frac{N}{2}\| v \|^2_{L^2(\Omega)}$$

where z_d is given in $L^6(\Omega)$ and where N is given >0.

Let U_{ad} be a closed convex subset of $L^2(\Omega)$. Then *there exist* optimal couples, i.e. couples $\{u,y\}$ such that

$$(1.8) \qquad \begin{cases} u \in U_{ad}, \ \{u,y\} \in L^2(\Omega) \times L^6(\Omega) \text{ and satisfy (1.2)} \\ \text{and} \\ J(u,y) = \min J(v,z), \ v \in U_{ad}, \ \{v,z\} \text{ satisfying (1.2).} \end{cases}$$

This type of problem has been called in J.L. LIONS, loc. cit., a *singular* distributed system, and the corresponding Optimality System (4) has been called "*Singular* Optimality System", in short S.O.S...

It has been proven in many cases, *with conditions* on U_{ad} (5), that if $\{u,y\}$ is an *optimal couple, there exists an adjoint state* p *such that*

$$(1.9) \qquad \begin{cases} -\Delta y - y^3 = m \\ -\Delta p - 3y^2 p = (y-z_d)^5, \\ y = p = 0 \text{ on } \Gamma, \end{cases}$$

(3) All functions considered throughout are real valued.
(4) Giving first order necessary conditions for $\{u,y\}$ to be an optimal couple.
(5) *For instance*, if there exists $\omega \subset \Omega$, ω open "arbitrarily small", such that U_{ad} contains *all* functions C^∞ *with support in* ω, then (1.9) (1.10) stands true.

$$\int_\Omega (p+Nu)(v-u)dx \geq 0 \ , \ \forall v \in \mathcal{U}_{ad} \ , \ u \in \mathcal{U}_{ad}. \tag{1.10}$$

[In(1.9)y belongs to the Sobolev space $H^2(\Omega)\cap H_0^1(\Omega)$ and p belongs to the Sobolev space $W^{2,6/5}(\Omega)$; we use here the notation:

$$W^{m,\lambda}(\Omega) = \{\phi \mid D^\alpha \phi \in L^\lambda(\Omega) \ , \ \forall \ \alpha, |\alpha| \leq m\} \tag{1.11}$$

and we set

$W^{m,2}(\Omega) = H^m(\Omega)$,

$H_0^m(\Omega)$ = closure in $H^m(\Omega)$ of smooth functions with compact support.]

It has also been given by M. RAMASWAMY [1] an example where $\mathcal{U}_{ad} = \{v_0\}$, $\Omega =]0,1[$, where p does not exist; this counter example is somewhat *unstable* with respect to choice of v_0, z_d, N.

It seems probable that *the result* (1.9)(1.10) is "*generally*" *true*.

Remark 1.1: For the numerical approximation of optimal couples, we refer to J.P. KERNEVEZ and J.L. LIONS [1].

Remark 1.2: A similar framework has been introduced by A.V. FURSIKOV [1] [2] for the study of Navier Stokes equations.

In the above formulation for *singular systems*, we think of (1.1) more as a *set of constraints* than as an *equation*. In this framework, it becomes possible to study the control of *non well set systems*. Let us give an example.

Let Ω be an open set in $\mathbb{R}^3$ (to fix ideas) with boundary $\partial\Omega = \Gamma = \Gamma_0 \cup \Gamma_1$ as in Fig. 1.

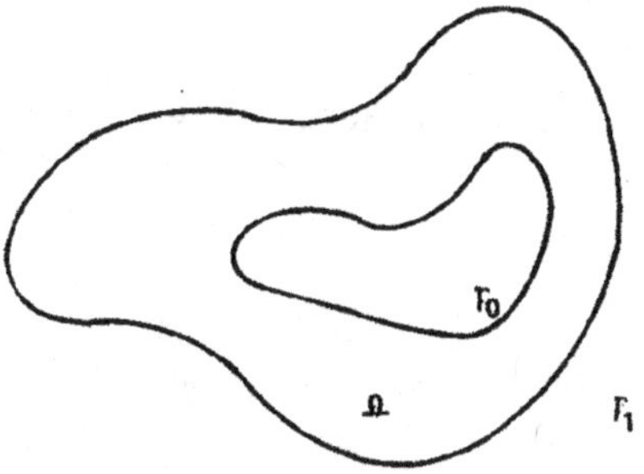

Fig. 1

We consider the "state equation"

$$(1.12) \qquad \begin{cases} \Delta z = 0 \quad \text{in} \quad \Omega, \\ z \in L^2(\Omega) \end{cases}$$

where z is subject to

$$(1.13) \qquad \begin{cases} z = v_0 \quad \text{on} \quad \Gamma_0, \\ \dfrac{\partial z}{\partial \nu} = v_1 \quad \text{on} \quad \Gamma_0; \end{cases} \quad (6)$$

(1.13) consists therefore of *Cauchy data* on Γ_0, and this is the most classical of non well set problems is Hadamard's terminology.

Remark 1.3: For $z \in L^2(\Omega)$, and $\Delta z \in L^2(\Omega)$ (in particular $\Delta z = 0$) *one can define* the traces of z and $\frac{\partial z}{\partial \nu}$ on Γ_0 (assuming Γ_0 to be smooth) (cf. J.L.

(6) $\frac{\partial}{\partial \nu}$ denotes the normal derivative to Γ_0, directed (to fix ideas) towards the exterior of Ω.

LIONS and E. MAGENES [1]); they belong to $H^{-1/2}(\Gamma_o)$ and $H^{-3/2}(\Gamma_o)$ respectively.

Let U^o_{ad} and U^1_{ad} be given closed convex subsets of $L^2(\Gamma_o)$. We define

$$v = \{v_o, v_1\}$$

and we consider the *set of couples* {v,z} such that

$$(1.14)\qquad \begin{cases} v \in U^o_{ad} \times U^1_{ad} \\ v,z \text{ satisfy } (1.12)(1.13). \end{cases}$$

This is the set of couples {control, state} admissible, and we *assume that this set is non empty*(7).

We can then consider the *cost function*

$$(1.15)\qquad J(v,z) = \frac{1}{2}\|z-z_d\|^2_{L^2(\Omega)} + \frac{N_o}{2}\|v_o\|^2_{L^2(\Gamma_o)} + \frac{N_1}{2}\|v_1\|^2_{L^2(\Gamma_o)}$$

where z_d is given in $L^2(\Omega)$ and where N_o, N_1 are given >0.

The problem

$$(1.16)\qquad \inf J(v,z)\ ,\ \{v,z\} \text{ satisfy } (1.14)$$

admits a unique solution {u,y}.

Formally the S.O.S. is given as follows:

(7) It can of course happen that this set is empty even with U^o_{ad} and U^1_{ad} non empty. This corresponds to the "non well posedness" of the state equation.

$$
(1.17)\quad
\begin{cases}
-\Delta y = 0, \\
-\Delta p = y - z_d \quad \text{in} \quad \Omega, \\
p = \dfrac{\partial p}{\partial \nu} = 0 \quad \text{on} \quad \Gamma_1, \\
\displaystyle\int_{\Gamma_o} \left(-\frac{\partial p}{\partial \nu} + N_o u_o\right)(v_o - u_o)\,d\Gamma_o \geq 0 \ , \ \forall v_o \in \mathcal{U}^o_{ad} \ , \ u_o \in \mathcal{U}^o_{ad}, \\
\displaystyle\int_{\Gamma_o} (p + N_1 u_1)(v_1 - u_1)\,d\Gamma_o \geq 0 \ , \ \forall v_1 \in \mathcal{U}^1_{ad} \ , \ u_1 \in \mathcal{U}^1_{ad}, \\
y = u_o \ , \ \dfrac{\partial y}{\partial \nu} = u_1 \quad \text{on} \quad \Gamma_o .
\end{cases}
$$

But in (1.17) p may be *an ultra distribution*, or *an analytic functional*.

Examples of such cases are given in J.L. LIONS [2] (we also refer to Yu. DUBINSKII [1], L.A. MEDEIROS [1], P.H. RIVERA and C.F. VASCONCELLOS [1] for related questions and in particular for functional spaces which are *not* spaces of distributions and which are needed for solving P.D.E.'s).

In J.L LIONS [3] we have shown that, *with special hypothesis* on $\mathcal{U}^o_{ad}$ or on $\mathcal{U}^1_{ad}$, then p becomes a *function*.

Still other families of problems where the above general framework turns out to be useful are *problems with uncomplete data*.

These are the problems we want to consider more in detail in this paper.

Let Ω be an open set as in Fig. 1. We consider the set of functions z such that

$$
(1.18)\qquad \Delta z = 0 \text{ in } \Omega \ , \ z \in L^2(\Omega),
$$

and for which we have *some* more information.

In Section 2 we consider the case where

$$(1.19)\quad \begin{cases} z|_{\Gamma_o} \ , \ z|_{\Gamma_1} \in K_o \times K_1, \\ K_i = \text{closed convex subset of } L^2(\Gamma_i),\ i=0,1. \end{cases}$$

We want to find

$$(1.20)\quad \begin{cases} \inf \ \|z-z_d\|^2_{L^2(\Omega)} \\ z \text{ satisfying}(1.18)(1.19), z_d \text{ given in } L^2(\Omega). \end{cases}$$

This is a simple situation, which serves as an introduction to the next situation, studied in Section 3. We consider the set of z's satisfying(1.18)and which satisfy

$$(1.21)\quad \begin{cases} z|_{\Gamma_1} \ , \ \frac{\partial z}{\partial \nu}|_{\Gamma_1} \in K \times M, \\ \text{where } K \text{ and } M \text{ are closed convex subsets of } L^2(\Gamma_1). \end{cases}$$

We assume that the set of z's which satisfy (1.18)(1.21) is non empty and we want to find

(1.22) $\inf\|z-z_d\|^2_{L^2(\Omega)}$ among functions z satisfying (1.18)(1.21).

Remark 1.4: If we set

$$(1.23)\quad z|_{\Gamma_1} = v_o \ , \ \frac{\partial z}{\partial \nu}|_{\Gamma_1} = v_1$$

and if we consider

$$(1.24)\quad \begin{cases} J_\alpha(v,z) = \frac{1}{2}\|z-z_d\|^2_{L^2(\Omega)} + \frac{\alpha}{2}\left[\|v_o\|^2_{L^2(\Gamma_1)} + \|v_1\|^2_{L^2(\Gamma_1)}\right], \\ \alpha > 0, \end{cases}$$

the problem $\inf J_\alpha(v,z)$, $v = \{v_o, v_1\} \in K \times M$, is similar to problem (1.15) .

Then problem (1.22) *corresponds to* (1.24) *with* $\alpha = 0$; it is therefore a singular problem with "*cheap control*".

In general we can expect the Lagrange multiplier (if it exists at all) to be in a space of general ultra distributions or of analytic functionals; in Section 3 we give some cases where it is a usual (L^2) function. Some extensions, along with open questions, are then given in Section 4 (for parabolic systems) and in Section 5 (for hyperbolic systems).

2. ELLIPTIC SYSTEM, «WELL SET» SITUATION

Let Ω be given as in Fig. 1. We consider the set of functions z such that

$$z \in L^2(\Omega)\ ,\ \Delta z = 0 \quad \text{in} \quad \Omega, \tag{2.1}$$

and which satisfy

$$\begin{cases} z|_{\Gamma_0} \in K_0 \ , \quad z|_{\Gamma_1} \in K_1, \\ K_i = \text{closed convex (non empty) subset of } L^2(\Gamma_i),\ i=0,1. \end{cases} \tag{2.2}$$

If z_d is given in $L^2(\Omega)$, we want to find

$$\begin{cases} \inf \|z-z_d\|^2_{L^2(\Omega)} \ , \\ z \text{ satisfying (2.1) (2.2).} \end{cases} \tag{2.3}$$

Since the set of z's which satisfy (2.1) (2.2) is a closed convex subset of $L^2(\Omega)$, problem (2.3) *admits a unique solution* y.

We want to find necessary conditions (and, in this case, necessary and sufficient conditions) for y to be the solution of (2.3).

We use the *classical penalty method*. We consider the set of z's such that

(2.4) $$z, \Delta z \in L^2(\Omega) \ , \ z|_{\Gamma_0} \in K_0, \ z|_{\Gamma_1} \in K_1$$

and we define, for $\varepsilon > 0$,

(2.5) $$J_\varepsilon(z) = \|z - z_d\|^2_{L^2(\Omega)} + \frac{1}{\varepsilon}\|\Delta z\|^2_{L^2(\Omega)} .$$

The problem

(2.6) $$\begin{cases} \inf J_\varepsilon(z) \\ z \text{ subject to (2.4)} \end{cases}$$

admits a unique solution, say y_ε.

It is a standard matter to show that

(2.7) $$y_\varepsilon \to y \text{ in } L^2(\Omega) \text{ as } \varepsilon \to 0,$$

where y is the solution of (2.3).

We write now the *Optimality System* (O.S.) for y_ε to be the solution of (2.5). If we set

(2.8) $$p_\varepsilon = \frac{1}{\varepsilon} \Delta y_\varepsilon$$

then

(2.9) $$(y_\varepsilon - z_d, z - y_\varepsilon) + (p_\varepsilon, \Delta(z - y_\varepsilon)) \geq 0 \ , \ \forall z \text{ subject to (2.4)},$$

where we have set

$$(f,g) = \int_\Omega fg \, dx.$$

We can choose in (2.9)

(2.10) $z = y_\varepsilon \pm \phi$, $\varphi \in \mathcal{D}(\Omega)$ (C^∞ functions with compact support in Ω);

it gives

(2.11) $$(p_\varepsilon, \Delta\phi) + (y_\varepsilon - z_d, \phi) = 0 \ , \ \forall \ \phi \in \mathcal{D}(\Omega),$$

i.e.

(2.12) $$-\Delta p_\varepsilon = y_\varepsilon - z_d \quad \text{in} \quad \Omega.$$

We can also in (2.10) take ϕ in $C^\infty(\bar{\Omega})$, subject to

$$\phi|_{\Gamma_0} = 0 \ , \ \phi|_{\Gamma_1} = 0.$$

Then $\frac{\partial\phi}{\partial\nu}$ is an arbitrary smooth function on $\Gamma = \Gamma_0 \cup \Gamma_1$, so that (2.11) gives

(2.13) $$p_\varepsilon = 0 \quad \text{on} \quad \Gamma_0 \cup \Gamma_1.$$

But (2.12) (2.13) imply that

(2.14) $$p_\varepsilon \in H^2(\Omega) \ , \ \|p_\varepsilon\|_{H^2(\Omega)} \leq C\|y_\varepsilon - z_d\|_{L^2(\Omega)} \leq C$$

(where the C's denote various constants, independant of ε).

Since p_ε belongs to $H^2(\Omega)$ we can in (2.9) perform the integrations by parts, and we obtain

(2.15) $$\int_{\Gamma_0 \cup \Gamma_1} \left(-\frac{\partial p_\varepsilon}{\partial\nu}\right) (z - y_\varepsilon) \, d\Gamma \geq 0.$$

In (2.15) one can take

$$\begin{cases} z = \{k_0, k_1\} & \text{on} \quad \Gamma_0 \times \Gamma_1 , \\ k_i \in K_i \ . \end{cases}$$

Passing to the limit as $\varepsilon \to 0$, *we obtain the O.S.* :

$$(2.16)\quad \begin{cases} -\Delta y = 0, \\ -\Delta p = y - z_d \quad \text{in } \Omega, \\ p = 0 \quad \text{on } \Gamma_0 \cup \Gamma_1, \end{cases}$$

$$(2.17)\quad \begin{cases} y|_{\Gamma_0} \in K_0, \\ \int_{\Gamma_0} (-\frac{\partial p}{\partial \nu})(k_0 - y)\,d\Gamma_0 \geq 0 , \ \forall k_0 \in K_0, \end{cases}$$

$$(2.18)\quad \begin{cases} y|_{\Gamma_1} \in K_1, \\ \int_{\Gamma_1} (-\frac{\partial p}{\partial \nu})(k_1 - y)\,d\Gamma_1 \geq 0 , \ \forall k_1 \in K_1. \end{cases}$$

Remark 2.1 The distinction of the two parts Γ_0, Γ_1 of Γ is irrelevant in this Section but will serve as an introduction to the questions considered in the next Section 3 .

Remark 2.2 In the above O.S. y belongs to $H^{1/2}(\Omega)$, $p \in H^2(\Omega)$. Therefore $\frac{\partial p}{\partial \nu} \in H^{1/2}(\Gamma)$ which is more than necessary for the integrals in (2.17) and in (2.18) to make sense.

Remark 2.3 Let us consider the state equation

$$(2.19)\quad \begin{cases} -\Delta y = 0 \quad \text{in } \Omega, \\ y = \{v_0, v_1\} \quad \text{on } \Gamma_0 \times \Gamma_1 , \ v_i \in K_i, \end{cases}$$

which admits a unique solution $y(v) \in H^{1/2}(\Omega)$.

We then define, for $\alpha > 0$,

$$(2.20)\quad J_\alpha(v) = \|y(v) - z_d\|^2_{L^2(\Omega)} + \alpha\left[\|v_0\|^2_{L^2(\Gamma_0)} + \|v_1\|^2_{L^2(\Gamma_1)}\right]$$

and we consider the problem

(2.21) $$\inf J_\alpha(v) \ , \ v \in K_o \times K_1 .$$

This problem admits a unique solution y_α; one has

(2.22) $$y_\alpha \to y \quad \text{in} \quad L^2(\Omega) \quad \text{as} \quad \alpha \to 0,$$

where y is the solution of (2.3).

The O.S. corresponding to (2.21) is given by:

(2.23) $$\begin{cases} -\Delta y_\alpha = 0, \\ -\Delta p_\alpha = y_\alpha - z_d \quad \text{in} \quad \Omega, \\ p_\alpha = 0 \quad \text{on} \quad \Gamma_o \cup \Gamma_1 , \end{cases}$$

(2.24) $$\begin{cases} y_\alpha = u_{o\alpha} \quad \text{on} \quad \Gamma_o, \\ \int_{\Gamma_o} \left(- \frac{\partial p_\alpha}{\partial \nu} + \alpha u_{o\alpha} \right) (v_o - u_{o\alpha}) d\Gamma_o \geq 0 \ , \ \forall v_o \in K_o , \end{cases}$$

(2.25) $$\begin{cases} y_\alpha = u_{1\alpha} \quad \text{on} \quad \Gamma_1, \\ \int_{\Gamma_1} \left(- \frac{\partial p_\alpha}{\partial \nu} + \alpha u_{1\alpha} \right) (v_1 - u_{1\alpha}) d\Gamma_1 \geq 0 \ , \ \forall v_1 \in K_1 . \end{cases}$$

When $\alpha \to 0$, $p_\alpha \to p$ in $H^2(\Omega)$ weakly, and one obtains (2.16) (2.17) (2.18).

3. ELLIPTIC SYSTEM, «NON WELL SET» SITUATION

3.1 Setting of the Problem

We consider again Ω as in Fig. 1 and we consider functions z such that

(3.1) $$z \in L^2(\Omega) \ , \ \Delta z = 0 \quad \text{in} \quad \Omega,$$

with the conditions

$$(3.2)\qquad \begin{cases} z|_{\Gamma_1} \in K, \quad \frac{\partial z}{\partial \nu}|_{\Gamma_1} \in M, \\ K, M \text{ closed convex subsets of } L^2(\Gamma_1). \end{cases}$$

Remark 3.1 We recall that if z satisfies (3.1), then, according to J.L. LIONS and E. MAGENES [1], $z|_{\Gamma_1}$ and $\frac{\partial z}{\partial \nu}|_{\Gamma_1}$ are well defined and belong to $H^{-1/2}(\Gamma_1)$ and to $H^{-3/2}(\Gamma_1)$ respectively. Therefore (3.2) makes sense.

Remark 3.2 Given $\{k,m\} \in K \times M$, the problem

$$(3.3)\qquad \begin{cases} \Delta z = 0 \quad \text{in} \quad \Omega, \\ z = k, \ \frac{\partial z}{\partial \nu} = m \quad \text{on} \quad \Gamma_1 \end{cases}$$

does not admit in general a solution. Therefore it is necessary to make *the hypothesis*

(3.4) *the set of* z's *which satisfy* (3.1) (3.2) *is not empty.*

Since of course (3.1) (3.2) do not in general uniquely define z, we are dealing with *a system with uncomplete data.*

Given z_d in $L^2(\Omega)$, we define

$$(3.5)\qquad J(z) = \|z-z_d\|^2_{L^2(\Omega)}$$

and we want to find

$$(3.6)\qquad \inf J(z)\ , \ z \text{ subject to (3.1) and (3.2)}$$

This problem admits a unique solution y:

$$(3.7)\qquad J(y) = \inf J(z).$$

We want now to find the S.O.S.

3.2 Penalty Approximation

As in Section 2, we consider the set of z's such that

(3.8) $$z, \Delta z \in L^2(\Omega), \ z|_{\Gamma_1} \in K, \ \frac{\partial z}{\partial \nu}\Big|_{\Gamma_1} \in M.$$

This set is certainly non empty if (3.4) holds true. We then consider

(3.9) $$J_\varepsilon(z) = \|z-z_d\|^2_{L^2(\Omega)} + \frac{1}{\varepsilon}\|\Delta z\|^2_{L^2(\Omega)}$$

and we look for

(3.10) $$\inf J_\varepsilon(z), \ z \text{ subject to (3.8).}$$

Let y_ε be the unique solution of (3.10). If we set

(3.11) $$p_\varepsilon = \frac{1}{\varepsilon} \Delta y_\varepsilon$$

then

(3.12) $$(y_\varepsilon - z_d, z-y_\varepsilon) + (p_\varepsilon, \Delta(z-y_\varepsilon)) \geq 0, \ \forall z \text{ subject to (3.8),}$$

(where we always set $(f,g) = \int_\Omega fg\, dx$).

We can take in (3.12)

$$z = y_\varepsilon \pm \phi,$$

where $\phi \in \mathcal{D}(\Omega)$ and more generally where $\phi \in C^\infty(\bar{\Omega})$ and $\phi = \frac{\partial \phi}{\partial \nu} = 0$ on Γ_1. It follows that

(3.13) $$-\Delta p_\varepsilon = y_\varepsilon - z_d \text{ in } \Omega$$

and that

(3.14) $$p_\varepsilon = \frac{\partial p_\varepsilon}{\partial \nu} = 0 \quad \text{on} \quad \Gamma_0 .$$

But (3.13) (3.14) do *not* imply that p_ε is bounded in a usual Lebesgue-Sobolev space, or even in a space of distributions. *One can only obtain that* p_ε *is bounded in a space of ultra distributions* (cf. J.L. LIONS [1] for questions of this type).

The problem we want to address now is: *when is it possible to obtain for* p_ε *an a priori estimate in a Sobolev's type space?*

A few remarks are in order.

Remark 3.3 If in (3.12) we perform the integration by parts *in a formal way*, we obtain

(3.15) $$\int_{\Gamma_1} p_\varepsilon \left(\frac{\partial z}{\partial \nu} - \frac{\partial y_\varepsilon}{\partial \nu}\right) d\Gamma_1 \geq 0,$$

(3.16) $$\int_{\Gamma_1} \left(- \frac{\partial p_\varepsilon}{\partial \nu}\right)(z-y_\varepsilon) d\Gamma_1 \geq 0.$$

These informations - when they can be made rigorous - will be used in order to answer the question, cf. Section 3.3 below.

Remark 3.4 *Towards the "weak form" of the* S.O.S. One can rewrite (3.12) in the form

(3.17) $$(y_\varepsilon - z_d, z) + (p_\varepsilon, \Delta z) \geq (y_\varepsilon - z_d, y_\varepsilon) + (p_\varepsilon, \Delta y_\varepsilon).$$

But according to (3.11), $(p_\varepsilon, \Lambda y_\varepsilon) > 0$ and therefore (3.17) implies

(3.18) $$(y_\varepsilon - z_d, z) + (p_\varepsilon, \Delta z) \geq (y_\varepsilon - z_d, y_\varepsilon) \ , \ \forall z \text{ subject to (3.8)}.$$

Remark 3.5 As usual, one has

(3.19) $$y_\varepsilon \to y \quad \text{in} \quad L^2(\Omega) \quad \text{as} \quad \varepsilon \to 0,$$

where y is the solution of (3.7). Therefore if we can find any space A (of distributions or of ultra distributions) such that

(3.20) $$p_\varepsilon \to p \quad \text{in} \quad A \quad \text{as} \quad \varepsilon \to 0$$

then (3.18) will give:

(3.21) $$\begin{cases} (y-z_d,z) + (p,\Delta z) \geq (y-z_d,y) \\ \forall z \text{ subject to (3.8) and such that } \Delta z \in A', \text{ the dual space of } A \\ \text{(assuming this set of z's non empty).} \end{cases}$$

This is the weak form of the S.O.S.

Remark 3.6 If $K = L^2(\Gamma_1)$ (*or* if $M = L^2(\Gamma_1)$) the problem is simple. Assume $K = L^2(\Gamma_1)$ to fix ideas, then (3.16), which can be easily justified in this case, gives

(3.22) $$\frac{\partial p_\varepsilon}{\partial \nu} = 0 \quad \text{on} \quad \Gamma_1.$$

Since, in particular, $p_\varepsilon = 0$ on Γ_0, it follows that

(3.23) $$p_\varepsilon \text{ remains in a bounded set of } H^2(\Omega) \text{ as } \varepsilon \to 0$$

and we conclude as in Section 2.

3.3 Study of a Particular Case

We consider now the case

(3.24) $$K = M = \{\phi \mid \phi \in L^2(\Gamma_1),\ \phi \geq 0 \text{ a.e. on } \Gamma_1\}.$$

Let ϕ_o, ϕ_1 be arbitrary smooth functions of K. We can then construct ϕ smooth in $\bar{\Omega}$, such that

$$\phi = \phi_o \quad , \quad \frac{\partial \phi}{\partial \nu} = \phi_1 \text{ on } \Gamma_1$$

and we can take $z = y_\varepsilon + \phi$ in (3.12). It comes

(3.25) $$(p_\varepsilon, \Delta\phi) + (y_\varepsilon - z_d, \phi) \geq 0.$$

We can integrate by parts in (3.25) since ϕ is smooth. We obtain

(3.26) $$\int_{\Gamma_1} p_\varepsilon \phi_1 \, d\Gamma_1 - \int_{\Gamma_1} \frac{\partial p_\varepsilon}{\partial \nu} \phi_o \, d\Gamma_1 \geq 0.$$

In (3.26) $p_\varepsilon \in H^{-1/2}(\Gamma_1)$, $\frac{\partial p_\varepsilon}{\partial \nu} \in H^{-3/2}(\Gamma_1)$ and the integrals express the duality between $H^{-s}(\Gamma_1)$ and $H^s(\Gamma_1)$, $s = 1/2$, $s = 3/2$ respectively. Since ϕ_o and ϕ_1 are *arbitrary* (smooth) positive functions, it follows that

(3.27) $$p_\varepsilon \geq 0, \; -\frac{\partial p_\varepsilon}{\partial \nu} \geq 0 \quad \text{on } \Gamma_- .$$

Let us proceed *formally* for a moment. If we multiply (3.13) by p_ε and if we integrate by parts, we obtain

(3.28) $$\int_\Omega |\nabla p_\varepsilon|^2 \, dx - \int_{\Gamma_1} p_\varepsilon \frac{\partial p_\varepsilon}{\partial \nu} \, d\Gamma_1 = (y_\varepsilon - z_d, p_\varepsilon).$$

But according to (3.27) $-\int_{\Gamma_1} p_\varepsilon \frac{\partial p_\varepsilon}{\partial \nu} \, d\Gamma_1 \geq 0$, so that (3.28) implies

(3.29) $$\int_\Omega |\nabla p_\varepsilon|^2 \, dx \leq (y_\varepsilon - z_d, p_\varepsilon).$$

But since $p_\varepsilon = 0$ on Γ_o, one has

$$\|p_\varepsilon\|_{L^2(\Omega)} \le C\left(\int_\Omega |\nabla p_\varepsilon|^2 \, dx\right)^{1/2}$$

and (3.29) "implies" that

(3.30) p_ε *remains in a bounded set of* $H^1(\Omega)$.

This procedure is formal, since a priori the integral $\int_{\Gamma_1} p_\varepsilon \frac{\partial p_\varepsilon}{\partial \nu} d\Gamma_1$ does not make sense.

But the result (3.30) will be justified below in Section 2.4.

The S.O.S. is then given as follows:

(3.31)
$$\begin{cases} -\Delta y = 0, \\ -\Delta p = y - z_d \quad \text{in } \Omega, \\ y \in L^2(\Omega), \; p \in H^1(\Omega), \end{cases}$$

(3.32)
$$p = \frac{\partial p}{\partial \nu} = 0 \quad \text{on } \Gamma_0,$$

(3.33)
$$\begin{cases} p \ge 0, \; -\frac{\partial p}{\partial \nu} \ge 0 \quad \text{on } \Gamma_1, \\ (y - z_d, y) \le 0, \; y \ge 0, \; \frac{\partial y}{\partial \nu} \ge 0 \quad \text{on } \Gamma_1. \end{cases}$$

Remark 3.7 If we perform *formal* integrations by parts,

$$(y - z_d, y) = (-\Delta p, y) = \int_{\Gamma_1} \left(-\frac{\partial p}{\partial \nu} y + p \frac{\partial y}{\partial \nu}\right) d\Gamma_1 .$$

Since $-\frac{\partial p}{\partial \nu}$, y, p, $\frac{\partial y}{\partial \nu}$ are ≥ 0 on Γ_1, we have $(y-z_d, y) \ge 0$ and therefore, using (3.33), $(y - z_d, y) = 0$, i.e.

(3.34)
$$p \frac{\partial y}{\partial \nu} = 0, \; y \frac{\partial p}{\partial \nu} = 0 \quad \text{on } \Gamma_1.$$

Conditions (3.33) *imply a weak form of* (3.34). The S.O.S. (3.31)(3.32)

(3.33) is in a weak form.

Remark 3.8 Open question. It would be interesting to obtain *other classes* of examples of sets K and M which imply that p_ε is bounded in, at least, $L^2(\Omega)$.

In order to show (3.31) (3.32) (3.33) we are going to introduce a new approximation of the problem, by regularization and penalty (so that we are not going to show exactly (3.30) but a variant of it, which also implies (3.31) (3.32) (3.33). It would be interesting to give a direct proof of (3.30)).

3.4 Approximation by Regularization and Penalty

We consider the set of functions z such that

$$(3.35)\qquad \begin{cases} z, \Delta z \in L^2(\Omega), \\ z|_{\Gamma_1} \in L^2(\Gamma_1)\ , \ \dfrac{\partial z}{\partial \nu} \in M \ (= K = \text{set of} \geq 0 \ \ L^2(\Gamma_1) \text{ functions on } \Gamma_1). \end{cases}$$

On the set (3.35) we define

$$(3.36)\qquad J_{\varepsilon\alpha}(z) = \|z-z_d\|^2_{L^2(\Omega)} + \alpha\|z\|^2_{L^2(\Gamma_1)} + \frac{1}{\varepsilon}\|\Delta z\|^2_{L^2(\Omega)} + \frac{1}{\varepsilon}\|z^-\|^2_{L^2(\Gamma_1)}$$

where $\alpha > 0$ and where $z^- = \sup(-z, 0)$.

The problem

$$(3.37)\qquad \inf J_{\varepsilon\alpha}(z)\ , \ z \text{ subject to } (3.35)$$

admits a unique solution $y_{\varepsilon\alpha}$.

When $\varepsilon \to 0$, $y_{\varepsilon\alpha} \to y_\alpha$, where y_α is the solution of

$$(3.38)\qquad J_\alpha(y_\alpha) = \inf J_\alpha(z)\ , \ z \text{ subject to } (3.8).$$

(3.39) $$J_\alpha(z) = \|z-z_d\|^2_{L^2(\Omega)} + \alpha\|z\|^2_{L^2(\Gamma_1)}.$$

Then, as $\alpha \to 0$,

(3.40) $$y_\alpha \to y,$$

where y is the solution of (3.7).

Let us introduce

(3.41) $$p_{\varepsilon\alpha} = \frac{1}{\varepsilon}\Delta y_{\varepsilon\alpha},\ q_{\varepsilon\alpha} = -\frac{1}{\varepsilon}y^-_{\varepsilon\alpha}.$$

The S.O.S. relative to (3.37) is given by

(3.42) $$\begin{cases} (y_{\varepsilon\alpha}-z_d, z-y_{\varepsilon\alpha}) + (p_{\varepsilon\alpha}, \Delta(z-y_{\varepsilon\alpha})) + (y_{\varepsilon\alpha}, z-y_{\varepsilon\alpha})_{L^2(\Gamma_1)} + \\ + (q_{\varepsilon\alpha}, z-y_{\varepsilon\alpha})_{L^2(\Gamma_1)} \geq 0 \quad \forall z \text{ subject to (3.35).} \end{cases}$$

It follows from (3.42) that

(3.43) $$\begin{cases} -\Delta p_{\varepsilon\alpha} = y_{\varepsilon\alpha} - z_d \text{ in } \Omega, \\ p_{\varepsilon\alpha} = \dfrac{\partial p_{\varepsilon\alpha}}{\partial\nu} = 0 \text{ on } \Gamma_0, \\ -\dfrac{\partial p_{\varepsilon\alpha}}{\partial\nu} + \alpha y_{\varepsilon\alpha} + q_{\varepsilon\alpha} = 0 \text{ on } \Gamma_1, \end{cases}$$

and since $q_{\varepsilon\alpha}, y_{\varepsilon\alpha}|_{\Gamma_1} \in L^2(\Gamma_1)$, it follows from (3.43) that

(3.44) $$p_{\varepsilon\alpha} \in H^{3/2}(\Omega).$$

Moreover (3.42) gives

$$(3.45) \qquad \left(p_{\varepsilon\alpha}, \frac{\partial z}{\partial \nu} - \frac{\partial y_{\varepsilon\alpha}}{\partial \nu}\right)_{L^2(\Gamma_1)} \geq 0$$

where $\frac{\partial z}{\partial \nu}$ can be taken an arbitrary (smooth) ≥ 0 function, so that

$$(3.46) \qquad p_{\varepsilon\alpha} \geq 0 \quad \text{on} \quad \Gamma_1.$$

If we now multiply (3.43) by $p_{\varepsilon\alpha}$ and if we integrate by parts, we obtain

$$(3.47) \qquad \int_\Omega |\nabla p_{\varepsilon\alpha}|^2 \, dx - \int_{\Gamma_1} (\alpha y_{\varepsilon\alpha} + q_{\varepsilon\alpha}) p_{\varepsilon\alpha} \, d\Gamma_1 = (y_{\varepsilon\alpha} - z_d, p_{\varepsilon\alpha}).$$

But $q_{\varepsilon\alpha} \leq 0$ (cf. (3.41)) so that, using (3.46) $-q_{\varepsilon\alpha} p_{\varepsilon\alpha} \geq 0$ and (3.47) implies

$$(3.48) \qquad \int_\Omega |\nabla p_{\varepsilon\alpha}|^2 \, dx - \alpha \int_{\Gamma_1} y_{\varepsilon\alpha} p_{\varepsilon\alpha} \, d\Gamma_1 \leq (y_{\varepsilon\alpha} - z_d, p_{\varepsilon\alpha}).$$

It follows from the structure of (3.36) that

$$(3.49) \qquad \| \sqrt{\alpha} \, y_{\varepsilon\alpha} \|_{L^2(\Gamma_1)} \leq C$$

so that, if we set $|||\phi||| = \left(\int_\Omega |\nabla \phi|^2 \, dx \right)^{1/2}$, (3.48) gives

$$(3.50) \qquad |||p_{\varepsilon\alpha}|||^2 \leq C \, \alpha \, \|p_{\varepsilon\alpha}\|_{L^2(\Gamma_1)} + C \|p_{\varepsilon\alpha}\|_{L^2(\Omega)}.$$

But $\|p_{\varepsilon\alpha}\|_{L^2(\Gamma_1)} \leq C |||p_{\varepsilon\alpha}|||$, $\|p_{\varepsilon\alpha}\|_{L^2(\Omega)} \leq C |||p_{\varepsilon\alpha}|||$ (since $p_{\varepsilon\alpha} = 0$ on Γ_0),

so that (3.50) implies

$$(3.51) \qquad |||p_{\varepsilon\alpha}||| \leq C.$$

All constants do not depend on ε and on α.

We now let $\varepsilon \to 0$. We extract a subsequence, still denoted by $p_{\varepsilon\alpha}$, such that

$$p_{\varepsilon\alpha} \to p_\alpha \quad \text{in } H^1(\Omega) \text{ weakly.} \tag{3.52}$$

We express $(3.43)_3$ as $-\frac{\partial p_{\varepsilon\alpha}}{\partial \nu} + \alpha y_{\varepsilon\alpha} \geq 0$, so that we can pass to the limit in (3.43):

$$\begin{cases} -\Delta p_\alpha = y_\alpha - z_d, \\ p_\alpha = \frac{\partial p_\alpha}{\partial \nu} = 0 \quad \text{on } \Gamma_0, \\ -\frac{\partial p_\alpha}{\partial \nu} + \alpha y_\alpha > 0 \quad \text{on } \Gamma_1; \end{cases} \tag{3.53}$$

(3.46) gives

$$p_\alpha \geq 0 \quad \text{on } \Gamma_1. \tag{3.54}$$

Taking $z=0$ in (3.42) gives

$$(y_{\varepsilon\alpha} - z_d, y_{\varepsilon\alpha}) + (p_{\varepsilon\alpha}, \Delta y_{\varepsilon\alpha}) + \alpha \| y_{\varepsilon\alpha} \|^2_{L^2(\Gamma_1)} + (q_{\varepsilon\alpha}, y_{\varepsilon\alpha})_{L^2(\Gamma_1)} \leq 0. \tag{3.55}$$

By virtue of (3.41), $(p_{\varepsilon\alpha}, \Delta y_{\varepsilon\alpha}) \geq 0$, $(q_{\varepsilon\alpha}, y_{\varepsilon\alpha})_{L^2(\Gamma_1)} \geq 0$ so that (3.55) implies

$$(y_{\varepsilon\alpha} - z_d, y_{\varepsilon\alpha}) + \alpha \| y_{\varepsilon\alpha} \|^2_{L^2(\Gamma_1)} \leq 0,$$

and passing to the limit

$$(y_\alpha - z_d, y_\alpha) + \alpha \| y_\alpha \|^2_{L^2(\Gamma_1)} \leq 0. \tag{3.56}$$

Since $|||p_\alpha||| \leq C$, we can now let $\alpha \to 0$, and we obtain (3.31) (3.32) (3.33) in the limit.

4. PARABOLIC SYSTEM

4.1 Setting of the Problem

Let Ω be given as in Fig. 1. We introduce:

$$Q = \Omega \times]0,T[\quad , \quad T > 0 \text{ given}$$

$$\Sigma_i = \Gamma_i \times]0,T[\quad , \quad i = 0,1,$$

and we consider the set of functions z such that

$$z \in L^2(Q) \quad , \quad \frac{\partial z}{\partial t} - \Delta z = 0 \text{ in } Q, \tag{4.1}$$

$$z(\cdot,o) \in K_o, \tag{4.2}$$

$$z|_{\Sigma_1} \in K \quad , \quad \frac{\partial z}{\partial \nu}|_{\Sigma_1} \in M \tag{4.3}$$

where

$$\begin{cases} K_o = \text{closed convex subset of } L^2(\Omega), \\ K \text{ (resp. } M) = \text{closed convex subset of } L^2(\Sigma_1). \end{cases} \tag{4.4}$$

Remark 4.1 If $z \in L^2(Q)$ such that $\frac{\partial z}{\partial t} - \Delta z = 0$, then

$$\frac{\partial z}{\partial t} = \Delta z \in L^2(0,T;H^{-2}(\Omega)) \tag{4.5}$$

so that z is (a.e. equal to) a continuous function from $[0,T] \to H^{-1}(\Omega)$.

Therefore $z|_{t=0} = z(\cdot,o)$ makes sense so that condition (4.2) is meaningful. We can "reverse" the argument so as to show that conditions (4.3) make sense.
Indeed

$$\Delta z = \frac{\partial z}{\partial t} \in H^{-1}(0,T;L^2(\Omega))$$

so that $z|_{\Sigma_1} \in H^{-1}(0,T;H^{-1/2}(\Gamma_1))$ and $\frac{\partial z}{\partial \nu}|_{\Sigma_1} \in H^{-1}(0,T;H^{-3/2}(\Gamma_1))$, so that conditions (4.3) are meaningful.

Remark 4.2 We cannot impose in general

(4.6) $$z = \mathbf{k} \, , \; \frac{\partial z}{\partial \nu} = m \text{ on } \Sigma_1 \; , \; k \in K \; , \; m \in M,$$

since these correspond to Cauchy data on Σ_1, i.e. to a non well set problem. Therefore we have to *assume* that

(4.7) *the set of* z's *which satisfy* (4.1) (4.2) (4.3) *is non empty.*

We consider now the problem (where z_d is given in $L^2(Q)$)

(4.8) $$\inf' \|z-z_d\|^2_{L^2(Q)} \; , \; z \text{ subject to (4.1) (4.2) (4.3).}$$

This problem *admits a unique solution* y and we want now to find the corresponding S.O.S. (if any).

4.2 Penalty Approximation

We consider the set of z's such that

(4.9) $$\begin{cases} z \in L^2(Q) \; , \; \frac{\partial z}{\partial t} - \Delta z \in L^2(Q), \\ z(\cdot,o) \in K_o \; , \; z|_{\Sigma_1} \in K \; , \; \frac{\partial z}{\partial \nu}|_{\Sigma_1} \in M \end{cases}$$

and we set

$$(4.10) \qquad J_\varepsilon(z) = ||z-z_d||^2_{L^2(Q)} + \frac{1}{\varepsilon} ||\frac{\partial z}{\partial t} - \Delta z||^2_{L^2(Q)} .$$

Let y_ε be the solution in the set (4.9) of

$$(4.11) \qquad J_\varepsilon(y_\varepsilon) = \inf J_\varepsilon(z) \text{ , } z \text{ subject to } (4.9).$$

The O.S. corresponding to (4.11) is given by (we set

$$(4.12) \qquad p_\varepsilon = -\frac{1}{\varepsilon}(\frac{\partial y_\varepsilon}{\partial t} - \Delta y_\varepsilon))$$

$$(4.13) \qquad \begin{cases} (y_\varepsilon - z_d, z-y_\varepsilon) - (p_\varepsilon, (\frac{\partial}{\partial t} - \Delta)(z-y_\varepsilon)) \geq 0 \\ \forall z \text{ subject to } (4.9) \text{ , where } (f,g) = \int_Q fg\, dx\, dt. \end{cases}$$

By arguments similar to those used in Section 3, one verifies that

$$(4.14) \qquad \begin{cases} -\frac{\partial p_\varepsilon}{\partial t} - \Delta p = y_\varepsilon - z_d \text{ in } Q \text{ , } p_\varepsilon \in L^2(Q), \\ p_\varepsilon(\cdot,T) = 0 \text{ in } \Omega, \\ p_\varepsilon = \frac{\partial p_\varepsilon}{\partial \nu} = 0 \text{ on } \Sigma_o. \end{cases}$$

The other conditions are obtained by *formal* integrations by parts in (4.13):

$$(4.15) \qquad \begin{cases} \int_{\Sigma_1} p_\varepsilon \frac{\partial(z-y_\varepsilon)}{\partial \nu} d\Sigma_1 \geq 0, \\ \int_{\Sigma_1} \left(-\frac{\partial p_\varepsilon}{\partial \nu}\right)(z-y_\varepsilon) d\Sigma_1 \geq 0, \\ \int_\Omega p_\varepsilon(\cdot,o)(z(\cdot,o) - y_\varepsilon(\cdot,o))dx \geq 0. \end{cases}$$

But (4.14) does *not* give a priori estimates on p_ε in usual spaces (but it does give estimates in spaces of ultra-distributions).

In order to pass to the limit we will use a *weak form* of (4.13), namely

$$(y_\varepsilon - z_d, z) - (p_\varepsilon, (\frac{\partial}{\partial t} - \Delta)z) \geq (y_\varepsilon - z_d, y_\varepsilon). \tag{4.16}$$

We consider now a particular case, analogous to the one considered in Section 3.3.

4.3 A Particular Case

We consider now the case

$$K = M = \{\phi \mid \phi \in L^2(\Sigma_1), \ \phi \geq 0 \text{ a.e. on } \Sigma_1\}. \tag{4.17}$$

Then *formally* it follows from (4.15) (and this can be justified as in Section 3) that

$$p_\varepsilon \geq 0, \ -\frac{\partial p_\varepsilon}{\partial \nu} \geq 0 \text{ on } \Sigma_1. \tag{4.18}$$

If we multiply $(4.14)_1$ by p_ε and if we integrate by parts in a formal fashion, we obtain:

$$-\frac{1}{2}\frac{d}{dt}\int_\Omega p_\varepsilon(x,t)^2\,dx - \int_{\Gamma_1} \frac{\partial p_\varepsilon}{\partial \nu} p_\varepsilon \, d\Gamma_1 + \int_\Omega |\nabla p_\varepsilon|^2\,dx = \tag{4.19}$$

$$= \int_\Omega (y_\varepsilon - z_d) p_\varepsilon \, dx$$

and by virtue of (4.18), (4.19) gives

$$-\frac{1}{2}\frac{d}{dt}\int_\Omega p_\varepsilon(x,t)^2\,dx + \int_\Omega |\nabla p_\varepsilon(x,t)|^2\,dx \leq \int_\Omega (y_\varepsilon - z_d) p_\varepsilon\,dx. \tag{4.20}$$

Since $p_\varepsilon(\cdot,T) = 0$, it "follows" that, as $\varepsilon \to 0$,

(4.21) $\quad p_\varepsilon$ remains in a bounded set of $L^\infty(0,T;L^2(\Omega)) \cap L^2(0,T;H^1(\Omega))$.

The S.O.S. is then given[8] by:

$$(4.22)\quad \begin{cases} \dfrac{\partial y}{\partial t} - \Delta y = 0 \;, \; y \in L^2(Q), \\ -\dfrac{\partial p}{\partial t} - \Delta p = y - z_d \text{ in } Q \;, \; p \in L^\infty(0,T;L^2(\Omega)) \cap L^2(0,T;H^1(\Omega)), \end{cases}$$

$$(4.23)\quad p(\cdot,T) = 0,$$

$$(4.24)\quad p = \frac{\partial p}{\partial \nu} = 0 \text{ on } \Sigma_0,$$

$$(4.25)\quad \begin{cases} y \geq 0, \; -\dfrac{\partial p}{\partial \nu} \geq 0 \text{ on } \Sigma_1, \\ \dfrac{\partial y}{\partial \nu} \geq 0 \;, \; p \geq 0 \text{ on } \Sigma_1, \\ (y - z_d, y) - \displaystyle\int_\Omega p(x,o)y(x,o)dx \leq 0,^{(9)} \end{cases}$$

$$(4.26)\quad \begin{cases} \displaystyle\int_\Omega p(x,o)(k_o - y(x,o))dx \geq 0 \;, \; \forall k_o \in K_o \\ y(\cdot,o) \in K_o. \end{cases}$$

Remark 4.3 If we multiply $(4.22)_2$ by y and if we *formally* integrate by parts, we find that

(8) A rigourous proof can be given along the same lines as in Section 3.4.
(9) Actually p is *continuous* from $[0,T] \to L^2(\Omega)$, so that $p(\cdot,o) \in L^2(\Omega)$.

$$(y-z_d, y) = -\int_{\Sigma_1} \frac{\partial p}{\partial \nu} y \, d\Sigma_1 + \int_{\Sigma_1} p \frac{\partial y}{\partial \nu} d\Sigma_1 + \int_\Omega p(x,o)y(x,o)dx$$

so that $(4.25)_3$ is a *weak form* of

$$\int_{\Sigma_1} \left(- \frac{\partial p}{\partial \nu} y + p \frac{\partial y}{\partial \nu}\right) d\Sigma_1 \le 0. \tag{4.27}$$

Since according to $(4.25)_1$ and $(4.25)_2$ all terms in (4.27) are ≥ 0, it follows that

$$p \frac{\partial y}{\partial \nu} = 0 \ , \ y \frac{\partial p}{\partial \nu} = 0 \text{ on } \Sigma_1. \tag{4.28}$$

Condition $(4.25)_3$ is a weak form of (4.28).

5. HYPERBOLIC SYSTEM

5.1 Setting of the Problem

We use the notations of Section 4. We consider the set of z's such that

$$z \in L^2(Q), \ \frac{\partial^2 z}{\partial t^2} - \Delta z = 0 \text{ in } Q, \tag{5.1}$$

where z is subject to(10)

$$\begin{cases} z(\cdot,o) \in K_o \ , \ \dfrac{\partial z}{\partial t} (\cdot,o) \in K_1, \\ K_i = \text{closed convex subset of } L^2(\Omega) , \end{cases} \tag{5.2}$$

(10) One verifies by arguments similar to those of Section 4.1 that these constraints make sense.

$$
(5.3)\qquad \begin{cases} z|_{\Sigma_1} \in K \ , \ \dfrac{\partial z}{\partial \nu}|_{\Sigma_1} \in M, \\ K \text{ (resp. } M) = \text{closed convex subset of } L^2(\Sigma_1). \end{cases}
$$

We assume that

(5.4) *the set of* z's *which satisfy* (5.1) (5.2) (5.3) *is not empty.*

We consider *the solution* y *of*

$$
(5.5)\qquad J(y) = \inf J(z) \ , \ z \text{ subject to } (5.1)\ (5.2)\ (5.3),
$$

where

$$
(5.6)\qquad J(z) = ||z-z_d||^2_{L^2(Q)} \ .
$$

We want to find the S.O.S. (if any) for this problem.

5.2 Penalty Approximation

We introduce the set of z's such that

$$
(5.7)\qquad \begin{cases} z, \ \dfrac{\partial^2 z}{\partial t^2} - \Delta z \in L^2(Q) \\ \text{and which satisfy } (5.2)\ (5.3). \end{cases}
$$

We consider next the penalized function

$$
(5.8)\qquad J_\varepsilon(z) = ||z-z_d||^2_{L^2(Q)} + \frac{1}{\varepsilon} \left|\left|\frac{\partial^2 z}{\partial t^2} - \Delta z\right|\right|^2_{L^2(Q)}
$$

and we consider

$$
(5.9)\qquad \inf J_\varepsilon(z) \ , \ z \text{ subject to } (5.7).
$$

Let y_ε be the solution of (5.9) and let us set

(5.10) $$p_\varepsilon = -\frac{1}{\varepsilon}\left(\frac{\partial^2}{\partial t^2} - \Delta\right) y_\varepsilon .$$

We have

(5.11) $$\begin{cases} (y_\varepsilon - z_d, z - y_\varepsilon) - \left(p_\varepsilon, \left(\frac{\partial^2}{\partial t^2} - \Delta\right)(z - y_\varepsilon)\right) \geq 0 \\ \forall z \text{ subject to } (5.7) \end{cases}$$

It follows from (5.11) that

(5.12) $$\begin{cases} \left(\frac{\partial^2}{\partial t^2} - \Delta\right) p_\varepsilon = y_\varepsilon - z_d \quad \text{in } Q, \\ p_\varepsilon(\cdot,T) = \frac{\partial p_\varepsilon}{\partial t}(\cdot,T) = 0 \quad \text{in } \Omega, \\ p_\varepsilon = \frac{\partial p_\varepsilon}{\partial \nu} = 0 \quad \text{on } \Sigma_0 . \end{cases}$$

Here again (5.12) does *not* give a priori estimates on p_ε, at least in "usual" spaces.

But we are now going to show, *in a formal fashion*, that "usual" a priori estimates can be obtained on p_ε if

(5.13) $$K = \{\phi \mid \phi \in L^2(\Sigma_1),\ \phi \geq 0 \text{ a.e. on } \Sigma_1\},$$

(5.14) $$M = \{\phi \mid \phi \in L^2(\Sigma_1),\ \int_0^t \phi(x,\sigma)d\sigma \leq 0 \text{ a.e. on } \Sigma_1\}.$$

Indeed (5.11) gives *formally* that

(5.15) $$\int_{\Sigma_1} p_\varepsilon \frac{\partial}{\partial \nu}(z - y_\varepsilon) d\Sigma_1 \geq 0,$$

(5.16) $$\int_{\Sigma_1} \left(-\frac{\partial p_\varepsilon}{\partial \nu}\right)(z - y_\varepsilon) d\Sigma_1 \geq 0.$$

Let θ be a smooth function on Σ_1, $\theta = 0$ near $t = T$, $\theta \leq 0$. Then we take in

(5.15)

$$\frac{\partial z}{\partial \nu} = \frac{\partial y_\varepsilon}{\partial \nu} + \frac{\partial \theta}{\partial t} ;$$

one has $\frac{\partial z}{\partial \nu} \in M$, so that this choice is valid and we obtain

$$\int_{\Sigma_1} p_\varepsilon \frac{\partial \theta}{\partial t} \, d\Sigma_1 \geq 0 ,$$

i.e.

$$\int_{\Sigma_1} \left(- \frac{\partial p_\varepsilon}{\partial t} \right) \theta \, d\Sigma_1 \geq 0 \tag{5.17}$$

(where the duality in (5.17) is taken in the sense of distributions). Since in (5.17) θ can be any (smooth) function which is ≤ 0, it follows that

$$\frac{\partial p_\varepsilon}{\partial t} \geq 0 \quad \text{on} \quad \Sigma_1 . \tag{5.18}$$

Using now (5.16), with the choice (5.13) for K, we obtain

$$- \frac{\partial p_\varepsilon}{\partial \nu} \geq 0 \quad \text{on} \quad \Sigma_1 . \tag{5.19}$$

If we now multiply $(5.12)_1$ by $\frac{\partial p_\varepsilon}{\partial t}$, we obtain, after setting

$$E(p_\varepsilon(t)) = \int_\Omega \left[\left(\frac{\partial p_\varepsilon}{\partial t} (x,t) \right)^2 + |\nabla_x p_\varepsilon(x,t)|^2 \right] dx: \tag{5.20}$$

$$\frac{d}{2dt} E(p_\varepsilon(t)) - \int_\Gamma \frac{\partial p_\varepsilon}{\partial \nu} \frac{\partial p_\varepsilon}{\partial t} \, d\Gamma = \int_\Omega (y_\varepsilon - z_d) \frac{\partial p_\varepsilon}{\partial t} (x,t) dx. \tag{5.21}$$

Using (5.18) (5.19), it follows that

$$\frac{1}{2} \frac{d}{dt} E(p_\varepsilon(t)) \leq \int_\Omega (y_\varepsilon - z_d) \frac{\partial p_\varepsilon}{\partial t} (x,t) dx. \tag{5.22}$$

This "energy inequality" together with $(5.12)_2$ implies that

(5.23) $$\begin{cases} p_\varepsilon \left(\text{resp. } \dfrac{\partial p_\varepsilon}{\partial t}\right) \text{ remains in a bounded set of} \\ L^\infty(0,T;H^1(\Omega)) \quad (\text{resp. } L^\infty(0,T;L^2(\Omega)). \end{cases}$$

We can then give a weak form of the S.O.S.

Remark 5.1 No attempt is made here to give a rigorous proof for (5.23). It is likely to work along lines similar to those of previous sections. It would be interesting to see *other cases* when similar estimates can be obtained.

REFERENCES

[1] Yu.Dubinskii. 'The algebra of pseudo differential operators with analytic symbols and its applications to Mathematical Physics'. *Russian Math. Surveys* 37, 5 (1982), p. 109-153.

[2] A.V.Fursikov. 'Control problems and Uniqueness results for tri-dimensional Navier-Stokes and Euler equations'. *Mat. Sbornik* 115 (157), (1981), p. 281-306.

[3] A.V.Fursikov. 'Properties of solutions of extrema problems connected with Navier-Stokes System' . *Mat. Sbornik*, 118 (160), (1982), p. 323-349.

[4] J.P.Kernevez and J.L.Lions. To appear

[5] J-L.Lions. 'Contrôle des Systèmes Distribués Singuliers'. *Gauthiers Villars* 1983. English translation. 1985.

[6] J-L.Lions. 'Some remarks on the optimal control of singular distributed systems'. *Summer Institute on Non Linear Analysis*. Berkeley Calif. 1983. To appear in A.M.S. publication.

[7] J-L.Lions. 'Some methods in the Mathematical Analysis of systems and their control'. *Science Press*, Beijing (1981).

[8] J-L.Lions and E.Magenes. 'Problèmes aux limites non homogènes et applications'. Dunod, Paris, Vol. 1 et 2, 1968. English translation, *Springer-Verlag*, 1970.

[9] L.A.Medeiros. 'Remarks on a non well posed problem'. To appear

[10] M.Ramaswamy. 'Thesis', Paris 1983.

[11] P.H.Rivera and C.F.Vasconcellos. 'Optimal control for a backward parabolic problem'. *SIAM J. Control*, 1985.

Chapter 13

Some Applications of Penalty Functions in Mathematical Programming

O. L. Mangasarian

ABSTRACT By using an exterior penalty function and recent boundedness and existence results for monotone complementarity problems, we give existence and boundedness results, for a pair of dual convex programs, of the following nature. If there exists a point which is feasible for the primal problem and which is interior to the constraints of the Wolfe dual, then the primal problem has a solution which is easily bounded in terms of the feasible point. Furthermore there exists no duality gap. We also show that by solving an exterior penalty problem for only two values of the penalty parameter we obtain an optimal point which is approximately feasible to any desired preassigned tolerance. This result is then employed to obtain an estimate of the perturbation parameter for a linear program which allows us to solve the linear program to any preassigned accuracy by an iterative scheme such as a successive over-relaxation (SOR) method.

AMS (MOS) Classification: 90C30, 90C25, 90C05

Key Words: Penalty functions, mathematical programming, duality, linear programs

---

Sponsored by the United States Army under Contract No. DAAG29-80-C-0041. This material is based on work sponsored by National Science Foundation Grant MCS-8200632.

1. INTRODUCTION

We consider in this work the constrained minimization problem

$$(1.1) \qquad \min_{x \in X} f(x), \quad X := X_o \cap X_1$$

where X_o and X_1 are subsets of the n-dimensional real space R^n which have a nonempty intersection X, and f: $X_o \to R$. Associated with the above problem is the classical exterior penalty problem [3,2,1]

$$(1.2) \qquad \min_{x \in X_o} P(x, \ ) := f(x) + \alpha Q(x)$$

where α is in R_+, the nonnegative real line, and $Q(x)$: $X_o \to R_+$ such that $Q(x) = 0$ for $x \in X$, else $Q(x) > 0$. We have two principal applications in mind regarding the penalty problem (1.2). The first application, which amploys in addition to (1.2) the recent boundedness and existence results for monotone complementarity problems [10]and which is described in Section 3 of the paper, gives existence and boundedness results for a convex program obtained from (1.1) and the associated dual problem. In particular we show in Theorem 3.1 that if there exists a point which is feasible for a primal convex program and is interior to the constraints of its Wolfe dual [12,5] , then the primal problem has a solution which is easily bounded in terms of the feasible point,and that there is no duality gap between the primal problem and its Wolfe dual. Theorem 3.2 shows that if there is a point which is interior to the constraints of a primal convex program which is also feasible for the associated Wolfe dual, then the Lagrangian dual [4,1] of the convex program has a nonempty solution set which is easily bounded by the feasible point,and in addition there is no duality gap between the primal problem and its Lagrangian dual. In section 4 our main concern is the recasting by means of an exterior penalty function of the standard linear programming problem as a quadratic minimization problem on the nonnegative orthant in the spirit of previous work [6,7,8]. The principal new result here is to show how to obtain a precise value of the penalty parameter which allows us to satisfy the Karush-Kuhn-Tucker optimality conditions [5] for the linear program to any preassigned degree of precision. Theorem 4.1 shows that this can be done by minimizing a convex function on the nonnegative orthant for only two values of the penalty parameter. Iterative methods developed in [6,7,8] can solve by this approach very large sparse linear programs which cannot be solved by a standard linear programming simplex package [8] .

Because of the key role played by exterior penalty functions in this work, we give in Section 2 some fundamental results regarding these functions in a form convenient for deriving our other results. Although some of these penalty results are known under more restrictive conditions [3, 2], some are new. For example, Theorem 2.3 shows that by solving only two exterior penalty function minimization problems, we can obtain an optimal point which is feasible to any preassigned feasibility tolerance. Theorem 2.8 shows that under rather mild assumptions each accumulation point of a sequence of solutions of penalty functions, corresponding to an increasing unbounded sequence of positive numbers, solves the associated constrained optimization problem. Furthermore the corresponding sequence of products of the panlty parameter and the penalty term tends to zero.

We briefly describe our notation now. Vectors will be column or row vectors depending on the context. For a vector x in the n-dimensional real space R^n, $\|x\|$ will denote an arbitrary norm, while $\|x\|_p$ will denote the p-norm

$$\|x\|_p := \left(\sum_{i=1}^{n} |x_i|^p\right)^{\frac{1}{p}} \quad \text{for} \quad 1 \le p < \infty$$

and

$$\|x\|_\infty := \max_{1\le i\le n} \quad |x_i| \ ,$$

where x_i is the i-th component of x; x_+ will denote the vector in R^n with components

$$(x_+)_i = \max\{x_i, 0\}, \quad i = 1,\ldots,n.$$

A vector of ones in any real space will be denoted by e. For a differentiable function $L: R^n \times R^m \to R$, $\nabla_x L(x,u)$ will denote the n-dimensional gradient vector $\frac{\partial L}{\partial x_i}(x,u)$, $i=1,\ldots,n$, while for $f: R^n \to R$, $\nabla f(x)$ will denote the n-dimensional gradient vector. The set of vectors in R^n with nonnegative components will be denoted by R^n_+.

2. SOME FUNDAMENTAL PROPERTIES OF EXTERIOR PENALTY FUNCTIONS

We collect in this section some fundamental properties of exterior penalty functions in a form convenient for our applications and under more general assumptions than usually given [3,11]. We begin with some elementary but important monotonicity properties for solutions of penalty problems.

Proposition 2.1
Let $x_i \in X_o$ be a solution of $\min_{x \in X_o} P(x,\alpha_i)$ for i = 1,2 with $\alpha_2 > \alpha_1 \geq 0$. Then

$$Q(x_2) \leq Q(x_1), \; f(x_1) \leq f(x_2), \; P(x_1,\alpha_1) \leq P(x_2,\alpha_2) \tag{2.1}$$

Proof: Addition of $P(x_2,\alpha_2) \leq P(x_1,\alpha_2)$ and $P(x_1,\alpha_1) \leq P(x_2,\alpha_1)$, gives, together with $\alpha_2 > \alpha_1$, the inequality $Q(x_2) \leq Q(x_1)$, which in turn together with $P(x_1,\alpha_1) \leq P(x_2,\alpha_1)$, and $\alpha_1 \geq 0$, gives $f(x_1) \leq f(x_2)$. We also have that

$$P(x_1,\alpha_1) \leq P(x_2,\alpha_1) \leq P(x_2,\alpha_2) \;. \qquad \#$$

Proposition 2.2

Let $\inf_{x \in X} f(x) > -\infty$, let $\alpha > 0$ and let $x(\alpha) \in X_o$ be such that $P(x(\alpha),\alpha) = \min_{x \in X_o} P(x,\alpha)$. Then

(2.2) $$f(x(\alpha)) \le \inf_{x \in X} f(x)$$

If $x(\alpha) \in X$ then

(2.3) $$f(x(\alpha)) = \min_{x \in X} f(x)$$

Proof: For any $\varepsilon > 0$ pick $x(\varepsilon) \in X$ such that

$$f(x(\varepsilon)) < \inf_{x \in X} f(x) + \varepsilon$$

Then

$$\varepsilon + \inf_{x \in X} f(x) > f(x(\varepsilon)) = P(x(\varepsilon),\alpha) \ge P(x(\alpha),\alpha) \ge f(x(\alpha))$$

Since $x(\alpha)$ does not depend on ε, (2.2) follows by letting ε approach zero. If $x(\alpha)$ is also in X, then (2.3) is obviously a consequence of (2.2). #

The following simple theorem shows how, for any desired feasibility tolerance $\delta > 0$, solving the penalty problem (1.2) for only two values of the penalty parameter α will yield a point $x_2 \in X_o$ such that $Q(x_2) \le \delta$ and $f(x_2) \le \inf_{x \in X} f(x)$. Hence if δ chosen sufficiently small, x_2 is an approximately feasible optimal solution for the minimization problem (1.1).

Theorem 2.3

Let $\delta > 0$, $\alpha_1 > 0$, let $\inf_{x \in X} f(x) > -\infty$, let $\hat{x} \in X$ and let $P(x_1,\alpha_1) = \min_{x \in X_o} P(x,\alpha_1)$. If $f(\hat{x}) \le f(x_1)$ then $\hat{x}$ solves $\min_{x \in X} f(x)$, else for

$$\alpha_2 > \alpha_1 \quad \text{and} \quad \alpha_2 \geq \frac{f(\hat{x}) - f(x_1)}{\delta} \tag{2.4}$$

it follows that

$$x_2 \in X_o,\ Q(x_2) \leq \delta,\ f(x_2) \leq \inf_{x \in X} f(x) \tag{2.5}$$

where

$$P(x_2,\alpha_2) = \min_{x \in X_o} P(x,\alpha_2),\ x_2 \in X_o$$

Proof: First note that if $f(\hat{x}) \leq f(x_1)$ then by (2.2) $\hat{x}$ solves $\min_{x \in X} f(x)$. Suppose now that $f(\hat{x}) > f(x_1)$ and (2.4) holds. Then

$$f(x_2) + \alpha_2 Q(x_2) \leq f(\hat{x}) + \alpha_2 Q(\hat{x}) = f(\hat{x}) \tag{2.6}$$

Hence by (2.4), (2.1) and (2.6) respectively it follows that

$$\delta \geq \frac{f(\hat{x}) - f(x_1)}{\alpha_2} \geq \frac{f(\hat{x}) - f(x_2)}{\alpha_2} \geq Q(x_2)$$

which establishes the first inequality of (2.5). The second inequality of (2.5) follows from (2.2). #

Remark 2.4 Theorem 2.3 can be applied to obtain an approximate solution of (1.1) in the sense of (2.5) as follows:

(a) Choose $\delta > 0$, $\alpha_1 > 0$, $\hat{x} \in X$.

(b) Compute $x_1 \in X_o$ such that: $P(x_1,\alpha_1) = \min_{x \in X_o} P(x,\alpha_1)$. If $f(\hat{x}) \leq f(x_1)$, stop, $\hat{x}$ solves (1.1).

(c) Choose α_2 such that $\alpha_2 > \alpha_1$ and $\alpha_2 \geq \frac{f(\hat{x}) - f(x_1)}{\delta}$.

(d) Compute $x_2 \in X_o$ such that: $P(x_2,\alpha_2) = \min_{x \in X_o} P(x,\alpha_2)$.

If α_2 of step (c) is too large, an $\bar{\alpha}_1$ such that $\alpha_1 < \bar{\alpha}_1 < \alpha_2$ can be chosen to replace α_1 and steps (a)-(b)-(c) are repeated. Also $\hat{x}$ may be replaced when possible by some $\tilde{x} \in [\hat{x},x_1] \cap X$ such that $f(\tilde{x}) < f(\hat{x})$.

The next result shows that for a sequence of solutions $\{x_i\}$ of the penalty problem (1.2) for an increasing unbounded sequence of penalty parameters $\{\alpha_i\}$, the sequence of penalties $\{Q(x_i)\}$ converges to 0 and the sequence $\{f(x_i)\}$ converges to a lower bound for $\inf_{x \in X} f(x)$, provided the latter is finite. We do not require that the sequence $\{x_i\}$ have an accumulation point here.

Theorem 2.5

Let $\inf_{x \in X} f(x) > -\infty$, let $\{\alpha_i\}$ be an increasing unbounded sequence of positive numbers, let $\{x_i\}$ be a corresponding sequence of point in X_0 not in X such that $P(x_i,\alpha_i) = \min_{x \in X_0} P(x,\alpha_i)$.
Then

$$\lim_{i \to \infty} Q(x_i) = 0 \quad \text{and} \quad \lim_{i \to \infty} f(x_i) \le \inf_{x \in X} f(x). \tag{2.7}$$

Proof: By (2.1), the sequence $\{Q(x_i)\}$ is nonincreasing and bounded below by 0 and hence converges to $\bar{Q} \ge 0$ and $Q(x_i) \ge \bar{Q}$, $i=1,2,\ldots$. If $\bar{Q}>0$ we get from (2.5) by picking i sufficiently large such that $\alpha_i \ge 2(f(\hat{x}) - f(x_1))/\bar{Q}$ where $\hat{x} \in X$, that $\bar{Q} \le Q(x_i) \le \bar{Q}/2$ which is a contradiction. Hence $\bar{Q} = 0$ and $\lim_{i \to \infty} Q(x_i) = 0$. Now again by (2.1), the sequence $f(x_i)$ is nondecreasing, and by (2.2) it is bounded above by $\inf_{x \in X} f(x)$. Hence $\{f(x_i)\}$ converges to $\bar{f}$ and

$$f(x_i) \le \bar{f} \le \inf_{x \in X} f(x) \qquad \#$$

To make the inequality in (2.7) an equality we need additional assumptions such as those given in the following corollary.

Corollary 2.6 If in addition to the assumptions of Theorem 2.5, f is Lipschitz continuous on X_o, that is for some $K > 0$

$$|f(y) - f(x)| \leq K\|y - x\|_2 \quad \text{for all } x, y \in X_o \tag{2.8}$$

and there exists a constant $\mu > 0$ such that for each $x \in X_o$ there exists an $\hat{x}(x) \in X$ such that

$$\|x - \hat{x}(x)\|_2 \leq \mu Q(x) \tag{2.9}$$

then

$$\lim_{i\to\infty} f(x_i) = \inf_{x \in X} f(x) \tag{2.10}$$

Proof: For each x_i there exists an $\hat{x}_i \in X$ such that

$$\|x_i - \hat{x}_i\|_2 \leq \mu Q(x_i)$$

Hence by (2.8) and (2.9)

$$0 \leq |f(x_i) - f(\hat{x}_i)| \leq K\|x_i - \hat{x}_i\|_2 \leq K\mu Q(x_i) \tag{2.11}$$

Since by (2.7) $\lim_{i\to\infty} Q(x_i) = 0$, it follows from (2.11) that

$$\lim_{i\to\infty} f(\hat{x}_i) = \lim_{i\to\infty} f(x_i) \tag{2.12}$$

From (2.11) and $\hat{x}_i \in X$ we get the inequalities

$$f(x_i) + K\mu Q(x_i) \geq f(\hat{x}_i) \geq \inf_{x \in X} f(x)$$

Taking the limit as $i \to \infty$ and using (2.7) gives

$$\inf_{x \in X} f(x) \geq \lim_{i\to\infty} f(x_i) \geq \inf_{x \in X} f(x)$$

Hence $\lim_{i\to\infty} f(x_i) = \inf_{x\in X} f(x)$. #

Remark 2.7 Condition (2.9) is satisfied if the feasible region X is convex and satisfies an appropriate constraint qualification [9, Theorem 2]. In particular (2.9) holds in the special case when $X_o = R^n$ and X_1 is defined by linear inequalities [9, Remark 2.2].

We observe that in both Theorem 2.5 and Corollary 2.6 the sequence $\{x_i\}$ need not have an accumulation point. A stronger result is obtained if $\{x_i\}$ has an accumulation point.

Theorem 2.8

Let $\inf_{x\in X} > -\infty$, and let $\{\alpha_i\}$ be an increasing unbounded sequence of positive numbers. Let $\{x_i\}$ be a corresponding sequence of points in X_o not in X such that $P(x_i,\alpha_i) = \min_{x\in X_o} P(x,\alpha_i)$ with an accumulation point $\bar{x} \in X_o$. If f and Q are lower semicontinuous at $\bar{x}$, then $Q(\bar{x}) = 0$ and $\bar{x}$ solves $\min_{x\in X} f(x)$. Furthermore

$$\lim_{j\to\infty} \alpha_{i_j} Q(x_{i_j}) = 0 \quad \text{for} \quad x_{i_j} \to \bar{x} \in X_o . \tag{2.13}$$

Proof: Let $x_{i_j} \to \bar{x} \in X_o$. From (2.7) and the lsc of Q we have

$$0 = \lim_{j\to\infty} Q(x_{i_j}) \geq Q(\bar{x}) \geq 0$$

Hence $Q(\bar{x}) = 0$ and $\bar{x} \in X$. From (2.7) and the lsc of f we have

$$f(\bar{x}) \leq \lim_{j\to\infty} f(x_{i_j}) \leq \inf_{x\in X} f(x)$$

Since $\bar{x} \in X$, it follows that $\bar{x}$ solves $\min_{x\in X} f(x)$. To establish (2.13) note that

$$0 \geq P(x_{i_j},\alpha_{i_j}) - P(\bar{x},\alpha_{i_j}) = f(x_{i_j}) - f(\bar{x}) + \alpha_{i_j} Q(x_{i_j})$$

Hence

$$f(\bar{x}) - f(x_{i_j}) \geq \alpha_{i_j} Q(x_{i_j}) \geq 0$$

By letting $j \to \infty$ and recalling that f is lsc at $\bar{x}$ it follows that $\lim_{j\to\infty} \alpha_{i_j} Q(x_{i_j}) = 0$. #

3. BOUNDS AND EXISTENCE FOR DUAL CONVEX PROGRAMS

We consider in this section the convex primal program

(3.1) $$\min_{x \in X} f(x), \quad X = \{x \mid x \in R^n_+, \ g(x) \leq 0\}$$

where $f\colon R^n \to R$, $g\colon R^n \to R^m$ are differentiable and convex on R^n. The Wolfe dual [12,5] associated with this problem is

(3.2) $$\max_{(x,u,v)\in Y} L(x,u) - vx ,$$

$$Y = \{(x,u,v) \mid x \in R^n, \ u \in R^m_+, \ v \in R^n_+, \ \nabla_x L(x,u) - v = 0\}$$

and the Lagrangian dual [4,1] is

(3.3) $$\max_{(u,v)\geq 0} \ \inf_{x \in R^n} L(x,u) - vx$$

where $L(x,u) := f(x) + ug(x)$ is the usual Lagrangian. Note that (3.2) is equivalent to

(3.2') $$\max_{(x,u)\in Z} L(x,u) - x\nabla_x L(x,u) ,$$

$$Z = \{(x,u) \mid x \in R^n, \ u \in R^m_+, \ \nabla_x L(x,u) \geq 0 \}$$

Note that (3.1) can be identified with problem (1.1) by setting $X_o = R^n_+$ and $X_1 = \{x \mid g(x) \le 0\}$.

Our primary objective here is to give simple conditions for the separate existence of a solution to each of primal and Lagrangian dual problems and to bound their solutions. Loosely speaking we shall establish existence of a solution and a bound for the primal (Lagrangian dual) problem under a primal and Wolfe-dual feasibility assumption together with a Wolfe-dual (primal) constraint interiority assumption . Our principal tools will be the recent boundedness and existence results for monotone complementarity problems and convex programs of [10] and the penalty function results outlined in the previous section. We begin with an existence and boundedness result for the primal problem (3.1).

Theorem 3.1

(Primal feasibility & Wolfe dual interior-feasibility ⇒ Primal solution existence-boundedness & zero duality gap with Wolfe dual). Let f and g be differentiable and convex on R^n and let $(\hat{x},\hat{u})$ satisfy

$$\hat{x} \in X, \quad (\hat{x},\hat{u}) \in Z, \quad \nabla_x L(\hat{x},\hat{u}) > 0$$

Then there exists a primal solution $\bar{x}$ to (3.1) which is bounded by

$$\|\bar{x}\|_1 \le \frac{-\hat{u}g(\hat{x}) + \hat{x}\nabla_x L(\hat{x},\hat{u})}{\min\limits_i (\nabla_x L(\hat{x},\hat{u}))_i} \tag{3.4}$$

In addition there exists no duality gap between the primal problem (3.1) and the Wolfe dual (3.2), that is:

$$\min_{x\in X} f(x) = f(\bar{x}) = \sup_{(x,u,v)\in Y} L(x,u) - vx \tag{3.5}$$

Proof: Consider the penalty function problem associated with (3.1)

$$\min_{x\ge 0} f(x) + \alpha e g(x)_+ \tag{3.6}$$

or equivalently

$$\min_{(x,z)\geq 0} f(x) + \alpha e z \quad \text{s.t.} \quad g(x) - z \leq 0 \tag{3.6'}$$

The Wolfe dual associated with (3.6') is

$$\max_{(x,z,u,v,w)} L(x,u) + z\,(\alpha e - u - w) - vx \tag{3.7}$$

$$\text{s.t.} \quad \nabla_x L(x,u) - v = 0, \quad \alpha e - u - w = 0\,, \quad u,v,w \geq 0$$

which is equivalent to

$$\max_{(x,u)} L(x,u) - x\nabla_x L(x,u) \quad \text{s.t.} \quad \nabla_x L(x,u) \geq 0, \quad \alpha e \geq u \geq 0 \tag{3.7'}$$

Note that the only difference between (3.7') and (3.2') is the constraint $\alpha e \geq u$. Now, for any $\varepsilon > 0$, the point $(\hat{x},\ \hat{z} := e\varepsilon,\ \hat{u})$ satisfies a "Slater" constraint qualification for the dual problems (3.6')-(3.7') for $\alpha >$ $> \|\hat{u}\|_\infty$. Hence these problems have equal extrema and a solution $(x(\alpha)$, $z(\alpha)$, $u(\alpha))$ such that $x(\alpha)$ is bounded by [10, Theorem 2.3]

$$\|x(\alpha)\|_1 \leq \frac{\hat{u}(-g(\hat{x}) + \varepsilon e) + \hat{x}\nabla_x L(\hat{x},\hat{u}) + \varepsilon e(\alpha e - \hat{u})}{\min_i\,(\nabla_x L(\hat{x},\hat{u}))_i} \tag{3.8}$$

Since the left side of (3.8) does not depend on ε, we can let $\varepsilon \to 0$ in (3.8) and we have

$$\|x(\alpha)\|_1 \leq \frac{-\hat{u}g(\hat{x}) + \hat{x}\nabla_x L(\hat{x},\hat{u})}{\min_i\,(\nabla_x L(\hat{x},\hat{u}))_i} \tag{3.9}$$

Note now that by the weak duality theorem [5] applied to (3.1) and (3.3) we have

$$\inf_{x\in X} f(x) \geq L(\hat{x},\hat{u}) - \hat{x}\nabla_x L(\hat{x},\hat{u}) > -\infty$$

Hence for an unbounded increasing sequence of positive numbers $\{\alpha_i\}$ exceeding $\|\bar{u}\|_\infty$, it follows [10, Theorem 2.3] that there exists a sequence of points $\{x(\alpha_i), u(\alpha_i)\}$ with $x(\alpha_i)$ bounded as in (3.9), such that each $x(\alpha_i)$ solves the penalty function problem (3.6) with $\alpha = \alpha_i$ and $(x(\alpha_i), u(\alpha_i))$ solves its dual (3.7'). Since $\{x(\alpha_i)\}$ is bounded it has an accumulation point $\bar{x}$ which is bounded by (3.9). Since $ez(\alpha_i)=e(g(x(\alpha_i)))_+$ is the penalty term for (3.6'), it follows by Theorem (2.8) that $e\bar{z} = eg(\bar{x})_+ = 0$, that $\bar{x}$ solves $\min_{x \in X} f(x)$ and that

$$(3.10) \qquad \lim_{j\to\infty} \alpha_{i_j} ez(\alpha_{i_j}) = \lim_{j\to\infty} \alpha_{i_j} e(g(x(\alpha_{i_j})))_+ = 0 \quad \text{for} \quad x(\alpha_{i_j}) \to \bar{x}$$

Now we establish the zero duality gap. Let $\{\varepsilon_i\}$ be any decreasing sequence of positive numbers converging to 0 and let $\{\alpha_i\}$ be an unbounded increasing sequence of positive numbers chosen as follows:

$$\infty > \sup_{(x,u)\in Z} L(x,u) - x\nabla_x L(x,u) - \varepsilon_i$$

(By weak duality theorem)

$$< L(x(\varepsilon_i), u(\varepsilon_i)) - x(\varepsilon_i)\nabla_x L(x(\varepsilon_i), u(\varepsilon_i))$$

(For some $(x(\varepsilon_i), u(\varepsilon_i)) \in Z$, by definition of sup)

$$\le L(x(\alpha_i), u(\alpha_i)) - x(\alpha_i)\nabla_x L(x(\alpha_i), u(\alpha_i))$$

(For α_i sufficiently large s.t. $\alpha_i \ge \|u(\varepsilon_i)\|_\infty$, because $(x(\alpha_i), u(\alpha_i))$ solves $\max L(x,u) - x\nabla_x L(x,u)$ s.t. $\nabla_x L(x,u) \ge 0$, $\alpha_i e \ge u \ge 0$)

$$= f(x(\alpha_i)) + \alpha_i ez(\alpha_i)$$

(By equality of primal-dual optimal objective functions of problem (3.6') and (3.7') with $\alpha = \alpha_i$)

$$= \sup_{(x,u)} \{L(x,u)-x\nabla_x L(x,u) \mid \nabla_x L(x,u) \ge 0,\ \alpha_i e \ge u \ge 0\} \le \sup_{(x,u)\in Z} L(x,u)-x\nabla_x L(x,u)$$

Since by (3.10), $\lim_{j\to\infty} \alpha_{i_j} ez(\alpha_{i_j}) = 0$ for $x(\alpha_{i_j}) \to \bar{x}$, it follows that

$$\sup_{(x,u)\in Z} L(x,u) - x\nabla_x L(x,u) = f(\bar{x}) = \min_{x\in X} f(x) \quad \#$$

We establish now an existence and boundedness result for the Lagrangian dual problem (3.3).

Theorem 3.2

(Wolfe-dual feasibility & primal interior-feasibility $\Rightarrow$ Lagrangian dual solution existence-boundedness & zero duality gap with primal). Let f and g be differentiable and convex on R^n and let $(\tilde{x},\tilde{u})$ satisfy:

(3.11) $$\tilde{x} \in X,\ (\tilde{x},\tilde{u}) \in Z,\ \tilde{x} > 0,\ g(\tilde{x}) < 0$$

There exists a dual optimal solution $(\bar{u},\bar{v})$ to the Lagrangian dual (3.3) which is bounded by

(3.12) $$\|\bar{u},\bar{v}\|_1 \le \frac{-\tilde{u}g(\tilde{x}) + \tilde{x}\nabla_x L(\tilde{x},\tilde{u})}{\min_{i,j} \{-g_i(\tilde{x}),\ \tilde{x}_j\}}$$

In addition there is no duality gap between the primal problem (3.1) and the Lagrangian dual (3.3), that is:

(3.13) $$\inf_{x\in X} f(x) = \max_{(u,v)\ge 0} \inf_{x\in R^n} L(x,u) - xv$$

Proof: For $\beta > 0$ consider the bounded version of (3.1)

(3.14) $$\min f(x) \quad \text{s.t.} \quad g(x) \le 0, \quad \beta e \ge x \ge 0$$

and its Wolfe dual

(3.15)
$$\max_{(x,u,v,w)} \quad L(x,u) - vx + w(x - \beta e)$$
$$\text{s.t.} \quad \nabla_x L(x,u) - v + w = 0, \quad u,v,w \geq 0$$

or equivalently

(3.15')
$$\max_{(x,u,w)} \quad L(x,u) - x\nabla_x L(x,u) - \beta ew$$
$$\text{s.t.} \quad \nabla_x L(x,u) + w \geq 0, \quad u,w \geq 0$$

which again is equivalent to

(3.15")
$$\max_{\substack{(x,u)\\ u\geq 0}} \; L(x,u) - x\nabla_x L(x,u) - \beta e(-\nabla_x L(x,u))_+$$

which is nothing other than an exterior penalty function formulation for the Wolfe dual (3.2') with penalty parameter β. Thus the bound β on the ∞-norm of the primal variable x becomes a penalty parameter on the Wolfe dual.

Now for any $\varepsilon > 0$, the point

$$(\tilde{x}, \tilde{u}, \tilde{w} := \varepsilon e)$$

satisfies a Slater constraint qualification for the dual problems (3.14)-(3.15') for $\beta > \|x\|_\infty$. Hence [10, Theoreme 2.3] there exists $(x(\beta), u(\beta), v(\beta), w(\beta))$ which solves the dual problems (3.14)-(3.15) with equal extrema. For any such solution, $(u(\beta), v(\beta))$ is bounded by [10, Theorem 2.2]

(3.16)
$$\|u(\beta), v(\beta)\|_1 \leq \frac{-\tilde{u}g(\tilde{x}) + \beta\varepsilon ee + \tilde{x}\nabla_x L(\tilde{x},\tilde{u})}{\min\limits_{i,j} \{-g_i(\tilde{x}), \tilde{x}_j\}}$$

Since the left side of (3.16) does not depend on ε, we can let $\varepsilon \to 0$ in (3.16) and we have

$$\|u(\beta),v(\beta)\|_1 \le \frac{-\tilde{u}g(\tilde{x}) + \tilde{x}\nabla_x L(\tilde{x},\tilde{u})}{\min\limits_{i,j} \{-g_i(\tilde{x}), \tilde{x}_j\}} \tag{3.17}$$

Define now

$$\phi(u,v) := \inf_{x \in R^n} L(x,u) - vx \tag{3.18}$$

$$\psi(u,v,w) := \inf_{x \in R^n} L(x,u) - vx + wx \tag{3.19}$$

Then

$$\phi(u,v) = \psi(u,v,0) \tag{3.20}$$

Note now that by the weak duality theorem [5]

$$\infty > f(x) \ge \sup_{(x,u)\in Z} L(x,u) - x\nabla_x L(x,u)$$

Hence for an unbounded increasing sequence of positive numbers $\{\beta_i\}$ exceeding $\|\tilde{x}\|_\infty$, it follows [10, Theorem 2.3] that there exists a sequence of points $\{x(\beta_i), u(\beta_i), v(\beta_i), w(\beta_i)\}$ which solve the dual pair (3.14)-(3.15) for $\beta = \beta_i$, giving equal extrema and such that $\{u(\beta_i), v(\beta_i)\}$ is bounded by (3.17). Since $ew(\beta_i) = e(-\nabla_x L(x(\beta_i), u(\beta_i)))_+$ constitutes the penalty term for (3.15"), it follows by (2.7) that $\{ew(\beta_i)\}$ converges to zero and since $w(\beta_i) \ge 0$, it follows that $\{w(\beta_i)\}$ also converges to $\bar{w} = 0$. Let $(\bar{u},\bar{v},0)$ be an accumulation point of the bounded sequence $\{u(\beta_i), v(\beta_i), w(\beta_i)\}$. Now we have

$c := L(\tilde{x},\tilde{u}) - \tilde{x}\nabla_x L(\tilde{x},\tilde{u}) \le \inf\limits_{x\in X} f(x)$ (By weak duality)

$\le f(x(\beta_i))$ (Since $x(\beta_i) \in X$)

$\le L(x(\beta_i), u(\beta_i)) - v(\beta_i)x(\beta_i) + w(\beta_i)x(\beta_i)$

(Since $u(\beta_i)g(x(\beta_i))=0, v(\beta_i)x(\beta_i) = 0$ and $w(\beta_i)x(\beta_i) \ge 0$)

$$\le L(x,u(\beta_i)) - v(\beta_i)x + w(\beta_i)x \quad \forall x \in R^n$$

$$\text{(Since } \nabla_x L(x(\beta_i),u(\beta_i)) - v(\beta_i) + w(\beta_i) = 0$$

$$L(x,u(\beta_i)) - v(\beta_i)x + w(\beta_i)x \text{ is convex in } x)$$

In the limit we have

$$c \le L(x,\bar{u}) - \bar{v}x + \bar{w}x \qquad \forall x \in R^n$$

and so (since $\bar{w} = 0$)

$$c \le \inf_{x\in R^n} L(x,\bar{u}) - \bar{v}x + \bar{w}x = \psi(\bar{u},\bar{v},\bar{w}) = \phi(\bar{u},\bar{v})$$

Since $\psi(\bar{u},\bar{v},\bar{w})$ is finite, it follows by Theorem A.1 of the Appendix,that $\psi(u,v,w)$ is upper semicontinuous at $(\bar{u},\bar{v},\bar{w})$ with respect to R_+^{m+2n}. Now let $\{\varepsilon_j\} \downarrow 0$. It follows by the upper semicontinuity of $\psi(u,v,w)$ at $(\bar{u},\bar{v},\bar{w})$ that there exists a subsequence $\{\beta_{i_j}\} \uparrow \infty$ of the unbounded increasing sequence $\{\beta_i\}$ such that $\{u(\beta_{i_j}), v(\beta_{i_j}), w(\beta_{i_j})\}$ converges to $(\bar{u},\bar{v},\bar{w}=0)$ and

(3.21) $$\phi(\bar{u},\bar{v}) + \varepsilon_j = \psi(\bar{u},\bar{v},\bar{w}) + \varepsilon_j$$

$$> \psi(u(\beta_{i_j}), v(\beta_{i_j}), w(\beta_{i_j}))$$

(By usc of ψ at $(\bar{u},\bar{v},\bar{w})$)

$$= \inf_x L(x,u(\beta_{i_j})) - v(\beta_{i_j})x + w(\beta_{i_j})x$$

(By definition of ψ)

$$= L(x(\beta_{i_j}),u(\beta_{i_j})) - v(\beta_{i_j})x(\beta_{i_j})+w(\beta_{i_j})x(\beta_{i_j})$$

(Since $x(\beta_{i_j})$ minimizes $L(x,u(\beta_{i_j}))-v(\beta_{i_j})x+w(\beta_{i_j})x$)

$$\ge f(x(\beta_{i_j}))$$

(Since $u(\beta_{i_j})g(x(\beta_{i_j}))=0$, $v(\beta_{i_j})x(\beta_{i_j}) = 0$ and $w(\beta_{i_j})x(\beta_{i_j}) \ge 0$)

$$\geq L(x(\beta_{i_j}), u(\beta_{i_j})) - vx(\beta_{i_j}) \quad \text{for} \quad (u,v) \geq 0$$

$$(\text{Since } g(x(\beta_{i_j}) \leq 0 \quad \text{and} \quad x(\beta_{i_j}) \geq 0)$$

$$\geq \phi(u,v) \qquad (\text{By definition of } \phi)$$

Note that for $\{\beta_{i_j}\} \uparrow \infty$, the sequence $\{f(x(\beta_{i_j}))\}$ of minima of (3.14) with $\beta = \beta_{i_j}$, constitutes a nonincreasing sequence bounded below by $\inf_{x \in X} f(x)$. Hence $\{f(x(\beta_{i_j})\}$ converges and

(3.22) $$\inf_{x \in X} f(x) \leq \lim_{j \to \infty} f(x(\beta_{i_j})$$

Letting $\varepsilon_j \to 0$ in the string of inequalities of (3.21) gives

$$\phi(\bar{u},\bar{v}) \geq \lim_{j \to \infty} f(x(\beta_{i_j})) \geq \phi(u,v) \qquad \forall (u,v) \geq 0$$

Hence

(3.23) $$\phi(\bar{u},\bar{v}) = \lim_{j \to \infty} f(x(\beta_{i_j})) = \max_{(u,v) \geq 0} \phi(u,v) = \max_{(u,v) \geq 0} \inf_{x \in R^n} L(x,u) - vx$$

and $(\bar{u},\bar{v})$ solves the Lagrangian dual problem (3.3). The bound (3.12) on $(\bar{u},\bar{v})$ follows from (3.17). To show a zero duality gap, just note that

$$\inf_{x \in X} f(x) \leq \lim_{j \to \infty} f(x(\beta_{i_j})) = \max_{(u,v) \geq 0} \phi(u,v) \leq \inf_{x \in X} f(x)$$

where the first inequality follows from (3.22), the equality from (3.23) and the last inequality from the weak duality thorem for the Lagrangian dual [4,1]. Hence

$$\inf_{x \in X} f(x) = \max_{(u,v) \geq 0} \phi(u,v) \qquad \#$$

We remark that the existence part of this theorem and the zero duality gap result can also be derived as a consequence of the strong duality theorem of Lagrangian duality (e.g. [4, Theorem 3]) which is based on the entirely different argument of a separating hyperplane. Our explicit bound on the dual optimal variables (3.12) however does not follow from Lagrangian duality and is based on the recent boundedness results of [10].

4. PENALTY FUNCTIONS IN LINEAR PROGRAMMING

In this final section we show how to use penalty function results to determine precisely the value of the parameter in the quadratic perturbation to a linear program [6,7,8] in order to obtain a solution to the perturbed problem which is dual feasible to within any preassigned tolerance. This is a practical and important issue which has not been completely resolved before in the iterative successive overrelaxation (SOR) methods for solving huge sparse linear programs [8] .

We consider the primal linear program

$$\max_{x} \; cx \quad \text{s.t.} \quad Ax \le b, \quad x \ge 0 \tag{4.1}$$

where A is given $m \times n$ real matrix, $c \in R^n$ and $b \in R^m$, and its dual

$$\min_{(u,v)} \; bu \quad \text{s.t.} \quad v = A^T u - c, \quad u,v \ge 0 \tag{4.2}$$

In [8] it has been shown that perturbed primal program

$$\max_{x} \; cx - \frac{\varepsilon}{2} xx \quad \text{s.t.} \quad Ax \le b, \quad x \ge 0 \tag{4.3}$$

is solvable for all $\varepsilon \in (0,\bar{\varepsilon}]$ for some $\bar{\varepsilon}$ if and only if (4.1) is solvable, in which case the unique solution $\bar{x}$ of (4.3) for $\varepsilon \in (0,\bar{\varepsilon}]$ is independent of ε and is the point in the solution set of (4.1) with least 2-norm. If we consider the Wolfe dual to (4.3) we obtain

$$\min_{(x,u,v)} \ bu + \frac{\varepsilon}{2}xx \quad \text{s.t.} \quad c - \varepsilon x - A^T u + v = 0, \quad u,v \geq 0 \tag{4.4}$$

Elimination of x through the constraint relation

$$x = \frac{1}{\varepsilon}(-A^T u + v + c) \tag{4.5}$$

gives

$$\min_{(u,v)\geq 0} \ bu + \frac{1}{2\varepsilon}\|-A^T u + v + c\|_2^2 \tag{4.6}$$

which is precisely the exterior penalty function associated with the dual linear program (4.2) with penalty parameter $\frac{1}{\varepsilon}$. Using standard exterior penalty function results, one needs that $\varepsilon \to 0$ in order for solutions $(u(\varepsilon), v(\varepsilon))$ of (4.6) to approach a solution of the dual linear problem (4.2). However by computing $\bar{x}$ from $(u(\varepsilon), v(\varepsilon))$ through the relation (4.5), it turns out [8] that for $\varepsilon \in (0,\bar{\varepsilon}]$, $\bar{x}$ is independent of ε and is the unique point in the solution set of (4.1) with least 2-norm. In [8] SOR methods were prescribed for solving (4.6) for ε sufficiently small and then computing x from (4.5). Very large sparse problems (n = = 20,000, m = 5,000) were solved by this technique, without knowing what $\bar{\varepsilon}$ is, but merely by decreasing ε until certain approximate optimality criteria were met. We would like to show here that by solving the penalty problem (4.6) for only *two* values of ε, we can satisfy the Karush-Kuhn-Tucker optimality conditions for the linear program to any preassigned tolerance. In fact such a solution will be primal feasible, satisfy the complementarity conditions between primal and dual linear programs, and satisfy dual feasibility to any required tolerance. More specifically we have the following.

Theorem 4.1

Let $\delta > 0$, $\varepsilon_1 > 0$, let $(\hat{u},\hat{v})$ be dual feasible, that is $\hat{v} = A^T\hat{u} - c \geq 0$, $\hat{u} \geq 0$, and let $(u(\varepsilon_1), v(\varepsilon_1))$ be a solution of (4.6) with $\varepsilon = \varepsilon_1$. If $b\hat{u} \leq bu(\varepsilon_1)$ then $(\hat{u},\hat{v})$ solves the dual problem (4.2), else for

$$\varepsilon_2 < \varepsilon_1 \quad \text{and} \quad \varepsilon_2 \leq \frac{\delta}{b\hat{u} - bu(\varepsilon_1)} \tag{4.7}$$

it follows that

$$\frac{1}{2}\|-A^Tu(\varepsilon_2)+v(\varepsilon_2)+c\|_2^2 \leq \delta, \quad bu(\varepsilon_2) \leq \min_u \{bu \mid A^Tu \geq c,\ u \geq 0\} \tag{4.8}$$

where $(u(\varepsilon_2), v(\varepsilon_2))$ is a solution of (4.6) with $\varepsilon = \varepsilon_2$. Furthermore for $x(\varepsilon_2)$ defined by

$$x(\varepsilon_2) := \frac{1}{\varepsilon_2}(-A^Tu(\varepsilon_2) + v(\varepsilon_2) + c) \tag{4.9}$$

we have that the Karush-Kuhn-Tucker conditions for the linear program (4.1) are satisfied to within a tolerance δ as follows

$$(4.10) \quad \left\langle \begin{array}{l} x(\varepsilon_2) \geq 0,\ Ax(\varepsilon_2) \leq b,\ u(\varepsilon_2) \geq 0,\ v(\varepsilon_2) \geq 0 \\ u(\varepsilon_2)(b - Ax(\varepsilon_2)) = 0,\ v(\varepsilon_2)x(\varepsilon_2) = 0 \\ \|-A^Tu(\varepsilon_2) + v(\varepsilon_2) + c\|_2 \leq (2\delta)^{\frac{1}{2}} \end{array} \right.$$

Proof: The first part of theorem, (4.7)-(4.8), follows directly from Theorem 2.3. The last part of the theorem (4.10) follows from (4.8) and from the Karush-Kuhn-Tucker optimality conditions for (4.6) with $\varepsilon = \varepsilon_2$, that is

$$
(4.11)\quad \begin{cases} b - \frac{1}{\varepsilon_2} A(-A^T u(\varepsilon_2) + v(\varepsilon_2) + c) \geq 0, \quad u(\varepsilon_2) \geq 0 \\ u(\varepsilon_2(b - \frac{1}{\varepsilon_2} A(-A^T u(\varepsilon_2) + v(\varepsilon_2) + c) = 0 \\ \frac{1}{\varepsilon_2} (-A^T u(\varepsilon_2) + v(\varepsilon_2) + c) \geq 0, \quad v(\varepsilon_2) \geq 0 \\ \frac{v(\varepsilon_2)}{\varepsilon_2} (-A^T u(\varepsilon_2) + v(\varepsilon_2) + c) = 0 \end{cases}
$$

These conditions together with (4.8) and the definition (4.9) imply (4.10). #

5. APPENDIX

Theorem A.1

Let $\psi(s) := \inf_{t \in T} h(s,t)$ where $h: S \times T \to R$, $\phi \neq S \subset R^k$, $\phi \neq T \subset R^n$ and h is upper semicontinuous on S with respect to S for each fixed $t \in T$. Then ψ is upper semicontinuous with respect to S at each $\bar{s} \in S$ for which $\psi(\bar{s}) > -\infty$.

Proof: Suppose ψ is not usc at $\bar{s}$ with respect to S. Then

$$
(A.1) \qquad \exists \varepsilon > 0: \forall \delta > 0 \quad \exists s(\delta) \in S: \|s(\delta) - \bar{s}\| < \delta, \ \psi(s(\delta)) - \psi(\bar{s}) \geq \varepsilon
$$

Let ε be fixed. Since $-\infty < \psi(\bar{s}) = \inf_{t \in T} h(\bar{s},t)$, there exists $t(\varepsilon) \in T$ such that

$$
(A.2) \qquad h(\bar{s}, t(\varepsilon)) < \psi(\bar{s}) + \varepsilon
$$

Combining (A.1), (A.2) and the definition of ψ gives

$$
(A.3) \qquad \begin{gathered} h(\bar{s}, t(\varepsilon)) < \psi(\bar{s}) + \varepsilon \leq \psi(s(\delta)) \leq h(s(\delta), t(\varepsilon)) \\ \forall \delta > 0, \text{ for some } s(\delta) \in S \text{ such that } \|s(\delta) - \bar{s}\| < \delta \end{gathered}
$$

Since $h(s,t(\varepsilon))$ is usc with respect to S at $\bar{s} \in S$ we have

(A.4) $\forall\gamma > 0,\ \exists\delta(\gamma) > 0:\ \forall s \in S\ \ \|s-\bar{s}\| < \delta(\gamma),\ h(s,t(\varepsilon)) < h(\bar{s},t(\varepsilon))+\gamma$

Combining (A.3) and (A.4) gives

(A.5) $$h(\bar{s},t(\varepsilon)) < \psi(\bar{s}) + \varepsilon < h(\bar{s},t(\varepsilon)) + \gamma \quad \forall\gamma > 0$$

Since $\bar{s}$ and ε do not depend on γ, (A.5) gives a contradiction by letting γ approach zero. Hence ψ is usc at $\bar{s}$ with respect to S. #

REFERENCES

[1] M.S.Bazaraa and C.M.Shetty. 'Nonlinear programming theory and algorithms'. *Wiley*, New York (1979).

[2] E.J.Beltrami. 'An algorithmic approach to nonlinear analysis and optimization'. *Academic Press*, New York (1970).

[3] A.V.Fiacco and G.P.McCormick. 'Nonlinear programming : Sequential unconstrained minimization techniques'. *Wiley*, New York (1968).

[4] A.M.Geoffrion. 'Duality in nonlinear programming: a simplified applications-oriented development'. *SIAM Review* 13, (1971) 1-37.

[5] O.L.Mangasarian. 'Nonlinear programming'. *McGraw-Hill*, New York (1969).

[6] O.L.Mangasarian. 'Iterative solution of linear programs'.*SIAM Journal of Numerical Analysis* 18, (1981) 606-614.

[7] O.L.Mangasarian. 'Sparsity-preserving SOR algorithms for separable quadratic and linear programming problems'. *Computers and Operations Research* 11, (1984) 105-112.

[8] O.L.Mangasarian. 'Normal solutions of linear programs',Mathematical Programming Study 22,(1984) 206-216.

[9] O.L.Mangasarian. 'A condition number for differentiable convex inequalities', *Mathematics of Operations Research* 10, (1985)175-179.

[10] O.L.Mangasarian and L.McLinden. 'Simple bounds for solutions of monotone complementarity problems and convex programs', *Mathematical Programming* 32, (1985) 32-40.

[11] G.P.McCormick. 'Nonlinear programming theory, algorithms and applications'. *Wiley*, New York (1983).

[12] P.Wolfe. 'A duality theorem for nonlinear programming'.*Quarterly of Applied Mathematics* 19 (1961) 239-244.

Chapter 14

Dual Properties of Sequential Gradient - Restoration Algorithms for Optimal Control Problems

A. Miele and T. Wang

Abstract. The problem of minimizing a functional, subject to differential constraints, nondifferential constraints, initial constraints, and final constraints, is considered in connection with sequential gradient-restoration algorithms (SGRA) for optimal control problems. The system of Lagrange multipliers associated with (i) the gradient phase of SGRA and (ii) the restoration phase of SGRA is examined. For each phase, it is shown that the Lagrange multipliers are endowed with a duality property: they minimize a special functional, quadratic in the multipliers, subject to the multiplier differential equations and boundary conditions, for given state, control, and parameter. These duality properties have considerable computational implications; they allow one to reduce the auxiliary optimal control problems associated with (i) and (ii) to mathematical programming problems involving a finite number of parameters as unknows.

Key Words. Numerical methods, computing methods, mathematical programming, optimal control, optimality properties, duality properties, restoration algorithms, gradient algorithms, gradient-restoration algorithms, sequential gradient-restoration algorithms, general boundary conditions, nondifferential constraints.

1. INTRODUCTION

Over the past several years, a successful family of first-order algorithms for the solution of optimal control problems involving differential constraints, nondifferential constraints, initial constraints, and final constraints has been developed at Rice University by Miele and his associates (Refs. 1-7). They are called gradient - restoration algorithms and have been designed for the solution of different classes of optimal control problems. They include sequential gradient-restoration algorithms (SGRA, Refs. 1-3) and combined gradient-restoration algorithms (CGRA, Refs. 4-7).

Sequential gradient-restoration algorithms (Refs. 1-3) involve a sequence of two-phase cycles, each cycle including a gradient phase and a restoration phase; in a complete gradient-restoration cycle, the value of the functional is decreased, while the constraints are satisfied to a predetermined accuracy; in the gradient phase, the value of the augmented functional is decreased, while avoiding excessive constraint violation; in the restoration phase, the constraint error is decreased, while avoiding excessive change in the value of the functional. On the other hand, combined gradient-restoration algorithms (Refs. 4-7) involve a sequence of single-phase cycles, each cycle including a combined gradient-restoration phase; in this phase, the value of the augmented functional is decreased simultaneously with the constraint error.

In this paper, we consider the system of Lagrange multipliers associated with (i) the gradient phase of SGRA and (ii) the restoration phase of SGRA. For each phase, we show that the Lagrange multipliers are endowed with a duality property: they minimize a special functional, quadratic in the multipliers, subject to the multiplier differential equations and boundary conditions, for given state, control, and parameter.

For previous studies of duality properties of the family of gradient-restoration algorithms for optimal control problems, see Refs. 8-10. Here, we focus our attention on the computational implications of the duality properties. With reference to SGRA, we show that the auxiliary optimal control problems associated with (i) and (ii) can be reduced to mathematical programming problems involving a finite number of parameters as unknowns. Hence, the algorithmic efficiency of SGRA can be enhanced.

Notations. Vector-matrix notation is used for conciseness. All vectors are column vectors.

Let t denote the independent variable, and let $x(t)$, $u(t)$, $v(t)$, π denote the dependent variables. The time t is a scalar; the state $x(t)$ is an n-vector; the control $u(t)$ is an m-vector; the control $v(t)$ is a c-vector; and the parameter π is a p-vector.

Let $f(x,u,v,\pi,t)$ denote a scalar function of the arguments x,u,v,π,t. The symbol f_x denotes the n-vector function whose components are the partial derivatives of the scalar function f with respect to the components of the vector x. Analogous definitions hold for the symbols f_u, f_v, f_π.

Let $S(x,u,v,\pi,t)$ denote a c-vector function of the arguments x,u,v,π,t. The symbol S_x denote the $n \times c$ matrix function whose elements are the partial derivatives of the components of the vector function S with respect to the components of the vector x. Analogous definitions hold for the symbols S_u, S_v, S_π.

The dot sign denotes derivative with respect to the time, that is, $\dot{x} = dx/dt$. The symbol T denotes transposition of vector or matrix. The subscript 0 denotes the initial point, and the subscript 1 denotes the final point.

The symbol $N(y) = y^T y$ denotes the quadratic norm of a vector y.

2. GENERAL PROBLEM

Problem P. We consider the problem of minimizing the functional

$$I = \int_0^1 f(x,u,v,\pi,t)dt + [h(x,\pi)]_0 + [g(x,\pi)]_1 , \tag{1}$$

with respect to the n-vector state $x(t)$, the m-vector control $u(t)$, the c-vector control $v(t)$, and the p-vector parameter π which satisfy the constraints

(2a) $$\dot{x} + \phi(x,u,v,\pi,t) = 0, \qquad 0 \le t \le 1,$$

(2b) $$S(x,u,v,\pi,t) = 0, \qquad 0 \le t \le 1,$$

(2c) $$[\omega(x,\pi)]_0 = 0,$$

(2d) $$[\psi(x,\pi)]_1 = 0.$$

In the above equations, f is a scalar; h is a scalar; g is a scalar; ϕ is an n-vector; S is a c-vector; ω is an a-vector, $a \le n$; and ψ is a b-vector, $b \le n$. We assume that the first and second derivatives of the functions f,h,g,ϕ,S,ω,ψ with respect to the vectors x,u,v,π exist and are continuous. We also assume that the $n \times a$ matrix ω_x has rank a at the initial point, that the $n \times b$ matrix ψ_x has rank b at the final point, that the $c \times c$ matrix S_v has rank c everywhere along the interval of integration, and that the constrained minimum exists.

From calculus of variations, it is known that Problem P is of the Bolza type. It can be recast as that of minimizing the augmented functional

(3) $$J = I + L,$$

subject to (2), where L denotes the Lagrangian functional

(4) $$L = \int_0^1 \lambda^T(\dot{x} + \phi)dt + \int_0^1 \rho^T S dt + (\sigma^T \omega)_0 + (\mu^T \psi)_1 .$$

In Eq. (4), the symbols $\lambda(t)$, $\rho(t)$, σ, μ denote Lagrange multipliers associated with the constraints (2). With this understanding, the first-order optimality conditions of Problem P take the form

(5a) $$\dot{\lambda} - f_x - \phi_x\lambda - S_x\rho = 0, \qquad 0 \le t \le 1,$$

(5b) $$f_u + \phi_u\lambda + S_u\rho = 0\,, \qquad 0 \le t \le 1,$$

(5c) $$f_v + \phi_v\lambda + S_v\rho = 0\,, \qquad 0 \le t \le 1,$$

(5d) $$\int_0^1 (f_\pi + \phi_\pi\lambda + S_\pi\rho)dt + (h_\pi + \omega_\pi\sigma)_0 + (g_\pi + \psi_\pi\mu)_1 = 0,$$

(5e) $$(-\lambda + h_x + \omega_x\sigma)_0 = 0\,,$$

(5f) $$(\lambda + g_x + \psi_x\mu)_1 = 0\,.$$

Summarizing, we seek the functions $x(t)$, $u(t)$, $v(t)$, π and the multipliers $\lambda(t)$, $\rho(t)$, σ, μ such that the feasibility equations (2) and the optimality conditions (5) are satisfied.

Alternative Form. Under the assumption that the matrix S_v is nonsingular, Eq. (5c) yields the following solution for the Lagrange multiplier associated with the nondifferential constraint (2b):

(6) $$\rho = -S_v^{-1}(f_v + \phi_v\lambda)\,.$$

As a consequence, the optimality conditions (5) can be rewritten in the following alternative form:

(7a) $$\dot{\lambda} - F_x - \Phi_x\lambda = 0\,, \qquad 0 \le t \le 1,$$

(7b) $$F_u + \Phi_u\lambda = 0\,, \qquad 0 \le t \le 1,$$

(7c) $$\int_0^1 (F_\pi + \Phi_\pi\lambda)dt + (h_\pi + \omega_\pi\sigma)_0 + (g_\pi + \psi_\pi\mu)_1 = 0\,,$$

(7d) $$(-\lambda + h_x + \omega_x\sigma)_0 = 0\,,$$

(7e) $$(\lambda + g_x + \psi_x\mu)_1 = 0\,.$$

Here, F_x, F_u, F_π denote the vectors

(8a) $$F_x = f_x - S_x S_v^{-1} f_v \ ,$$

(8b) $$F_u = f_u - S_u S_v^{-1} f_v \ ,$$

(8c) $$F_\pi = f_\pi - S_\pi S_v^{-1} f_v \ ,$$

and Φ_x, Φ_u, Φ_π denote the matrices

(9a) $$\Phi_x = \phi_x - S_x S_v^{-1} \phi_v \ ,$$

(9b) $$\Phi_u = \phi_u - S_u S_v^{-1} \phi_v \ ,$$

(9c) $$\Phi_\pi = \phi_\pi - S_\pi S_v^{-1} \phi_v \ .$$

Summarizing, we seek the functions $x(t)$, $u(t)$, $v(t)$, π and the multipliers $\lambda(t)$, σ, μ such that the feasibility equations (2) and the optimality conditions (7) are satisfied.

In the following sections, we refer to the form (7) of the optimality conditions.

Performance Indexes. The form of Eqs. (2) and (7) suggests that the following scalar performance indexes are useful in computational work:

(10a) $$P = \int_0^1 N(\dot{x} + \phi)dt + \int_0^1 N(S)dt + N(\omega)_0 + N(\psi)_1 ,$$

(10b) $$Q = \int_0^1 N(\dot{\lambda} - F_x - \Phi_x\lambda)dt + \int_0^1 N(F_u + \Phi_u\lambda)dt$$
$$+ N\left[\int_0^1 (F_\pi + \Phi_\pi\lambda)dt + (h_\pi + \omega_\pi\sigma)_0 + (g_\pi + \psi_\pi\mu)_1\right]$$
$$+ N(-\lambda + h_x + \omega_x\sigma)_0 + N(\lambda + g_x + \psi_x\mu)_1 \ .$$

Here, P denotes the error in the constraints and Q the error in the optimality conditions. Therefore, numerical convergence can be characterized by the relations

$$P \le \varepsilon_1 \ , \quad Q \le \varepsilon_2 \ , \tag{11}$$

where ε_1, ε_2 are preselected, small, positive numbers.

3. GRADIENT PHASE, PRIMAL FORMULATION

The gradient phase of the sequential gradient-restoration algorithm involves a single iteration and is designed to decrease the augmented functional, while avoiding excessive constraint violation. The gradient iteration is started whenever Ineq. (11-1) is satisfied.

Let $x(t)$, $u(t)$, $v(t)$, π denote the nominal functions. Let $\tilde{x}(t)$, $\tilde{u}(t)$, $\tilde{v}(t)$, $\tilde{\pi}$ denote the varied functions. Let $\Delta x(t)$, $\Delta u(t)$, $\Delta v(t)$, $\Delta\pi$ denote the displacements leading from the nominal functions to the varied functions. By definition, the following relations hold:

$$\tilde{x}(t) = x(t) + \Delta x(t) \ , \tag{12a}$$

$$\tilde{u}(t) = u(t) + \Delta u(t) \ , \tag{12b}$$

$$\tilde{v}(t) = v(t) + \Delta v(t) \ , \tag{12c}$$

$$\tilde{\pi} = \pi + \Delta\pi \ . \tag{12d}$$

The displacements $\Delta x(t)$, $\Delta u(t)$, $\Delta v(t)$, $\Delta\pi$ are computed by solving the following auxiliary minimization problem.

Problem G1. Minimize the first variation of the functional (1), with respect to the vectors $\Delta x(t)$, $\Delta u(t)$, $\Delta v(t)$, $\Delta\pi$ which satisfy the linearized form of the constraints (2) plus a quadratic isoperimetric constraint imposed on the vectors $\Delta u(t)$, $\Delta\pi$, $\Delta x(0)$.

Therefore, we minimize the functional

$$(13)\qquad I_1 = \int_0^1 (f_x^T\Delta x+f_u^T\Delta u+f_v^T\Delta v+f_\pi^T\Delta\pi)dt+(h_x^T\Delta x+h_\pi^T\Delta\pi)_0+(g_x^T\Delta x+g_\pi^T\Delta\pi)_1 ,$$

with respect to the vectors $\Delta x(t)$, $\Delta u(t)$, $\Delta v(t)$, $\Delta\pi$ which satisfy the constraints

$$(14a)\qquad \Delta\dot{x} + \phi_x^T\Delta x + \phi_u^T\Delta u + \phi_v^T\Delta v + \phi_\pi^T\Delta\pi = 0 , \qquad 0 \le t \le 1 ,$$

$$(14b)\qquad S_x^T\Delta x + S_u^T\Delta u + S_v^T\Delta v + S_\pi^T\Delta\pi = 0 , \qquad 0 \le t \le 1 ,$$

$$(14c)\qquad (\omega_x^T\Delta x + \omega_\pi^T\Delta\pi)_0 = 0 ,$$

$$(14d)\qquad (\psi_x^T\Delta x + \psi_\pi^T\Delta\pi)_1 = 0 ,$$

and

$$(15)\qquad \int_0^1 \Delta u^T\Delta u dt + \Delta\pi^T\Delta\pi + (\Delta x^T\Delta x)_0 - K = C .$$

Alternative Form. Under the assumption that the matrix S_v is non-singular, Eq. (14b) yields the following solution for the dependent control displacement:

$$(16)\qquad \Delta v = -(S_v^{-1})^T(S_x^T\Delta x + S_u^T\Delta u + S_\pi^T\Delta\pi) .$$

As a consequence, upon eliminating Δv from Eqs. (13)-(14) and upon recalling the definitions (8)-(9), Problem G1 is reformulated as follows: Minimize the functional

$$(17)\qquad I_1 = \int_0^1 (F_x^T\Delta x+F_u^T\Delta u+F_\pi^T\Delta\pi)dt + (h_x^T\Delta x+h_\pi^T\Delta\pi)_0 + (g_x^T\Delta x+g_\pi^T\Delta\pi)_1 ,$$

with respect to the vectors $\Delta x(t)$, $\Delta u(t)$, $\Delta\pi$ which satisfy the constraints

(18a) $$\Delta\dot{x} + \phi_x^T \Delta x + \phi_u^T \Delta u + \phi_\pi^T \Delta\pi = 0, \qquad 0 \le t \le 1,$$

(18b) $$(\omega_x^T \Delta x + \omega_\pi^T \Delta\pi)_0 = 0,$$

(18c) $$(\psi_x^T \Delta x + \psi_\pi^T \Delta\pi)_1 = 0,$$

and

(19) $$\int_0^1 \Delta u^T \Delta u dt + \Delta\pi^T \Delta\pi + (\Delta x^T \Delta x)_0 - K = 0.$$

Problem G1 is a Bolza problem. Let $\lambda(t)$, σ, μ, $1/2\alpha$ denote Lagrange multipliers associated with the constraints (18)-(19). With this understanding, the first-order optimality conditions of Problem G1 take the form

(20a) $$\dot{\lambda} - F_x - \phi_x \lambda = 0, \qquad 0 \le t \le 1,$$

(20b) $$F_u + \phi_u \lambda + \Delta u/\alpha = 0, \qquad 0 \le t \le 1,$$

(20c) $$\int_0^1 (F_\pi + \phi_\pi \lambda) dt + (h_\pi + \omega_\pi \sigma)_0 + (g_\pi + \psi_\pi \mu)_1 + \Delta\pi/\alpha = 0,$$

(20d) $$(-\lambda + h_x + \omega_x \sigma + \Delta x/\alpha)_0 = 0,$$

(20e) $$(\lambda + g_x + \psi_x \mu)_1 = 0.$$

Summarizing, we seek the functions $\Delta x(t)$, $\Delta u(t)$, $\Delta\pi$ and the multipliers $\lambda(t)$, σ, μ, $1/2\alpha$ such that the feasibility equations (18)-(19) and the optimality conditions (20) are satisfied. After Eqs. (18)-(20) are solved, $\Delta v(t)$ is obtained with (16).

4. GRADIENT PHASE, DUAL FORMULATION

Let y(t), z, w denote the vectors defined by

(21a) $$F_u + \Phi_u\lambda + y = 0, \quad 0 \le t \le 1,$$

(21b) $$\int_0^1 (F_\pi + \Phi_\pi\lambda)dt + (h_\pi + \omega_\pi\sigma)_0 + (g_\pi + \psi_\pi\mu)_1 + z = 0,$$

(21c) $$(-\lambda + h_x + \omega_x\sigma)_0 + w = 0.$$

It is interesting to note that the vectors $\lambda(t)$, σ, μ and $y(t)$, z, w can also be obtained by solving the following auxiliary minimization problem.

Problem G2. Minimize the functional

(22) $$I_2 = (1/2)\left[\int_0^1 y^T y dt + z^T z + w^T w\right],$$

with respect to the vectors $\lambda(t)$, σ, μ and $y(t)$, z, w which satisfy the constraints

(23a) $$\dot{\lambda} - F_x - \Phi_x\lambda = 0, \quad 0 \le t \le 1,$$

(23b) $$F_u + \Phi_u\lambda + y = 0, \quad 0 \le t \le 1,$$

(23c) $$\int_0^1 (F_\pi + \Phi_\pi\lambda)dt + (h_\pi + \omega_\pi\sigma)_0 + (g_\pi + \psi_\pi\mu)_1 + z = 0,$$

(23d) $$(-\lambda + h_x + \omega_x\sigma)_0 + w = 0,$$

(23e) $$(\lambda + g_x + \psi_x\mu)_1 = 0.$$

Problem G2 is a Bolza problem. Let $\lambda_*(t)$, $-y_*(t)$, $-z_*$, $-w_*$, $-\nu_*$

denote Lagrange multipliers associated with the constraints (23). With this understanding, the first-order optimality conditions of Problem G2 take the form

(24a) $$\dot{\lambda}_* + \phi_x^T \lambda_* + \phi_u^T y_* + \phi_\pi^T z_* = 0, \qquad 0 \le t \le 1,$$

(24b) $$(\omega_x^T w_* + \omega_\pi^T z_*)_0 = 0,$$

(24c) $$(\psi_x^T \nu_* + \psi_\pi^T z_*)_1 = 0,$$

and

(25a) $$y - y_* = 0, \qquad 0 \le t \le 1,$$

(25b) $$z - z_* = 0,$$

(25c) $$w - w_* = 0,$$

(25d) $$(\lambda_* - w_*)_0 = 0,$$

(25e) $$(\lambda_* - \nu_*)_1 = 0.$$

Let the following substitutions be employed:

(26a) $$\lambda_* = \Delta x(t)/\alpha, \qquad 0 \le t \le 1,$$

(26b) $$y_* = y = \Delta u(t)/\alpha, \qquad 0 \le t \le 1,$$

(26c) $$z_* = z = \Delta\pi/\alpha,$$

(26d) $$w_* = w = \Delta x(0)/\alpha,$$

(26e) $$\nu_* = \Delta x(1)/\alpha.$$

Then, one can readily verify that Eqs. (23) of Problem G2 reduce to Eqs. (20) of Problem G1, that Eqs. (24) of Problem G2 reduce to Eqs.(18) of Problem G1, and that Eqs. (25) become identities.

Clearly, after the transformation (26) is applied, the solution of Problem G2 yields the solution of Problem G1, and viceversa. This means that the multipliers $\lambda(t)$, σ, μ associated with the gradient phase of SGRA are endowed with a duality property: they also minimize the quadratic functional (22), subject to (23), for given state, control, and parameter.

5. GRADIENT PHASE, COMPUTATIONAL IMPLICATIONS

In this section, we exploit the previously established duality property and we show that the execution of a gradient iteration can be reduced to solving a mathematical programming problem involving a finite number of parameters as unknowns. Hence, the algorithmic efficiency of the gradient phase of SGRA can be enhanced.

First, we consider Eqs. (23a) and (23e). We observe that, if μ is assigned, $\lambda(1)$ can be computed with (23e) and $\lambda(t)$ can be computed by backward integration of (23a).

Next, we execute b + 1 backward integrations, using Eqs. (23a)and (23e) in combination with the following choices for the multiplier μ:

(27a) $$\mu_1 = \delta_1, \quad \mu_2 = \delta_2, \ldots, \mu_b = \delta_b,$$

(27b) $$\mu_{b+1} = 0.$$

In (27a), δ_1, $\delta_2, \ldots,$ δ_b denote the vectors correspondings to the columns of the identity matrix of order b; in (27b), 0 denotes the null vector of dimension b.

Let $\lambda_1(t)$, $\lambda_2(t), \ldots,$ $\lambda_b(t)$, $\lambda_{b+1}(t)$ denote the particular solutions of Eqs. (23a) and (23e), corresponding to the choices (27) for the multiplier μ. Let $\tilde{\mu}$ denote the $b \times (b+1)$ matrix[1]

(28a) $$\tilde{\mu} = [\mu_1, \mu_2, \ldots, \mu_b, \mu_{b+1}] ;$$

let $\tilde{\lambda}(t)$denote the $n \times (b+1)$ matrix

(28b) $$\tilde{\lambda}(t) = [\lambda_1(t), \lambda_2(t), \ldots, \lambda_b(t), \lambda_{b+1}(t)];$$

and let k denote the $(b+1)$-vector

(28c) $$k = [k_1, k_2, \ldots, k_b, k_{b+1}]^T .$$

If the method of particular solution is employed (Refs. 11-13),the general solution of Eqs. (23a) and (23e) can be written in the form

(29) $$\mu = \tilde{\mu}k, \qquad \lambda(t) = \tilde{\lambda}(t)k,$$

with the following understanding: the components of the vector k must satisfy the normalization condition

(30) $$U^T k = 1,$$

where

(31) $$U = [1, 1, \ldots, 1, 1]^T$$

denotes the $(b+1)$-vector whose components are all equal to one.

Next, we combine Eqs. (23b), (23c), (23d) with Eqs. (29) and obtain the relations

(32a) $$y = -F_u - \Phi_u \tilde{\lambda} k, \qquad 0 \le t \le 1 ,$$

(1) Clearly, $\tilde{\mu} = [I, 0]$, where I denotes the identity matrix of order b.

(32b) $$z = -\left[\int_0^1 F_\pi dt + (h_\pi)_0 + (g_\pi)_1\right] - \left[\int_0^1 \phi_\pi \tilde{\lambda} dt + (\psi_\pi)_1 \tilde{\mu}\right] k - (\omega_\pi)_0 \sigma,$$

(32c) $$w = -(h_x)_0 + (\tilde{\lambda})_0 k - (\omega_x)_0 \sigma.$$

These relations show that $y(t)$, z, w depend only on the parameters k, σ.

Finally, upon combining (22) and (32), we obtain the following quadratic function of k, σ:

(33) $$I_2 = (1/2)k^T A k + (1/2)\sigma^T B \sigma + k^T C \sigma + D^T k + E^T \sigma + (1/2)H,$$

where

(34) $$U^T k - 1 = 0.$$

Here, the matrices A, B, C, the vectors D, E, and the scalar H are known. They are defined by

(35a) $$A = \int_0^1 (\phi_u \tilde{\lambda})^T (\phi_u \tilde{\lambda}) dt + (\tilde{\lambda}^T \tilde{\lambda})_0 + \left[\int_0^1 \phi_\pi \tilde{\lambda} dt + (\psi_\pi)_1 \tilde{\mu}\right]^T \left[\int_0^1 \phi_\pi \tilde{\lambda} dt + (\psi_\pi)_1 \tilde{\mu}\right],$$

(35b) $$B = (\omega_x^T \omega_x + \omega_\pi^T \omega_\pi)_0,$$

(35c) $$C = \left[\int_0^1 \phi_\pi \tilde{\lambda} dt + (\psi_\pi)_1 \tilde{\mu}\right]^T (\omega_\pi)_0 - (\tilde{\lambda}^T \omega_x)_0,$$

(35d) $$D = \left[\int_0^1 \phi_\pi \tilde{\lambda} dt + (\psi_\pi)_1 \tilde{\mu}\right]^T \left[\int_0^1 F_\pi dt + (h_\pi)_0 + (g_\pi)_1\right] + \int_0^1 (\phi_u \tilde{\lambda})^T F_u dt - (\tilde{\lambda}^T h_x)_0,$$

(35e) $$E = (\omega_\pi^T)_0 \left[\int_0^1 F_\pi dt + (h_\pi)_0 + (g_\pi)_1 \right] + (\omega_x^T h_x)_0,$$

(35f) $$H = \left[\int_0^1 F_\pi dt + (h_\pi)_0 + (g_\pi)_1 \right]^T \left[\int_0^1 F_\pi dt + (h_\pi)_0 + (g_\pi)_1 \right]$$
$$+ \int_0^1 F_u^T F_u dt + (h_x^T h_x)_0 .$$

Because of the duality property, the parameters k, σ can be obtained by minimizing (33), subject to (34). Clearly, this auxiliary minimization problem is a mathematical programming problem.

Let β denote a scalar Lagrange multiplier associated with the constraint (34). Let G denote the augmented function

(36) $$G = (1/2)k^T Ak + (1/2)\sigma^T B\sigma + k^T C\sigma + D^T k + E^T \sigma + (1/2)H + \beta(U^T k - 1).$$

With this understanding, the first-order optimality conditions of the auxiliary minimization problem take the form

(37) $$G_k = 0, \quad G_\sigma = 0.$$

Hence, the values of k, σ, β are determined by solving the following linear algebraic system:

(38a) $$Ak + C\sigma + U\beta + D = 0,$$

(38b) $$C^T k + B\sigma + E = 0,$$

(38c) $$U^T k - 1 = 0,$$

whose dimension is $a + b + 2$. Once k, σ, β are known, the multipliers μ and $\lambda(t)$ are determined with Eqs. (29). Then, $\Delta u(t)/\alpha$, $\Delta\pi/\alpha$, $\Delta x(0)/\alpha$ are obtained with Eqs. (20b), (20c), (20d). Finally, $\Delta x(t)/\alpha$ is determined by forward integration of (18a).

We note that, except for the determination of the stepsize , the iteration is completed. The stepsize α can be determined a posteriori using some suitable search procedure (Refs. 1-3).

6. RESTORATION PHASE, PRIMAL FORMULATION

The restoration phase of the sequential gradient-restoration algorithm involves one or more iterations and is designed to decrease the constraint error, while avoiding excessive change in the value of the functional. A restorative iteration is started whenever Ineq.(11-1) is violated.

Let $x(t)$, $u(t)$, $v(t)$, π denote the nominal functions. Let $\tilde{x}(t)$,$\tilde{u}(t)$, $\tilde{v}(t)$, $\tilde{\pi}$ denote the varied functions. Let $\Delta x(t)$, $\Delta u(t)$, $\Delta v(t)$,$\Delta\pi$ denote the displacements leading from the nominal functions to the varied functions. By definition, the following relations hold:

(39a) $$\tilde{x}(t) = x(t) + \Delta x(t),$$

(39b) $$\tilde{u}(t) = u(t) + \Delta u(t),$$

(39c) $$\tilde{v}(t) = v(t) + \Delta v(t),$$

(39d) $$\tilde{\pi} = \pi + \Delta\pi.$$

The displacements $\Delta x(t)$, $\Delta u(t)$, $\Delta v(t)$, $\Delta\pi$ are computed by solving the following auxiliary minimization problem.

Problem R1. Minimize the norm squared of the vectors $\Delta u(t)$,$\Delta\pi$,$\Delta x(0)$, with respect to the vectors $\Delta x(t)$, $\Delta u(t)$, $\Delta v(t)$, $\Delta\pi$ which satisfy the linearized form of the constraints (2).

Therefore, we minimize the functional

(40) $$I_1 = (1/2\alpha)\left[\int_0^1 \Delta u^T \Delta u dt + \Delta\pi^T\Delta\pi + (\Delta x^T\Delta x)_0\right],$$

with respect to the vectors $\Delta x(t)$, $\Delta u(t)$, $\Delta v(t)$, $\Delta\pi$ which satisfy the the constraints

(41a) $$\Delta\dot{x} + \phi_x^T\Delta x + \phi_u^T\Delta u + \phi_v^T\Delta v + \phi_\pi^T\Delta\pi + \alpha(\dot{x} + \phi) = 0, \qquad 0 \le t \le 1,$$

(41b) $$S_x^T\Delta x + S_u^T\Delta u + S_v^T\Delta v + S_\pi^T\Delta\pi + \alpha S = 0, \qquad 0 \le t \le 1,$$

(41c) $$(\omega_x^T\Delta x + \omega_\pi^T\Delta\pi + \alpha\omega)_0 = 0,$$

(41d) $$(\psi_x^T\Delta x + \psi_\pi^T\Delta\pi + \alpha\psi)_1 = 0.$$

The symbol α in (40)-(41) denotes the restoration stepsize, $0 < \alpha \le 1$.

Alternative Form. Under the assumption that the matrix S_v is nonsingular, Eq. (41b) yields the following solution for the dependent control displacement:

(42) $$\Delta v = -(S_v^{-1})^T(S_x^T\Delta x + S_u^T\Delta u + S_\pi^T\Delta\pi + \alpha S).$$

Let the following definitions be introduced[2]:

(43) $$\hat{\phi} = \phi - (S_v^{-1}\phi_v)^T S$$

and

(44a) $$\Phi_x = \phi_x - S_x S_v^{-1}\phi_v,$$

(44b) $$\Phi_u = \phi_u - S_u S_v^{-1}\phi_v,$$

(2) Note that the definitions (44) are the same as the definitions (9).

(44c) $$\tilde{\phi}_\pi = \phi_\pi - S_\pi S_v^{-1}\phi_v.$$

As a consequence, upon eliminating Δv from Eqs. (41) and upon recalling the definitions (43)-(44), Problem R1 is reformulated as follows: Minimize the functional

(45) $$I_1 = (1/2\alpha)\left[\int_0^1 \Delta u^T \Delta u dt + \Delta\pi^T\Delta\pi + (\Delta x^T \Delta x)_0\right],$$

with respect to the vectors $\Delta x(t)$, $\Delta u(t)$, $\Delta\pi$ which satisfy the constraints

(46a) $$\Delta\dot{x} + \phi_x^T\Delta x + \phi_u^T\Delta u + \phi_\pi^T\Delta\pi + \alpha(\dot{x} + \hat{\phi}) = 0, \qquad 0 \le t \le 1,$$

(46b) $$(\omega_x^T\Delta x + \omega_\pi^T\Delta\pi + \alpha\omega)_0 = 0,$$

(46c) $$(\psi_x^T\Delta x + \psi_\pi^T\Delta\pi + \alpha\psi)_1 = 0.$$

Problem R1 is a Bolza problem. Let $\lambda(t)$, σ, μ denote Lagrange multipliers associated with the constraints (46). With this understanding, the first-order optimality conditions of Problem R1 take the form

(47a) $$\dot{\lambda} - \phi_x\lambda = 0, \qquad 0 \le t \le 1,$$

(47b) $$\phi_u\lambda + \Delta u/\alpha = 0, \qquad 0 \le t \le 1,$$

(47c) $$\int_0^1 \phi_\pi\lambda dt + (\omega_\pi\sigma)_0 + (\psi_\pi\mu)_1 + \Delta\pi/\alpha = 0,$$

(47d) $$(-\lambda + \omega_x\sigma + \Delta x/\alpha)_0 = 0,$$

(47e) $$(\lambda + \psi_x\mu)_1 = 0.$$

Summarizing, we seek the functions $\Delta x(t)$, $\Delta u(t)$, $\Delta\pi$ and the multipliers

$\lambda(t)$, σ, μ such that the feasibility equations (46) and the optimality conditions (47) are satisfied. After Eqs. (46)-(47) are solved, $\Delta v(t)$ is obtained with (42).

7. RESTORATION PHASE, DUAL FORMULATION

Let $y(t)$, z, w denote the vectors defined by

(48a) $$\Phi_u\lambda + y = 0, \qquad 0 \le t \le 1\ ,$$

(48b) $$\int_0^1 \Phi_\pi\lambda dt + (\omega_\pi\sigma)_0 + (\psi_\pi\mu)_1 + z = 0,$$

(48c) $$(-\lambda + \omega_x\sigma)_0 + w = 0.$$

It is interesting to note the vectors $\lambda(t)$, σ, μ and $y(t)$, z, w can also be obtained by solving the following auxiliary minimization problem.

Problem R2. Minimize the functional

(49) $$I_2 = (1/2)\left[\int_0^1 y^T y dt + z^T z + w^T w\right] - \left[\int_0^1 \lambda^T(\dot{x} + \hat{\phi})dt + (\sigma^T\omega)_0 + (\mu^T\psi)_1\right],$$

with respect to the vectors $\lambda(t)$, σ, μ and $y(t)$, z, w which satisfy the constraints

(50a) $$\dot{\lambda} - \Phi_x\lambda = 0, \qquad 0 \le t \le 1,$$

(50b) $$\Phi_u\lambda + y = 0, \qquad 0 \le t \le 1,$$

(50c) $$\int_0^1 \Phi_\pi\lambda dt + (\omega_\pi\sigma)_0 + (\psi_\pi\mu)_1 + z = 0,$$

(50d) $$(-\lambda + \omega_x \sigma)_0 + w = 0.$$

(50e) $$(\lambda + \psi_x \mu)_1 = 0.$$

Problem R2 is a Bolza problem. Let $\lambda_*(t)$, $-y_*(t)$, $-z_*$, $-w_*$, $-\nu_*$ denote Lagrange multipliers associated with the constraints (50). With this understanding, the first-order optimality conditions of Problem R2 take the form

(51a) $$\dot{\lambda}_* + \phi_x^T \lambda_* + \phi_u^T y_* + \phi_\pi^T z_* + (\dot{x} + \hat{\phi}) = 0, \qquad 0 \le t \le 1,$$

(51b) $$(\omega_x^T w_* + \omega_\pi^T z_* + \omega)_0 = 0,$$

(51c) $$(\psi_x^T \nu_* + \psi_\pi^T z_* + \psi)_1 = 0,$$

and

(52a) $$y - y_* = 0, \qquad 0 \le t \le 1,$$

(52b) $$z - z_* = 0,$$

(52c) $$w - w_* = 0,$$

(52d) $$(\lambda_* - w_*)_0 = 0,$$

(52e) $$(\lambda_* - \nu_*)_1 = 0.$$

Let the following substitutions be employed:

(53a) $$\lambda_* = \Delta x(t)/\alpha, \qquad 0 \le t \le 1,$$

(53b) $$y_* = y = \Delta u(t)/\alpha, \qquad 0 \le t \le 1,$$

(53c) $$z_* = z = \Delta \pi/\alpha,$$

(53d) $$w_* = w = \Delta x(0)/\alpha,$$

(53e) $$\nu_* = \Delta x(1)/\alpha.$$

Then, one can readily verify that Eqs. (50) of Problem R2 reduce to Eqs. (47) of Problem R1, that Eqs. (51) of Problem R2 reduce to Eqs. (46) of Problem R1, and that Eqs. (52) become identities.

Clearly, after the transformation (53) is applied, the solution of Problem R2 yields the solution of Problem R1, and viceversa. This means that the multipliers $\lambda(t)$, σ, μ associated with the restoration phase of SGRA are endowed with a duality property: they also minimize the quadratic functional (49), subject to (50), for given state, control, and parameter.

8. RESTORATION PHASE, COMPUTATIONAL IMPLICATIONS

In this section, we exploit the previously established duality property and we show that the execution of a restorative iteration can be reduced to solving a mathematical programming problem involving a finite number of parameters as unknowns. Hence, the algorithmic efficiency of the restoration phase of SGRA can be enhanced.

First, we consider Eqs. (50a) and (50e). We observe that, if μ is assigned, $\lambda(1)$ can be computed with (50e) and $\lambda(t)$ can be computed by backward integration of (50a).

Next, we execute b + 1 backward integrations, using Eqs. (50a) and (50e) in combination with the following choices for the multiplier μ:

(54a) $$\mu_1 = \delta_1, \qquad \mu_2 = \delta_2, \ldots, \mu_b = \delta_b,$$

(54b) $$\mu_{b+1} = 0.$$

In (54a), δ_1, $\delta_2, \ldots$, δ_b denote the vectors corresponding to the columns of the identity matrix of order b; in (54b), 0 denotes the null vector of dimension b.

Let $\lambda_1(t)$, $\lambda_2(t),\ldots,\lambda_b(t)$, $\lambda_{b+1}(t)$ denote the particular solutions of Eqs. (50a) and (50e), corresponding to the choices (54) for the multiplier μ. Let $\tilde{\mu}$ denote be $b \times (b+1)$ matrix[3]

(55a) $$\tilde{\mu} = [\mu_1,\mu_2,\ldots,\mu_b,\mu_{b+1}];$$

let $\tilde{\lambda}(t)$ denote the $n \times (b+1)$ matrix

(55b) $$\tilde{\lambda}(t) = [\lambda_1(t),\lambda_2(t),\ldots,\lambda_b(t),\lambda_{b+1}(t)];$$

and let k denote the $(b+1)$-vector

(55c) $$k = [k_1,k_2,\ldots,k_b,k_{b+1}]^T.$$

If the method of particular solution is employed (Refs. 11-13), the general solution of Eqs. (50a) and (50e) can be written in the form

(56) $$\mu = \tilde{\mu}k, \quad \lambda(t) = \tilde{\lambda}(t)k,$$

with the following understanding: the components of the vector k must satisfy the normalization condition

(57) $$U^T k = 1,$$

where

(58) $$U = [1,1,\ldots,1,1]^T$$

([3]) Clearly, $\tilde{\mu} = [I,0]$, where I denotes the identity matrix of order b.

denotes the (b + 1)-vector whose components are all equal to one.

Next, we combine Eqs. (50b), (50c), (50d) with Eqs. (56) and obtain the relations

(59a) $$y = -\phi_u \tilde{\lambda} k, \qquad 0 \le t \le 1,$$

(59b) $$z = -\left[\int_0^1 \phi_\pi \tilde{\lambda} dt + (\psi_\pi)_1 \tilde{\mu}\right] k - (\omega_\pi)_0 \sigma,$$

(59c) $$w = (\tilde{\lambda})_0 k - (\omega_x)_0 \sigma.$$

These relations show that $y(t)$, z, w depend only on the parameters k,σ.

Finally, upon combining (49) and (59), we obtain the following quadratic function of k, σ:

(60) $$I_2 = (1/2)k^T A k + (1/2)\sigma^T B \sigma + k^T C \sigma + D^T k + E^T \sigma,$$

where

(61) $$U^T k - 1 = 0.$$

Here, the matrices A, B, C and the vectors D, E are known. They are defined by

(62a) $$A = \int_0^1 (\phi_u \tilde{\lambda})^T (\phi_u \tilde{\lambda}) dt + (\tilde{\lambda}^T \tilde{\lambda})_0 + \left[\int_0^1 \phi_\pi \tilde{\lambda} dt + (\psi_\pi)_1 \tilde{\mu}\right]^T \left[\int_0^1 \phi_\pi \tilde{\lambda} dt + (\psi_\pi)_1 \tilde{\mu}\right],$$

(62b) $$B = (\omega_x^T \omega_x + \omega_\pi^T \omega_\pi)_0,$$

(62c) $$C = \left[\int_0^1 \phi_\pi \tilde{\lambda} dt + (\psi_\pi)_1 \tilde{\mu}\right]^T (\omega_\pi)_0 - (\tilde{\lambda}^T \omega_x)_0,$$

(62d) $$D = -\int_0^1 \tilde{\lambda}^T (\dot{x} + \hat{\phi}) dt - (\tilde{\mu}^T \psi)_1,$$

$$E = -(\omega)_0. \tag{62e}$$

Because of the duality property, the parameters k, σ can be obtained by minimizing (60), subject to (61). Clearly, this auxiliary minimization problem is a mathematical programming problem.

Let β denote a scalar Lagrange multiplier associated with the constraint (61). Let G denote the augmented function

$$G = (1/2)k^TAk + (1/2)\sigma^TB\sigma + k^TC\sigma + D^Tk + E^T\sigma + \beta(U^Tk - 1). \tag{63}$$

With this understanding, the first-order optimality conditions of the auxiliary minimization problem take the form

$$G_k = 0, \quad G_\sigma = 0. \tag{64}$$

Hence, the values of k, σ, β are determined by solving the following linear algebraic system:

$$Ak + C\sigma + U\beta + D = 0, \tag{65a}$$

$$C^Tk + B\sigma + E = 0, \tag{65b}$$

$$U^Tk - 1 = 0, \tag{65c}$$

whose dimension is $a + b + 2$. Once k, σ, β are known, the multipliers μ and $\lambda(t)$ are determined with Eqs. (56). Then, $\Delta u(t)/\alpha$, $\Delta\pi/\alpha$, $\Delta x(0)/\alpha$ are obtained with Eqs. (47b), (47c), (47d). Finally, $\Delta x(t)/\alpha$ is determined by forward integration of (46a).

We note that, except for the determination of the stepsize, the iteration is completed. The stepsize α can be determined a posteriori using some suitable search procedure (Refs. 1-3).

9. COMMENTS

In sections 3-8, primal-dual properties were derived employing only first-order conditions. It can be verified readily that both the Legendre-Clebsch condition and the Weierstrass condition are satisfied for all of the auxiliary minimization problems studied in connection with both the primal formulation and the dual formulation. It can also be verified that the solution of each auxiliary minimization problem is unique. These topics are omitted for the sake of brevity.

It is clear that some simple relations hold between the values of the functionals associated with the primal formulation and the values of the functionals associated with the dual formulation. These relations are stated below without proof.

For the gradient phase, the functionals I_1 and I_2 associated with Problems G1 and G2 satisfy the relation

(66) $$I_1 + 2\alpha I_2 = 0,$$

where α is the stepsize of the gradient phase.

For the restoration phase, the functionals I_1 and I_2 associated with Problems R1 and R2 satisfy the relation

(67) $$I_1 + \alpha I_2 = 0,$$

where α is the stepsize of the restoration phase.

10. CONCLUSIONS

The problem of minimizing a functional, subject to differential constraints, nondifferential constraints, initial constraints, and final constraints, is considered in connection with sequential gradient-restoration algorithms (SGRA) for optimal control problems. The system of

Lagrange multipliers associated with (i) the gradient phase of SGRA and (ii) the restoration phase of SGRA is examined. For each phase, it is shown that the Lagrange multipliers are endowed with a duality property: they minimize a special functional, quadratic in the multipliers, subject to the multiplier differential equations and boundary conditions, for given state, control, and parameter. These duality properties have considerable computational implications: they allow one to reduce the auxiliary optimal control problems associated with (i) and (ii) to mathematical programming problems involving a finite number of parameters as unknowns.

REFERENCES

[1] A.Miele, R.E.Pritchard, and J.N.Damoulakis. 'Sequential gradient - restoration algorithm for optimal control problems'. *Journal of Optimization Theory and Applications*, vol.5, n° 4, (1970), 235-282.

[2] A.Miele, J.N.Damoulakis, J.R.Cloutier, and J.L.Tietze. 'Sequential gradient-restoration algorithm for optimal control problems with non-differential constraints'. *Journal of Optimization Theory and Applications*, vol.13, n°2, (1974), 218-255.

[3] S.Gonzales, and A.Miele. 'Sequential gradient-restoration algorithm for optimal control problems with general boundary conditions'. *Journal of Optimization Theory and Applications*, vol.26, n°3, (1978) , 395-425.

[4] A.Miele. 'Gradient methods in control theory, Part 6, Combined gradient-restoration algorithm'. Rice University , Aero-Astronautics Report n°74, (1970).

[5] A.Miele. 'Combined gradient-restoration algorithm for optimal control problems'. Rice University , Aero-Astronautics Report n°91, (1971).

[6] A.Miele, V.K.Basapur, and E.M.Coker. 'Combined gradient-restoration algorithm for optimal control problems with nondifferential constraints and general boundary conditions, Part 1, Theory'.Rice University,Aero-Astronautics Report n°163, (1983).

[7] A.Miele, V.K.Basapur, and E.M.Coker. 'Combined gradient-restoration algorithm for optimal control problems with nondifferential constraints and general boundary conditions, Part 2, Examples'. Rice University, Aero-Astronautics Report n°164, (1983).

[8] A.Miele, and C.T.Liu.'Supplementary optimality properties of gradient-restoration algorithms for optimal control problems'. *Journal of Optimization Theory and Applications*, vol.32, n°4, (1980),577-593.

[9] A.Miele, and T.Wang. 'Supplementary optimality properties of the restoration phase of sequential gradient-restoration algorithms for optimal control problems'. *Journal of Optimization Theory and Applications*, vol.41, n°1, (1983),169-184.

[10] A.Miele, and T.Wang. 'Supplementary optimality properties of the family of gradient-restoration algorithms for optimal control problems'. Proceedings of the 4th IFAC Workshop on Applications of Nonlinear Programming to Optimization and Control, Edited by H.E.Rauch, San Francisco, California, (1983),109-120.

[11] A.Miele. 'Method of particular solutions for linear, two-point boundary- value problems'. *Journal of Optimization Theory and Applications*, vol.2, n°4, (1968),260-273.

[12] A.Miele, and R.R.Iyer. 'Modified quasilinearization method for solving nonlinear, two-point boundary-value problems'. *Journal of Mathematical Analysis and Applications*, vol.36, n°3, (1971), 674-692.

[13] A.Miele, and R.R.Iyer. 'General technique for solving nonlinear , two-point boundary-value problems via the method of particular solutions'. *Journal of Optimization Theory and Applications*, vol.5, n°5, (1970),382-399.

[14] A.Miele, and C.T.Liu. 'Supplementary optimality properties of gradient-restoration algorithms'. Rice-University, Aero-Astronautics Report n°149, (1979).

[15] A.Miele, and T.Wang. 'Supplementary optimality properties of the restoration phase of sequential gradient-restoration algorithms for optimal control problems'. Rice-University, Aero-Astronautics Report n°165, (1983).

[16] A.Miele, and T.Wang. 'Supplementary optimality properties of combined gradient-restoration algorithms for optimal control problems'. Rice University, Aero-Astronautics Report n°166, (1983).

[17] A.Miele, and T.Wang. 'Supplementary optimality properties of the family of gradient-restoration algorithms for optimal control problems'. Rice University, Aero-Astronautics Report n°171, (1983).

Chapter 15

Stochastic Homogenization and Ergodic Theory

L. Modica

One of the most studied problems in infinite dimensional optimization has been recently the mathematical theory of homogenization. This theory relies on an asymptotic analysis of classical minimization problems for integral functionals of Calculus of Variations depending on a small parameter $\varepsilon \to 0+$.

Our aim here is to give a report about some recent results in non-linear stochastic homogenization obtained in collaboration with G. Dal Maso. Linear stochastic homogenization has been deeply studied by V.V. Yurinskij, S.M. Kozlov, G.C. Papanicolaou and S.R.S.Varadhan and others.

The main feature of our approach is to consider an optimization problem

$$\text{(P)} \qquad \min_{u \in X} F(u)$$

as function of F and to endow the space of all functionals F in which we are interested by a metric such that the map

$$F \to \min_{u \in X} F(u)$$

is continuous. This metric is related, in our case, to Γ-convergence. Then the convergence of a family (P_ε) of optimization problems as $\varepsilon \to 0+$ to a limit problem (P_0) becomes a result of convergence in this metric of (F_ε) to F_0. Moreover, if we have stochastic optimization problems $(P_\varepsilon(\omega))$, with ω random parameter, then convergence in probability of $(P_\varepsilon(\omega))$ becomes simply convergence in probability of $(F_\varepsilon(\omega))$ to $F_0(\omega)$.

By this approach and by elementary Probability Theory we first obtained a result of convergence in probability (proposition 3) for an equicoercive stochastic homogenization process. Recently, we have improved this result (proposition 5) by using a powerful result in Ergodic Theory due to M.A. Ackoglu and U. Krengel and we have faced the problem of the non-coercive case by a generalization of the classical compactness result for the weak* convergence of measures (proposition 7).

To fix the ideas, let us consider in the unitary cube Q of the three-dimensional space $\mathbb{R}^3$ a dielectric material built up by N^3 small cubes with edge $1/N$ of two different homogeneous and isotropic substances, with dielectric constants $\lambda>0$ and $\Lambda>0$, disposed in a three-dimensional chessboard structure. If the boundary of the cube is held to a given (even non-constant) electric potential u_0, the energy of the electric field in Q is given by

$$E_N = \min_u \left\{ \iiint_Q a_N \, |\text{grad } u|^2 \, dx\, dy\, dz \; : \; u = u_0 \text{ on } \partial Q \right\}$$

where $a_N : Q \to \mathbb{R}$ is the function which describes the dielectric constant in Q and holds λ or Λ.

The homogenization problem in this case is to study the behavior of E_N, and of the corresponding potential u_N which realizes the minimum, as $N \to +\infty$, in particular to check whether

$$(1) \qquad E_\infty = \lim_{N\to+\infty} E_N = \min_u \left\{ \iiint_Q a_\infty |Du|^2 \, dx\, dy\, dz \; : \; u = u_0 \text{ on } \partial Q \right\}$$

for a suitable a_∞, which would be the dielectric constant of the homogenized material.

Of course, this ia only a model for the general theory of homoge -

nization. Anyway, we want to recall that (1) holds but, to my knowledge, no explicit formula for a_∞ in terms of λ and Λ is known. The situation is more favorable for the analogous one and two-dimensional structures in which a_∞ is respectively the harmonic and the geometric mean of λ and Λ.

The general formulation of the homogenization problem in which we are interested is as follows. Let us define here "integrand" any function $f\colon \mathbb{R}^n \times \mathbb{R}^n \to \mathbb{R}$ such that $f(x,p)$ is Lebesgue measurable in x, convex in p and

$$c_1|p|^\alpha \le f(x,p) \le c_2(1+|p|^\alpha) \qquad \forall (x,p) \in \mathbb{R}^n \times \mathbb{R}^n$$

with $c_1 > 0$, $c_2 > 0$, $\alpha > 1$ fixed real constants. If $f(x,p)$ is periodic in x, we say that f is a periodic integrand. To each integrand f we associate an integral functional $F\colon L^\alpha_{loc}(\mathbb{R}^n) \times A_0 \to \overline{\mathbb{R}}$ ($\overline{\mathbb{R}} = \mathbb{R} \cup \{+\infty, -\infty\}$) given by

$$F(u,A) = \begin{cases} \int_A f(x,Du(x))\,dx & \text{if } u_{|A} \in W^{1,\alpha}(A) \\ +\infty & \text{otherwise} \end{cases}$$

where A_0 is the family of all open bounded subsets of $\mathbb{R}^n$ and $W^{1,\alpha}(A)$ is the usual Sobolev space of the functions of $L^\infty(A)$ whose first partial derivatives belong to $L^\infty(A)$. Hereafter Du will denote the gradient of u.

For such a functional F we may consider, for any $A \in A_0$ and $u_0 \in W^{1,\alpha}(A)$ fixed, the Dirichlet problem

$$P(F,u_0,A) : \min_u \{F(u,A) : u-u_0 \in W^{1,\alpha}_0(A)\}$$

which has a solution because of the classical results of Calculus of Variations: we shall denote by $m(F,u_0,A)$ the minimum value of $P(F,u_0,A)$.

Now, let f be a periodic integrand and define

$$F_\varepsilon(u,A) = \begin{cases} \int_A f(x/\varepsilon,\ Du(x))\,dx & \text{if } u_{|A} \in W^{1,\alpha}(A) \\ +\infty & \text{otherwise.} \end{cases}$$

The homogenization problem is for us to study

$$\lim_{\varepsilon \to 0+} m(F_\varepsilon,\ u_0,\ A)\ .$$

(It is not difficult, by choosing $n=3$, $A=Q$, $\varepsilon=1/N$, $c_1=\lambda$, $c_2=\Lambda$, $\alpha=2$ and

$$f(x,p) = \begin{cases} \lambda|p|^2 & \text{if } \sum_{i=1}^{3} [x_i] \text{ is even} \\ \Lambda|p|^2 & \text{otherwise} \end{cases}$$

where $[x_i]$ is the integer part of the i-th component of $x \in \mathbb{R}^3$, to recover our initial model problem).

A way for attacking this problem is to endow the set $F=F(c_1,c_2,\alpha)$ of all the integral functionals associated to our integrands (even non-periodic) with a metric d such that the map $F \to m(F,u_0,A)$ is continuous in (F,d) for every $A \in A_0$ and $u_0 \in W^{1,\alpha}(A)$. If this metric exists, the homogenization problem is reduced to the study of the limit in (F,d) of (F_ε) as $\varepsilon \to 0+$. The answer to this first problem is positive, as the following proposition shows.

Proposition 1 (E.De Giorgi, G.Dal Maso-L.Modica).
There exists a metric d on F such that the map $F \to m(F,u_0,A)$ is continuous in (F,d) for every $A \in A_0$ and $u_0 \in W^{1,\alpha}(A)$; the space (F,d) is compact; all the compact metrics which satisfy the above continuity property are topologically equivalent ; the convergence of a sequence in (F,d) is equivalent to the $\Gamma(L^{\alpha-})$-convergence.

In this setting the homogenization problem has a very general solution.

Proposition 2 (P.Marcellini, G.Dal Maso-L.Modica)
If f is a periodic integrand, then the associated family $(F_\varepsilon)_{\varepsilon>0}$ converges in (F,d) as $\varepsilon \to 0+$ to an integral functional F_o whose integrand $f_o(x,p)$ does not depend on x and is given by

$$f_o(p) = \lim_{t\to+\infty} |Q_t|^{-1} m(F,l_p,Q_t)$$

where $F = F_1$ is the functional associated to f, $Q = \{x \in \mathbb{R}^n : |x_i|<t,\ i=1,..,n\}$ is a cube, $|Q_t| = (2t)^n$ is its measure and $l_p = p.x$ is the linear function on $\mathbb{R}^n$ with gradient p.

This formulation of the homogenization problem has another advantage: it permits to attack directly the stochastic homogenization. Returning for a moment to our initial model problem, the stochastic case would correspond to a random distribution in Q of the small cubes of the two different substances. In general, by stochastic homogenization process we mean a family $(F_\varepsilon)_{\varepsilon>0}$ of random variables defined on a probability space (Ω,T,P) with values in (F,d) (endowed with the Borel σ-algebra) with the following properties:

(a) there exists a random functional $F: \Omega \to F$ such that F_ε has the same distribution law of the random functionals $\rho_\varepsilon F$ given by

$$[(\rho_\varepsilon F)(\omega)](u,A) = \begin{cases} \int_A f(\omega,x/\varepsilon,Du(x))\,dx & \text{if } u_{|A} \in W^{1,\alpha}(A) \\ +\infty & \text{otherwise} \end{cases}$$

where $f(\omega,x,p)$ is the integrand of $F(\omega)$, that is

$$P\{\omega \in \Omega: F_\varepsilon(\omega) \in S\} = P\{\omega \in \Omega: \rho_\varepsilon F(\omega) \in S\}$$

for every open subset S of (F,d);

(b) the random functional F is stochastically periodic, that is F has the same distribution law of the random functional $\tau_z F$ given by

$$(\tau_z F)(\omega)\ (u,A) = \begin{cases} \int_A f(\omega,x+z,Du(x))\, dx & \text{if } u_{|A}\ W^{1,\alpha}(A) \\ +\infty & \text{otherwise} \end{cases}$$

for every $z \in \mathbb{Z}^n$.

Then we have the following stowhastic form of the previous proposition.

Proposition 3 (G.Dal Maso-L.Modica)
Let (F_ε) be a stochastic homogenization process. Suppose that there exists $M > 0$ such that the two families

$$\{F(\cdot)(u,A) \quad : \quad u \in L^\alpha_{loc}(\mathbb{R}^n)\}$$

$$\{F(\cdot)(u,B) \quad : \quad u \in L^\alpha_{loc}(\mathbb{R}^n)\}$$

of real extended random variables are independent for any pair $A,B \in \mathcal{A}_0$ with $\operatorname{dist}(A,B) \geq M$. Then there exists an almost everywhere constant random functional $F_0 : \Omega \to \mathcal{F}$ such that (F_ε) converges to F_0 in probability as $\varepsilon \to 0+$ (i.e.

$$\lim_{\varepsilon \to 0+} P\,\{\omega \in \Omega : d(F_\varepsilon(\omega),F_0(\omega)) \geq \eta\} = 0 \qquad \forall \eta > 0)$$

and whose integrand $f_0(\omega,x,p)$ for P-almost all $\omega \in \Omega$ depends only on p and is given by

$$f_0(p) = \lim_{t \to +\infty} |Q_t|^{-1} \int_\Omega m(F(\omega),1_p,Q_t)\, dP(\omega)\ .$$

It is obvious that, when $F(\omega)$ is constant in ω, $\rho_\varepsilon F=F_\varepsilon$ and $\tau_z F=F$, this proposition agrees with the previous one.

Convergence in probability obtained in the last proposition cannot be improved in P-almost everywhere convergence, because the hypotheses on F_ε and F are given in terms of their distribution laws. Then, the question arises whether there is P-almost everywhere convergence by assuming $F_\varepsilon = \rho_\varepsilon F$.

The answer is positive, but the proof requires some more subtle arguments and relies on recent results in Ergodic Theory.

Let us begin by recalling a subadditive ergodic theorem. We say that a function $\mu : \mathcal{A}_o \to \mathbb{R}$ is dominated and subadditive if

$$0 \le \mu(A) \le C\,|A| \qquad \forall A \in \mathcal{A}_o$$

with C real constant and

$$\mu(A) \le \sum_{k\in K} \mu(A_k)$$

for every finite and disjointed family $(A_k)_{k\in K}$ and for every $A \in \mathcal{A}_o$ such that $|A \setminus \bigcup_{k\in K} A_k| = 0$.

Let us denote by $\mathcal{M}$ the family of all dominated and subadditive functions $\mu : \mathcal{A}_o \to \mathbb{R}$, endowed by the trace of the product σ-algebra of $\mathbb{R}^{\mathcal{A}_o}$. The group $\mathbb{Z}^n$ of the translation in $\mathbb{R}^n$ acts obviously on $\mathcal{M}$: indeed we may define

$$(\tau_z\mu)(A) = \mu(\tau_z A) \qquad \forall z \in \mathbb{Z}^n \;,\; A \in \mathcal{A}_o$$

where $\tau_z A = \{x \in \mathbb{R}^n : x-z \in A\}$. A measurable subset E of $\mathcal{M}$ is said to be invariant if

$$\mu \in E \iff \tau_z\mu \in E$$

for every $z \in \mathbb{Z}^n$.

Proposition 4 (M.A.Akcoglu-U.Krengel)

Let $\mu: \Omega \to M$ be a measurable function stochastically periodic and ergodic in the sense that $\mu(\cdot)$ and $\tau_z\mu(\cdot)$ have the same distribution law for every $z \in \mathbb{Z}^n$ and $P\{\omega \in \Omega: \mu(\omega) \in E\} = 0$ or 1 for every invariant subset E of M. Then there exists

$$\lim_{\substack{h\to+\infty \\ h\in N}} |hQ|^{-1} [\mu(\omega)](hQ)$$

for P-almost all $\omega \in \Omega$ and for every $Q \in \mathcal{Q}$ where $\mathcal{Q}$ is the family of all cubes in $\mathbb{R}^n$ with vertex in $\mathbb{Z}^n$.

Our interest in this subadditibe ergodic theorem depends on the fact that, if $F: \Omega \to \mathcal{F}$ is a stochastically periodic random functional and $p \in \mathbb{R}^n$, then the function $\mu: \Omega \to R^{A_0}$ defined by

$$\mu(\omega)(A) = m(F(\omega),1_p,A) = \min_u \{F(\omega)(u,A): u-1_p \in W_0^{1,\alpha}(A)\}$$

is dominated and subadditive, as it is easy to check. The crucial property is that the sum of the minima of an integral functional on two disjoint open sets is less or equal than the minimum of the integral functional on their union.

Moreover, if F is ergodic (with respect to $\mathbb{Z}^n$ in the obvious sense), then μ is ergodic. In this case the previous proposition gives that there exists

$$\lim_{h\to+\infty} |hQ|^{-1} m(F(\omega),1_p,hQ)$$

for every $Q \in \mathcal{Q}$. Let us recall that, in proposition 3, the limit

$$\lim_{h\to+\infty} |hQ|^{-1} \int_\Omega m(F(\omega),1_p,hQ)\, dP(\omega)$$

was the integrand $f_0(p)$ of the limit functional. Then the following proposition is somewhat natural even if the proof requires some technicalities.

Proposition 5 (G.Dal Maso-L.Modica)
Let (F_ε) be a stochastic homogenization process such that $F_\varepsilon = \rho_\varepsilon F$ with F stochastically periodic and ergodic. Then $(F_\varepsilon(\omega))$ converges for P-almost all $\omega \in \Omega$ to a single functional $F_o \in F$, whose integrand f_o depends only on p and is given by

$$f_o(p) = \lim_{\substack{t\to+\infty \\ t\in\mathbb{R}}} |Q_t|^{-1}\, m(F(\omega), l_p, Q_t)$$

(the limit is P-almost everywhere constant in ω).

Let us remark that the same technique of the proof of the last proposition allows us to obtain an analogous result in the non-ergodic case, in which obviously the limit functional is a random functional, i.e. depends on ω. Moreover, the hypothesis of "independence at large distances" of proposition 3 is of "mixing" type and, in particular, implies ergodicity of F, hence proposition 5 contains proposition 3.

In the previous results the equicoerciveness hypothesis $c_1 > 0$ and the dependence of the integrands on x and Du but not on u play an important role. The possibility of extending our results to the non-equicoercive case $c_1 = 0$ (physically, a material with holes) has been very recently exploited by studying a general concept of weak* convergence of measures.

Indeed, the integral functionals of the class F calculated on a fixed $A \in A_o$ are lower semicontinuous functions between $L^\alpha(A)$ and $\overline{\mathbb{R}}$ and the topology on them given by Γ-convergence is compact and metrizable because of equicoerciveness, so the probability measures associated to a stochastic homogenization process are sequentially weak* compact.

If we want to treat in the same way the non-equicoercive case, without loosing the compactness, we may enlarge the class of the functionals by considering the class S of all real extended lower semicontinuous functions on $L^{\alpha}(A)$, which is sequentially compact with respect to the Γ-convergence and may be endowed with the topology whose closed subsets are the Γ-sequentially closed subset of S. Unfortunately, as recently proved by D.Dal Maso (Dal Maso senior), this topology is pathological because the only real continuous functions on S are the constants,hence the ordinary weak* convergence of measures is not meaningful on S.

On the other hand, there are many upper and lower semicontinuous functions on S; for example, the two important inequalities in Γ-convergence theory

$$F_h \xrightarrow{\Gamma(X^-)} F_\infty \Rightarrow \quad \text{(i)} \quad \inf_{x\in O} F_\infty(x) \geq \limsup_{h\to+\infty} \inf_{x\in O} F_h(x) \quad \forall\, O \text{ open in } X$$

$$\text{(ii)} \quad \inf_{x\in K} F_\infty(x) \leq \liminf_{h\to+\infty} \inf_{x\in K} F_h(x) \quad \forall\, K \text{ compact in } X$$

tell us that the function on S

$$F \to \inf_{x\in O} F(x) \qquad (O \text{ open in } L^{\alpha}(A))$$

$$F \to \inf_{x\in K} F(x) \qquad (K \text{ compact in } L^{\alpha}(A))$$

are respectively upper and lower semicontinuous. Therefore, it seems useful to base a weak* Γ-convergence of measures on S upon the semicontinuous functions on S.

The first meaningful fact in this direction is the possibility of determining a measure on S by some semicontinuous functions.

Proposition 6 (G.Dal Maso-E.De Giorgi-L.Modica)
If μ_1, μ_2 are two measures on the Borel subsets of S such that

$$\int_S \varphi \, d\mu_1 = \int_S \varphi \, d\mu_2$$

for every bounded, non-decreasing, lower semicontinuous function $\varphi : S \to R$, then $\mu_1 = \mu_2$.

The second one is the following compactness result.

Proposition 7 (G.Dal Maso-E.De Giorgi-L.Modica)
Let (μ_h) be a sequence of measures on the Borel subsets of S such that $\mu_h(S) \leq p < +\infty$ for every $h \in \mathbb{N}$. Then there exist a subsequence (μ_{h_k}) and a measure μ_∞ on the Borel subsets of S such that

$$\text{(i)} \quad \limsup_{h \to +\infty} \int_S \psi d\mu_{h_k} \leq \int_S \psi d\mu_\infty$$

$$\liminf_{k \to +\infty} \int_S \varphi d\mu_{h_k} \geq \int_S \varphi d\mu_\infty$$

for every bounded upper semicontinuous function $\psi : S \to \mathbb{R}$ and for every bounded lower semicontinuous function $\varphi : S \to \mathbb{R}$;

$$\text{(ii)} \quad \int_S \varphi d\mu_\infty \leq$$

$$\sup \left\{ \liminf_{k \to +\infty} \int_S \psi d\mu_{h_k} : \psi : S \to \mathbb{R} \text{ bounded non-decreasing upper semicontinuous such that } \psi \leq \varphi \right\}$$

for every bounded non-decreasing lower semicontinuous function $\varphi : S \to \mathbb{R}$.

REFERENCES

[1] D.Dal Maso. 'Questioni di topologia legate alla Γ-convergenza'. *Ricerche di Matematica*, 32(1983), 135-162.
[2] G.Dal Maso, L.Modica. 'Nonlinear Stochastic Homogenization'. (To appear on Annali di Matematica Pura e Applicata).
[3] G.Dal Maso, L.Modica. 'Integral Functionals determined by their Minima'. (To appear on Rendiconti del Seminario Matematico di Padova).

[4] G.Dal Maso, L.Modica. 'Nonlinear Stochastic Homogenization and Ergodic Theory'. (To appear).
[5] G.Dal Maso, E.De Giorgi, L.Modica. (In preparation).
[6] E.De Giorgi. 'Convergenza in energia di operatori ellittici'.Contributi del Centro Linceo Interdisciplinare e loro Applicazioni.Accademia Nazionale dei Lincei, n.16, Roma, (1976).
[7] S.M.Kozlov. 'Averaging of Random Operators'. *Math. USSR Sbornik* 37 (1980).
[8] P.Marcellini. 'Periodic Solutions and Homogenization of Nonlinear Variational Problems'. *Ann. Mat. Pura Appl.* (4) 117 (1978),139-152.
[9] G.C.Papanicolaou, S.R.S.Varadhan. 'Boundary Value Problems with Rapidly Oscillating Random Coefficients'. Proceed. of Colloq. on Random Fields,Rigorous Results in statistical mechanics and quantum fields, ed. by J.Fritz, J.L.Lebowitz, D.Szasz. *Colloquia Mathematica Societ. Janos Bolyai*, 10, North Holland, Amsterdam, (1979), 835-873.
[10] V.V.Yurinskij. 'Averaging an Elliptic-Value Problem with Random Coefficients'. *Siberian Math. Journal*, 21 (1980), 470-482.

Chapter 16

On the Hypo-Convergence of Probability Measures

G. Salinetti and R. Wets

1. INTRODUCTION

Weak, or more precisely weak* , convergence of probability measures provides a topological structure for the study of the convergence in distribution of random variables, stochastic processes,etc.. Here we introduce a direct approach based on the correspondence between probability measures and certain increasing set functions that have specific topological properties. Convergence questions are then addressed in terms of the convergence of these (semicontinuous) functions in the framework provided by the theory of epi-convergence (that has been developed in Optimization Theory to study the limits of semicontinuous functions.

This work was motivated initially by our study of the convergence of stochastic infima (extremal processes) and the connections to questions of robustness in statistics and to the general theory of convergence of stochastic processes [1] . This was elaborated in the first presentation of these results in a communication to the European Meeting of Statistician at Palermo (Sicily) in September 1982. The connections to problems in stochastic homogenization, and related questions in the Calculus of Variations, was brought to our attention by the recent article of De Giorgi [2] whose research is following a path, that in some way, is parallel to ours; for example the results presented here essentially settle Conjecture 2 formulated in [2].

Basically, we work with the underlying space (domain of definition of the probability measures) locally compact and separated, this allows us to exploit to the fullest the compactness of the space of semicontinuous functions equipped with the epi- or hypo-topology [3, Section 4]. This is not a standard assumption in probability theory in that it would exclude a large number of interesting functional spaces and thus preclude the application of this theory to stochastic processes. However, that is only the case if we restrict ourselves to a "classical" functional view of stochastic processes. If, instead, we approach the theory of stochastic processes as in [1, Section 3] or [4] where paths are viewed as elements of a space of semicontinuous functions which we then equip with the epi-topology, the underlying (functional) space is compact and thus we are in the setting considered here. A more specific example of this construction is to be found in the work of Dal Maso and Modica [5] on the stochastic homogenization problem, here the space of integral functionals on $L^p_{loc}(R^n)$ is given the epi-topology which renders it compact. Another approach to the "compactification" of the space of stochastic processes is by identifying the paths of a process $(X_t(\cdot))_{t\in T}$ with their graphs, and define on the space of graphs the topology of set convergence, as done in [1, Section 6]: we are again dealing with a compact space as domain for the probability measures. Of course if we are dealing with a finite number of random variables, then the underlying space is finite dimensional and we are in the locally compact case.

After a brief review of some background material about the convergence of sets and semicontinuous function (Section 2), we show in Section 3 that to each probability measure one can associate an upper semicontinuous function defined on the (hyper) space of closed sets. This increasing set function can viewed as a "topological" version of the original probability measure or as a distribution function on the hyperspace of closed sets. It is this latter point of view that we emphasize (by our choice of terminology) in this note. The structural compactness of this space of upper semicontinuous functions with the hypo-topology is the key to a number of convergence results for collections of probability measures (defined on locally compact spaces). In

Section 4, we also exhibit the connection with Prohorov's theorem.

2. CONVERGENCE OF SETS AND SEMICONTINUOUS FUNCTIONS

We review briefly the results about the convergence of closed sets and of semicontinuous functions that are needed in this setting;we rely mostly on [3] changing simply the emphasis from the space of lower semicontinuous functions and the epi-topology to the space of upper semicontinuous functions and the hypo-topology. The *epigraph* of a function $f: X \to \bar{R} = [-\infty,\infty]$, denoted by epi f, is the set

$$\text{epi } f := \{(x,\alpha) \in X \times R \mid f(x) \leq \alpha\},$$

whereas its *hypograph*, denoted by hypo f, is the set

$$\text{hypo } f := \{(x,\alpha) \in X \times R \mid f(x) \geq \alpha\},$$

i.e. it is the set of points on and below the graph of f.

Let (E,τ) be a locally compact and separated (Hausdorff) topological space. By F, or by $F(E)$, we denote the *hyperspace of τ-closed subsets of* E .We equip F with the topology T of set convergence. Recall that

$$F = \tau\text{-}\lim_{\nu} F^{\nu} = \tau\text{-}\lim_{\nu \in (N,H)} F^{\nu}$$

where $\{F^{\nu}, \forall \in (N,H)\}$ is a filtered family of closed sets in E with the filter on the index space N, if

(2.1) $$\tau\text{-}\lim\sup_{\nu} F^{\nu} = F = \tau\text{-}\lim\inf_{\nu} F^{\nu},$$

and

(2.2) $$\tau\text{-}\lim\sup_{\nu} F^{\nu} := \bigcap_{H \in H} \tau\text{-cl}\left(\bigcap_{\nu \in H} F^{\nu}\right)$$
$$= \{x = \tau\text{-cluster } \{x^{\nu}\} \mid \text{for all } \nu, x^{\nu} \in F^{\nu}\}$$

and

(2.3)
$$\tau\text{-}\lim\inf_{\nu} F^{\nu} := \bigcap_{H \in \mathcal{H}^{\#}} \tau\text{-cl}\left(\bigcup_{\nu \in H} F^{\nu}\right)$$
$$= \{x = \tau\text{-}\lim x^{\nu} \mid \text{for all } \nu,\ x^{\nu} \in F^{\nu}\},$$

with

$$\mathcal{H}^{\#} := \{H' \subset N \mid H' \cap H \neq \emptyset \quad \text{for all } H \in \mathcal{H}\},$$

the grill of $\mathcal{H}$. For example, when $N = \mathbb{N}$, the natural numbers, and $\mathcal{H}$ is the Fréchet filter (all sets of finite co-cardinality) then $\mathcal{H}^{\#}$ consists of all countable subsets of $\mathbb{N}$. The corresponding topology $\mathcal{T}$ on $\mathcal{F}$ can be generated from the subbase of open sets

(2.4)
$$\{\mathcal{F}^{K},\ K \in \mathcal{K}\} \quad \text{and} \quad \{\mathcal{F}_{G},\ G \in \mathcal{G}\}$$

with

$\mathcal{K}$:= the hyperspace of τ-compact subsets of E,

$\mathcal{G}$:= the hyperspace of τ-open subsets of E,

and where for any $D \subset E$

$$\mathcal{F}^{D} = \{F \in \mathcal{F} \mid F \cap D = \emptyset\},$$

$$\mathcal{F}_{D} = \{F \in \mathcal{F} \mid F \cap D \neq \emptyset\}.$$

Choquet [6] is responsable for this characterization of the topology of set convergence; for more about this, including many other characterizationsof set-convergence, consult [7] and the references given there. Note that $\emptyset \in \mathcal{F}$ and it could be an accumulation point of a filtered family whose members "escape" from every compact subset of E. This is enough to guarantee the compactness of $(\mathcal{F}, \mathcal{T})$, a property that plays an important role in the ensuing development.

Theorem 2.1 ([8], [3, Proposition 3.2]).
If (E,τ) is locally compact and separated, then (F,T) is regular and compact. Moreover, if (E,τ) is separable, then T has a countable base.

Now, a function is lower semicontinuous if and only if its epigraph is a closed subset of $E \times R$, and since

$$E := \text{all epigraphs of lower semicontinuous functions on } X$$

is a T-closed subset of $F(X \times R)$, E with the T-relative topology is compact and regular. This construction leads to a topology for the space of lower semicontinuous functions that corresponds to convergenee in terms of epigraphs (and for upper semicontinuous functions in terms of the hypographs).

Let

$$SC^l(X) := \{f: X \to \bar{R} \mid f \text{ is lower semicontinuous}\}$$

and let us denote by *epi* the *epi-topology* induced by the T-relative topology when $SC^l(X)$ is identified with E through the relation: $f \leftrightarrow \text{epi } f$. Similarly, we denote by *hypo* the *hypo-topology* on

$$SC^u(X) := \{f: X \to \bar{R} \mid f \text{ is upper semicontinuous}\},$$

induced by the T-relative topology on the hyperspace of hypographs included in $F(X \times R)$. As a direct consequence of Theorem 2.1, and the fact that the hyperspaces of epigraphs and hypographs are T-closed subsets of $F(X \times R)$, we obtain:

Theorem 2.2 [3, Corollary 4.2]
If X is separated and locally compact, then $(SC^l(X), \text{epi})$ and $(SC^u(X), \text{hypo})$ are regular and compact topological spaces. Moreover if X is separable, then $(SC^l(X), \text{epi})$ and $(SC^u(X), \text{hypo})$ admit countable bases.

Moreover, it is easy to see that:

Corollary 2.3 [3, Corollary 4.4]. Any subset of bounded functions in $SC^u(X)$ is pre-compact with respect to hypo-topology. Similarly, any subset of bounded functions of $SC^l(X)$ is pre-compact with respect to the epi-topology.

Convergence of upper semicontinuous functions in the hypo-topology means set convergence of the hypographs as subsets of $X \times R$, and similarly for lower semicontinuous functions and epigraphs. If the space X admits a countable base then we have the very useful characterization of hypo-convergence provided by the next theorem [3, Section 1]:

Theorem 2.4
Suppose (X,τ) is a locally compact, separated space that admits a countable base, and suppose that f, and the functions of the filtered family $\{f^\nu, \nu \in N\}$ are τ-upper semicontinuous. Then

$$f = \underset{\nu \in N}{\text{hypo-lim}_\tau}\, f^\nu$$

if and only if for all x in X,

$$(2.5) \qquad \limsup_{\nu \in N} f^\nu(x^\nu) \le f(x) \quad \text{whenever} \quad x = \lim_{\nu \in N} x^\nu$$

and

$$(2.6) \qquad \text{there exists } \{x^\nu, \nu \in N\} \text{ with } x = \lim_{\nu \in N} x^\nu \text{ and } \liminf_{\nu \in N} f^\nu(x^\nu) \ge f(x).$$

And similarly for τ-lower semicontinuous functions, replacing hypo by epi, lim inf by lim sup and vice-versa, $\le$ by $\ge$ and vice-versa in (2.5) and (2.6).

In order to carry out certain measure theoretic arguments, we rely also on the following notion of limits for families of sets. Given a space E and $\{D^\nu, \nu \in (N,\mathcal{H})\}$ a filtered family of subsets of E, we define

(2.7)
$$\liminf_{\nu\in N} D^\nu := \bigcap_{H\in\mathcal{H}^\#} \bigcup_{\nu\in H} D^\nu ,$$

and

(2.8)
$$\limsup_{\nu\in N} D^\nu := \bigcap_{H\in\mathcal{H}} \bigcup_{\nu\in H} D^\nu .$$

We see that these limits correspond to those defined by (2.3) and (2.2) if we take τ to be the discrete topology on E, one refers sometimes to (2.7) and (2.8) as the set theoretic limits. Of course for any topology τ on E, and any family $\{F^\nu, \nu\in N\}$ of closed subsets of E:

(2.9)
$$\liminf_{\nu\in N} F^\nu \subset \tau\text{-}\liminf_{\nu\in N} F^\nu,$$

and

(2.10)
$$\limsup_{\nu\in N} F^\nu \subset \tau\text{-}\limsup_{\nu\in N} F^\nu.$$

3. SEMICONTINUOUS MEASURES ON F

Probability measures will be defined on the Borel field $\mathcal{B}(\Omega)$, generated by the τ-open subsets, of the topological space (Ω,τ) that we take to be *locally compact, separated and separable*. We show that every probability measure P defined on $\mathcal{B}$ determines and is uniquely determined by an increasing set-function D on $F = F(\Omega)$, the hyperspace of τ-closed subsets of Ω. This will leads us to the study of the convergence of probability measures in the framework laid out in Section 2.

For any sequence $\{F^\nu \in F, \nu = 1,\dots\}$ and P a probability measure on $(\Omega,\mathcal{B})$, we have

$$\text{(3.1)} \qquad \limsup_{\nu\to\infty} P(F^\nu) \le P(\limsup_{\nu\to\infty} F^\nu) \le P(\tau\text{-}\limsup_{\nu\to\infty} F^\nu)$$

where the last inequality follows from (2.10). Since (F,T) has a countable base (Theorem 2.1), and for any filtered family $\{F^\nu \subset F, \nu \in (N,H)\}$, τ-$\limsup_{\nu\in N} F^\nu$ is closed and thus measurable, we also have that

$$\text{(3.2)} \qquad \limsup_{\nu\in N} P(F^\nu) \le P(\tau\text{-}\limsup_{\nu\in N} F^\nu) \; .$$

Since $F = \tau\text{-}\lim_{\nu\in N} F^\nu$ implies $F = \tau\text{-}\limsup_{\nu\in N} F^\nu$, we have shown:

Proposition 3.1

Given (Ω,τ) a locally compact, separated and separable topological space. The resctriction to $F(\Omega)$ of a probability measure $P: B(\Omega)\to[0,1]$ is upper semicontinuous with respect to T, the topology of set convergence on F.

This suggests associating to P, the family of probability measures on (Ω,B), a family D of upper semicontinuous functions on (F,T) that can be analyzed in purely topological terms. We introduce the follo - ing concept:

Definition 3.2: A (set-)function $D: F\to[0,1]$ is a probability semicontinuous measure (sc-measure) on F if

(3.3) i. $D(\emptyset) = 0$, $D(\Omega) = 1$;

(3.4) ii. $D(F_1) \le D(F_2)$ whenever $F_1 \subset F_2$;

(3.5) iii. $D(F_1 \cup F_2)+D(F_1 \cap F_2)=D(F_1)+D(F_2)$ for all F_1, F_2 ;

(3.6) iv. D is T-upper semicontinuous.

Calling D a "distribution function" can be justified by analogy, its properties are similar to those of distribution functions defined on the real line. Indeed given $V: R \to [0,1]$, a distribution function on R, we can think of it as a set-function on the class $\{(-\infty,z],\ z \in R\}$ that plays role F in Definition 3.2. Property i. corresponds to having

$$\lim_{z \downarrow -\infty} V(z) = 0 \quad , \quad \lim_{z \uparrow \infty} V(z) = 1 \ ;$$

property ii., which makes D an increasing (set-)function corresponds to having V monotone nondecreasing; property iv. is connected to the right-semicontinuity of V on R. Property iii. has no counterpart in the 1-dimensional case-it would correspond to the tautology $V(z_1)+V(z_2)=V(z_1) + V(z_2)$ for any pair (z_1,z_2)- but if V is defined on R^n then iii. corresponds to the condition that the sum and differences of certain (unbounded) rectangles must stay nonnegative.

Proposition 3.3

Suppose (Ω,τ) is a locally compact, separated and separable topological space. Then every probability measure $P: \mathcal{B}(\Omega) \to [0,1]$ determines a probability sc-measure D on F through the relation $D(F) := P(F)$ for all $F \in F$. Conversely, every probability sc-measure D on F admits an extension to $\mathcal{B}$ that generates a unique probability measure P on $\mathcal{B}$ such that $P = D$ on F.

Proof: The restriction of P to F satisfies of course the conditions (3.3) - (3.5) of the definition of a probability sc-measure, whereas (3.6) is guaranteed by Proposition 3.1. On the other hand, if D is a probability sc-measure on F, setting $P = D$ on F and $P(G) = 1-D(\Omega\backslash G)$ for every open set G in G, we see that condition (3.5) determines uniquely P on A, the algebra (field) consisting of the finite unions of pairwise disjoint sets in F and G. The standard arguments [9, Theorem 1.3.10] can now be used to extend P to $\mathcal{B}$, the sigma-field generated by A. This extension is clearly unique since (Ω,τ) is metrizable and thus for every $A \in \mathcal{B}$, $P(A) = \sup \{P(F) \mid F \subset A\}$. #

Let $\mathcal{P}$ denote the class of probability measures on $(\Omega,\mathcal{B})$ and $\mathcal{D}$ the associated class of probability sc-measures on the topological space $(\mathcal{F}(\Omega), \mathcal{T})$. By Proposition 3.1,

$$\mathcal{D} \subset SC_b^u(\mathcal{F}) \subset SC^u(\mathcal{F}) ,$$

where upper semicontinuity is with respect to the $\mathcal{T}$-topology and

(3.7) $$SC_b^u(\mathcal{F}) := \text{space of } \mathcal{T}\text{-upper semicontinuous functions on } \mathcal{F} \text{ with values in } [0,1].$$

Since $(\mathcal{F}(\Omega), \mathcal{T})$ is a compact, regular topological space we know from Corollary 2.3, and the fact that any hypo-limit of a collection of functions bounded by 0 and 1 is itself bounded by 0 and 1, that

(3.8) $$(SC_b^u(\mathcal{F}), \text{hypo}) \quad \text{is regular and compact.}$$

The space $\mathcal{D}$ is thus pre-compact, and every collection contains a sub-collection that hypo-converges to a function in $SC_b^u(\mathcal{F})$. This hypo-limit is "almost" a distribution function as is demonstrated next.

Lemma 3.4 Suppose $\{D^\nu, \nu \in (N,\mathcal{H})\}$ is a filtered family of probability sc-measures on $\mathcal{F}(\Omega)$ with

(3.9) $$D := \text{hypo-}\lim_{\nu \in N} D^\nu .$$

Then $D: \mathcal{F} \to [0,1]$ satisfies conditions (3.4), (3.6) with $D(\Omega) = 1$, and for any pair $F_1, F_2 \in \mathcal{F}$,

(3.10) $$D(F_1 \cup F_2) + D(F_1 \cap F_2) \geq D(F_1) + D(F_2)$$

If, in addition $D(\emptyset) = 0$, then D is a probability sc-measure.

Proof: From the arguments that precede we already know that $D \in SC_b^u(F)$ and this implies that D is $\mathcal{T}$-upper semicontinuous, and clearly D takes its values in $[0,1]$. Moreover, from (2.5) it follows that

$$\limsup_{\nu \in N} D^\nu(\Omega) \leq D(\Omega)$$

and this implies that $D(\Omega) = 1$.

To show that D is an increasing set function on F, i.e. to establish (3.4), let us take F_1, F_2 in F with $F_1 \subset F_2$, we have to show that $D(F_1) \leq D(F_2)$. From (2.6) with (2.5) we know that there exists a filtered collection $\{F^\nu, \nu \in N\}$ converging to F_1 such that

$$D(F_1) = \lim_{\nu \in N} D^\nu(F^\nu) \quad .$$

Now observe that the sequence $\{F_2 \cup F^\nu, \nu = 1, \ldots\}$ converges to F_2. Indeed we have

$$F_2 \subset \tau\text{-}\liminf_{\nu \in N} (F_2 \cup F^\nu) \subset \tau\text{-}\limsup_{\nu \in N} (F_2 \cup F^\nu) \subset F_2;$$

all the inclusions being obvious except possibly the last one, which is obtained by observing that any collection $\{x^\nu \in F_2 \cup F^\nu, \nu \in N\}$ has its cluster points in $F_2 \cup F_1 = F_2$. From this it follows that

$$D(F_1) = \lim_{\nu \in N} D^\nu(F^\nu) \leq \liminf_{\nu \in N} D^\nu(F_2 \cup F^\nu) \leq D(F_2),$$

the first inequality coming from the fact that the D^ν are probability sc-measures on F and thus by (3.4)

$$D^\nu(F^\nu) \leq D^\nu(F_2 \cup F^\nu) \ ,$$

and the last one comes from (2.5) since D is the hypo-limit of the D^ν.

To obtain (3.10), i.e. that for any pair F_1, $F_2 \in F$,

$$D(F_1 \cup F_2) + D(F_1 \cap F_2) \geq D(F_1) + D(F_2),$$

we again appeal to Theorem 2.4, combining (2.6) with (2.5); to obtain the existence of filtered families $\{F^{1\nu}, \nu \in N\}$ and $\{F^{2\nu}, \nu \in N\}$ such that

$$F_1 = \tau\text{-}\lim_{\nu \in N} F^{1\nu}, \quad \text{and} \quad F_2 = \tau\text{-}\lim_{\nu \in N} F^{2\nu}$$

and

$$D(F_1) = \lim_{\nu \in N} D^\nu(F^{1\nu}), \quad \text{and} \quad D(F_2) = \lim_{\nu \in N} D^\nu(F^{2\nu}).$$

For such a family, arguing as above, we have

$$F_1 \cup F_2 \subset \tau\text{-}\lim\inf_{\nu \in N} (F^{1\nu} \cup F^{2\nu}) \subset \tau\text{-}\lim\sup_{\nu \in N} (F^{1\nu} \cup F^{2\nu}) \subset F_1 \cup F_2,$$

and

$$F_1 \cap F_2 \subset \tau\text{-}\lim\inf_{\nu \in N} ((F^{1\nu} \cap F^{2\nu}) \cup (F_1 \cap F_2)) \subset \tau\text{-}\lim\sup_{\nu \in N} ((F^{1\nu} \cap F^{2\nu}) \cup (F_1 \cap F_2)) \subset$$
$$\subset F_1 \cap F_2.$$

By relying on (3.5) and (3.6) for each D^ν individually, we obtain

$$\begin{aligned} D(F_1) + D(F_2) &= \lim_{\nu \in N} [D^\nu(F^{1\nu}) + D^\nu(F^{2\nu})] \\ &\leq \lim\sup_{\nu \in N} [D^\nu(F^{1\nu} \cup F^{2\nu}) + D^\nu(F^{1\nu} \cap F^{2\nu})] \\ &\leq \lim\sup_{\nu \in N} [D^\nu(F^{1\nu} \cup F^{2\nu}) + D^\nu((F^{1\nu} \cap F^{2\nu}) \cup (F_1 \cap F_2))] \\ &\leq D(F_1 \cup F_2) + D(F_1 \cap F_2), \end{aligned}$$

where the last inequality follows from (2.5).

To complete the proof there remains only to show that in addition

$D(\emptyset) = 0$, then actually (3.5) holds, or equivalently

$$D(F_1 \cup F_2) + D(F_1 \cap F_2) \leq D(F_1) + D(F_2). \tag{3.11}$$

Let us assume that F_1 and F_2 are nonempty since otherwise (3.11) is trivially satisfied. Let

$$\varepsilon F_2 := \{\omega \in \Omega \mid \text{dist}(\omega, F_2) \leq \varepsilon\}$$

be the (closed) ε-*enlargement* of F_2 where dist is any metric compatible with the τ-topology. Note that $F_2 = \tau\text{-}\lim_{\varepsilon \to 0} \varepsilon F_2$, which in view of the upper semicontinuity of D implies that to any $\delta > 0$ there corresponds $\varepsilon > 0$ such that

$$D(\varepsilon F_2) < D(F_2) + \delta/2 . \tag{3.12}$$

Since $D = \text{hypo-}\lim_{\nu \to \infty} D_\nu$, from (2.6) and (2.5) it follows that there exists sequences $\{R^\nu, \nu = 1, \ldots\}$ and $\{S^\nu, \nu = 1, \ldots\}$ such that

$$F_1 \cup F_2 = \tau\text{-}\lim_{\nu \to \infty} R^\nu \quad , \quad \text{and} \quad F_1 \cap F_2 = \tau\text{-}\lim_{\nu \to \infty} S^\nu$$

such that for $\delta > 0$ and ν sufficiently large,

$$D(F_1 \cup F_2) + D(F_1 \cap F_2) < D^\nu(R^\nu) + D^\nu(S^\nu) + \delta/2. \tag{3.13}$$

Because D^ν is a probability sc-measure, with

$$\bar{\varepsilon} F_2^c := \text{cl}(\Omega \setminus \varepsilon F_2)$$

in view of (3.4) and (3.5), we have

$$D^\nu(R^\nu) \leq D^\nu(R^\nu \cap \varepsilon F_2) + D^\nu(R^\nu \cap \bar{\varepsilon} F_2^c),$$

$$D^\nu(R^\nu \cap \varepsilon F_2) \le D^\nu(\varepsilon F_2),$$

$$D^\nu(R^\nu \cap \bar{\varepsilon}F_2^c) + D^\nu(S^\nu) = D^\nu((R^\nu \cap \bar{\varepsilon}F_2^c) \cup S^\nu) + D^\nu(R^\nu \cap \bar{\varepsilon}F_2^c \cap S^\nu),$$

and

$$D^\nu((R^\nu \cap \bar{\varepsilon}F_2^c) \cup S^\nu) \le D^\nu((R^\nu \cap \bar{\varepsilon}F_2^c) \cup S^\nu \cup F_1) .$$

Substituting in (3.13), for ν sufficiently large, this yields

$$(3.14) \qquad \begin{aligned} D(F_1 \cup F_2) + D(F_1 \cap F_2) &< D^\nu(\varepsilon F_2) + D^\nu((R^\nu \cap \bar{\varepsilon}F_2^c) \cup S^\nu \cup F_1) \\ &\quad + D^\nu(R^\nu \cap \bar{\varepsilon}F_2^c \cap S^\nu) + \delta/2 . \end{aligned}$$

We have that

$$(3.15) \qquad F_1 = \tau\text{-}\lim_{\nu \in N} (R^\nu \cap \bar{\varepsilon}F_2^c) \cup S^\nu \cup F_1;$$

to prove this observe that [10, § 25.II],

$$\begin{aligned} F_1 &= \tau\text{-}\lim_{\nu\in N} \inf(F_1 \cup [(R^\nu \cap \bar{\varepsilon}F_2^c) \cup S^\nu] \subset \tau\text{-}\lim_{\nu\in N} \sup(F_1 \cup (R^\nu \cap \bar{\varepsilon}F_2^c) \cup S^\nu) = \\ &= F_1 \cup \tau\text{-}\lim_{\nu\in N} \sup S^\nu \cup \tau\text{-}\lim_{\nu\to\infty} \sup(R^\nu \cap \bar{\varepsilon}F_2^c) \\ &= F_1 \cup \tau\text{-}\lim_{\nu\in N} \sup(R^\nu \cap \bar{\varepsilon}F_2^c) \subset F_1 \cup (\bar{\varepsilon}F_2^c \cap \tau\text{-}\lim_{\nu\in N} \sup R^\nu) \\ &= F_1 \cup (\bar{\varepsilon}F_2^c \cap (F_1 \cup F_2)) \subset F_1 \cup ((F_1 \cup F_2) \setminus F_2) = F_1 . \end{aligned}$$

Also, note that:

$$(3.16) \qquad \emptyset = \tau\text{-}\lim_{\nu\in N} (R^\nu \cap S^\nu \cap \bar{\varepsilon}F_2^c)$$

as follows from the following inclusions:

$$\emptyset \subset \tau\text{-}\liminf_{\nu \in N} (R^\nu \cap S^\nu \cap \bar{\varepsilon}F_2^c) \subset \tau\text{-}\limsup_{\nu \in N} (R^\nu \cap S^\nu \cap \bar{\varepsilon}F_2^c)$$

$$\subset \tau\text{-}\limsup_{\nu \in N} R^\nu \cap \tau\text{-}\limsup_{\nu \in N} S^\nu \cap \bar{\varepsilon}F_2^c \subset (F_1 \cup F_2) \cap (F_1 \cap F_2) \cap \bar{\varepsilon}F_2^c$$

$$\subset F_1 \cap F_2 \cap \bar{\varepsilon}F_2^c = \emptyset.$$

Now returning to (3.14) and taking lim sup on both sides, with (3.12), (3.15) and (2.5), (3.16) and the assumption that $D(\emptyset)=0$ we obtain:

$$D(F_1 \cup F_2) + D(F_1 \cap F_2) \le D(F_1) + D(F_2) + \delta .$$

Since this holds for every $\delta > 0$, it yields (3.11), and this completes the proof of the lemma. #

As a consequence of this lemma and Corollary 2.3 we have the following result which can be viewed as a *generalization of Helly's Theorem.*

Corollary 3.5 Given $\{D^\nu, \nu \in (N,H)\}$ a filtered collection probability sc-measures on $F(\Omega)$. Then there always exists a subcollection $\{D^\nu, \nu \in (N',H'), N' \in H\}$ (where H' is the restriction of H to N') and a function D on F such that

i. $D: F \to [0,1]$, $D(\Omega) = 1$;

ii. $D(F_1) \le D(F_2)$ whenever $F_1 \subset F_2$;

iii. $D(F_1 \cup F_2) + D(F_1 \cap F_2) \ge D(F_1) + D(F_2)$, for all $F_1, F_2 \in F$;

iv. D is T-upper semicontinuous.

To have $D(\emptyset) = 0$ when $D = \text{hypo-lim } D^\nu$ is intimately connected to the notion of tightness. As for probability measures, *a collection of probability sc-measures* $\{D^\nu, \nu \in (N,H)\}$ *on* $F(\Omega)$ *is* tight *if for every* $\varepsilon > 0$ *there exists a compact set* $K \subset \Omega$ *such that for all* $\nu \in N$

$$D^\nu(K) > 1 - \varepsilon . \tag{3.17}$$

Theorem 3.6

Suppose $\{D^\nu, \nu\in(N,H)\}$ is a filtered collection of probability sc-measures on F, and

$$D := \text{hypo-}\lim_{\nu\in N} D^\nu .$$

Then D is a probability sc-measure on F if and only if the collection $\{D^\nu, \nu\in N\}$ is tight.

Proof: To begin with let us assume that the collection $\{D^\nu, \nu\in(N,H)\}$ is tight. Let $\varepsilon > 0$ and K_ε be the compact set such that for all ν,

$$D^\nu(K_\varepsilon) > 1 - \varepsilon .$$

Since $D = \text{hypo-}\lim D^\nu$, there exists $\{F^\nu \in F, \nu\in(N,H)\}$ such that

$$\tau\text{-}\lim_{\nu\in N} F^\nu = \emptyset$$

and

$$D(\emptyset) = \tau\text{-}\lim_{\nu\in N} D^\nu(F^\nu) ;$$

we use here (2.6) and (2.5). Since $\emptyset = \lim_{\nu\in N} F^\nu$, it means that there exists $H\in H$ such that for all $\nu\in H$

$$F^\nu \cap K_\varepsilon = \emptyset ,$$

as follows from the definition of τ-lim sup, cf.(2.2). On the other hand for every $\nu\in H$

$$D^\nu(F^\nu) + D^\nu(K_\varepsilon) = D^\nu(F^\nu \cup K_\varepsilon) \le D^\nu(\Omega) = 1$$

and consequently for all $\nu\in H$

$$D^\nu(F^\nu) < \varepsilon .$$

Hence

$$D(\emptyset) = \lim_{\nu \in N} D^\nu(F^\nu) < \varepsilon .$$

and since this holds for every $\varepsilon > 0$, it yields $D(\emptyset) = 0$, which in view of Lemma 3.4 implies that D is a probability sc-measure on F.

To prove the "only if" part, let us proceed by contradiction. Suppose that $D(\emptyset) = 0$ but the collection $\{D^\nu, \nu \in N\}$ is not tight. Then for some $\varepsilon > 0$, and every compact K there exists $H \in \mathcal{H}$ such that for all $\nu \in H$

$$D^\nu(K) < 1 - \varepsilon.$$

Since

$$D^\nu(K) + D^\nu(\mathrm{cl}(\Omega \backslash K)) \geq D^\nu(\Omega) = 1,$$

it follows that for all $\nu \in H$

$$D^\nu(\mathrm{cl}(\Omega \backslash K)) \geq \varepsilon .$$

The definition of D as the hypo-limit of the $\{D^\nu, \nu \in N\}$, in particular implies that

$$\limsup_{\nu \in N} D^\nu(\mathrm{cl}(\Omega \backslash K)) \leq D(\mathrm{cl}(\Omega \backslash K)).$$

which with the preceding inequality yields

$$D(\mathrm{cl}(\Omega \ K)) \geq \varepsilon \tag{3.18}$$

for all compact $K \subset \Omega$. Consider now a collection $\{K^\nu, \nu = 1, \ldots\}$ of compact sets such that $\bigcup_{\nu=1}^{\infty} K^\nu = \Omega$. Such a collection exists since (Ω, τ) is locally compact and separable. Then, the sequence

$$\{K_u^\nu = \bigcup_{l=1}^{\nu} K^l , \nu = 1, \ldots\}$$

is an increasing sequence such that $\bigcup_{\nu=1}^{\infty} K_u^\nu = \Omega$. Then

$$\{cl(\Omega \setminus K_u^\nu),\quad \nu = 1,\dots\}$$

is a sequence that converges to $\emptyset$. Indeed, to any compact set $S \subset \Omega$ there corresponds ν_S such that for all $\nu \geq \nu_S$,

$$\delta S = \{x \in \Omega \mid dist(x,S) \leq \delta\} \subset K_u^\nu$$

where dist is any metric compatible with the topology τ. This means that for all $\nu \geq \nu_S$

$$cl(\Omega \setminus K_u^\nu) \cap S = \emptyset ,$$

and since such a statement holds for any compact set S, it follows that [7, Lemma]

$$\tau\text{-}\lim_{\nu \to \infty} (cl(\Omega \setminus K_u^\nu)) = \emptyset \quad .$$

Moreover D is upper semicontinuous on F (with respect to set convergence), thus

$$\limsup_{\nu \to \infty} D(cl(\Omega \setminus K_u^\nu)) \leq D(\emptyset) = 0 \ .$$

But from (3.18) it would follow that

$$\limsup_{\nu \to \infty} D(cl(\Omega \setminus K_u^\nu) \geq \varepsilon > 0 \quad .$$

in clear contradiction with the above. Hence the collection $\{D^\nu, \nu \in (N,H)\}$ must be tight. #

If instead of building probability sc-measures on F, we used as domain of definition the hyperspace G of open subsets of Ω, we obtain a number of related results that are merely the mirror image of those for probability sc-measures on F . It is however useful to record them here since they provide additional criteria for weak* convergence of probability measures as we shall see in the next section. We equip G with the "complementary" topology T_g to T, namely, for a filtered collection $\{G^\nu, \nu \in (N,H)\}$, we write

$$G = T\text{-}\lim_{\nu \in N} G^\nu \quad \text{if} \quad (\Omega \backslash G) = T\text{-}\lim_{\nu \in N} (\Omega \backslash G^\nu).$$

Definition 3.7: A (set-)function $D_g: G(\Omega) \to [0,1]$ is a probability sc-measure on G if

(3.19) i. $D_g(\emptyset) = 0$, $D_g(\Omega) = 1$;

(3.20) ii. $D_g(G_1) \le D_g(G_2)$ whenever $G_1 \subset G_2$;

(3.21) iii. $D_g(G_1 \cup G_2) + D_g(G_1 \cap G_2) = D_g(G_1) + D_g(G_2)$ for all G_1, G_2;

(3.22) iv. D_g is T_g-lower semicontinuous.

This time, the analogy is with a distribution function V on R with V(z) taking on the probability mass associated to the open interval $(-\infty, z)$. As for probability sc-measures on F, there is one-to-one relation between probability measures defined on $B(\Omega)$ and probability sc-measures on $G(\Omega)$. Given P we define

$$D_g(G) = P(G) \quad \text{for all open sets } G$$

and given D_g, one can show that there is a unique extension of D_g on B that generates a probability measure on B.

Lemma 3.8 Suppose $\{D_g^\nu, \nu \in (N,H)\}$ is a filtered family of probability sc-measures on $G(\Omega)$ with

$$D_g = \text{epi-}\lim_{\nu \in N} D_g^\nu .$$

Then, $D_g : G \to [0,1]$ satisfies conditions (3.20), (3.22) of the Definition 3.7; moreover $D(\emptyset) = 0$, and for any pair $G_1, G_2 \in G$

$$D_g(G_1 \cup G_2) + D_g(G_1 \cap G_2) \le D_g(G_1) + D_g(G_2) . \tag{3.23}$$

If in addition $D_g(\Omega) = 1$, then D_g is a distribution function.

A collection $\{D_g^\nu, \nu \in N\}$ of probability sc-measures on $G(\Omega)$ is tight if for every $\varepsilon > 0$ there exists a compact set $K \subset \Omega$ such that for all $\nu \in N$

$$D_g^\nu(\Omega \setminus K) < \varepsilon .$$

Theorem 3.9

Suppose $\{D_g^\nu, \nu \in (N,H)\}$ is a filtered collection of probability sc-measures on G, and

$$D_g := \operatorname*{epi-lim}_{\nu \in N} D_g^\nu$$

Then D_g is a probability sc-measure on G if and only if the collection $\{D_g^\nu, \nu \in N\}$ is tight.

4. HYPO-CONVERGENCE OF PROBABILITY SC-MEASURES ON *F* AND PROHOROV'S THEOREM

In this section we study the relationship between the weak*-convergence of probability measures and the hypo-convergence of the associated probability sc-measures on F.

We work in the same setting as in Section 3: (Ω,τ) is locally compact, separated and separable, which also means that it is metrizable. We assume that a metric has been selected so that we can think of, and refer to (Ω,τ) as a metric space. Recall that weak or more precisely weak*, convergence of a filtered family of probability measures

$$\{P^\nu : \mathcal{B}(\Omega) \to [0,1] \ , \ \nu \in (N, \mathcal{H})\}$$

to a probability measure P means that

(4.1) $$\int g \, dP = \lim_{\nu \in N} \int g \, dP^\nu$$

for every bounded continuous function $g: \Omega \to R$, or equivalently [11,Theorem 2.1]

(4.2) $$\limsup_{\nu \in N} P^\nu(F) \le P(F) \quad \text{for every } F \in \mathcal{F}(\Omega),$$

(4.3) $$\liminf_{\nu \in N} P^\nu(G) \ge P(G) \quad \text{for every } G \in \mathcal{G}(\Omega),$$

and

(4.4) $$\lim_{\nu \in N} P^\nu(A) = P(A) \quad \text{for all } \ A \in \mathcal{B}(\Omega),\ P(\text{bdy } A) = 0,$$

where bdy A is the boundary of A.

Theorem 4.1

Suppose $\{P;\ P^\nu,\ \nu \in (N,\mathcal{H})\}$ is a filtered collection of probability measures defined on $\mathcal{B}(\Omega)$ where $\mathcal{B}$ is the Borel field on (Ω,τ) a locally compact, separable, complete metric space(Ω,τ), and $\{D;\ D^\nu,\ \nu \in (N,\mathcal{H})\}$ the corresponding collection of probability sc-measures on $\mathcal{F}(\Omega)$. Then

$$P = \text{weak}^*\text{-}\lim_{\nu \in N} P^\nu,$$

if and only only if

$$D = \text{hypo-}\lim_{\nu \in N} D^\nu \ .$$

Proof: If $D = \text{hypo-}\lim_{\nu \in N} D^\nu$ then (4.2) follows directly from (2.5), and hence $P = \text{weak}^*\text{-}\lim_{\nu \in N} P^\nu$.

Now let us show the converse. To begin with we show that to every $F \in F$, there corresponds $\{F^\nu, \nu \in (N,H)\}$ such that $F = \tau\text{-}\lim_{\nu \in N} F^\nu$ that satiasfies (2.6), i.e. such that

$$D(F) \leq \liminf_{\nu \in N} D^\nu(F^\nu) .$$

Let assume that F is nonempty, otherwise the inequality is trivially satisfied. For $\varepsilon > 0$, let

$$\varepsilon^\circ F := \{x \mid \text{dist}(x,F) < \varepsilon\}$$

denote the (open) ε-enlargement of F, where dist is the metric on Ω.We have that $\text{cl}(\varepsilon^\circ F) = \varepsilon F$, the closed ε-enlargement of F; also

$$\tau\text{-}\lim_{\varepsilon \downarrow 0} \varepsilon F = F ,$$

and whenever $\varepsilon_1 < \varepsilon_2$,

$$\varepsilon_1 F \subset \varepsilon_2^\circ F \subset \varepsilon_2 F \quad .$$

Since by (4.3) for every $\varepsilon > 0$

$$P(\varepsilon^\circ F) \leq \liminf_{\nu \in N} P^\nu(\varepsilon^\circ F) ,$$

then to every $\delta > 0$, there corresponds $H_\delta \in H$ such that for all $\nu \in H_\delta$,

$$D(F) \leq P(\varepsilon^\circ F) \leq P^\nu(\varepsilon^\circ F) + \delta \leq P^\nu(\varepsilon F) + \delta = D^\nu(\varepsilon F) + \delta .$$

In particular, to every $k = 1,\ldots,$ there corresponds $H_k \in H$ such that for all $\nu \in H_k$,

$$D(F) \leq D^\nu(k^{-1}F) + k^{-1} .$$

Then with $\bigcap_{\ell=1}^{k} H_\ell = H'_k \in H$, $F^\nu = k^{-1}F$ for all $\nu \in H'_{k-1}\setminus H'_k$ where $H'_o = N$, we have that $F = \tau\text{-}\lim_{\nu\in N} F^\nu = \tau\text{-}\lim_{k\to\infty} k^{-1}F$, and moreover, with $\delta_\nu = k^{-1}$ whenever $F^\nu = k^{-1}F$ we obtain

$$D(F) \le \liminf_{\nu\in N}(D^\nu(F^\nu)+\delta_\nu) = \liminf_{\nu\to\infty} D^\nu(F^\nu).$$

There remains only to show that for any filtered collection $\{F_\nu, \nu\in \epsilon(N,H)\}$ converging to F,

$$\limsup_{\nu\in N} D^\nu(F^\nu) \le D(F) \quad . \tag{4.5}$$

Since $F = \tau\text{-}\lim_{\nu\in N} F^\nu$, to every pair $\varepsilon > 0$, $K \in K(\Omega)$, there corresponds $H \in H$ such that for all $\nu \in H$ [7, Theorem vi_b]

$$F_\nu \cap K \subset \varepsilon^o F \subset \varepsilon F \quad . \tag{4.6}$$

The probability measure P is tight, it is defined on a complete separable metric space. Thus to every $\varepsilon > 0$ we can associate a compact set $K_\varepsilon \in K$ such that

$$P(cl(\Omega\setminus K_\varepsilon) \le \varepsilon \; .$$

For such a compact set K_ε , by (4.6) and (4.2), we have

$$\begin{aligned}\limsup_{\nu\in N} D^\nu(F^\nu) &\le \limsup_{\nu\in N} P^\nu(F^\nu \cap K_\varepsilon) + \limsup_{\nu\in N} P^\nu(cl(\Omega\setminus K_\varepsilon))\\ &\le \limsup_{\nu\in N} P^\nu(\varepsilon F) + \limsup_{\nu\in N} P^\nu(cl(\Omega\setminus K_\varepsilon))\\ &\le P(\varepsilon F) + P(cl(\Omega\setminus K_\varepsilon)) \le D(\varepsilon F) + \varepsilon \; .\end{aligned}$$

Since the above holds for all $\varepsilon > 0$, $F = \tau\text{-}\lim_{\varepsilon\to 0} \varepsilon F$ and D is τ-upper semicontinuous at F, this implies (4.5). #

Theorem 4.1 gives a further characterization of weak convergence of probability measures on metric spaces that are complete, separable and locally compact. It also generalizes the well-known fact that on R^k, there is equivalence between weak convergence of the probability measures and the convergence of the distribution functions. This convergence of distribution functions can be characterized, as is often done, in terms of "pointwise" conditions, we can then see that tightness is related to an equi-upper semicontinuity [3] condition. This will be done in a subsequent publication.

As a direct consequence of Theorem 4.1, we obtain the well known Theorems of Prohorov and Varadajaran [11, Section 6] both direct and inverse parts. Recall that a family $\mathcal{P}$ of probability measures in $\mathcal{B}(\Omega)$ is *tight* if for every $\varepsilon > 0$ there exists a compact set $K_\varepsilon \subset \Omega$ such that for all $P \in \mathcal{P}$,

$$P(K_\varepsilon) > 1 - \varepsilon .$$

It is said to be *relatively precompact* if

$$P \in \text{weak}^*\text{-cl}\, \mathcal{P} \text{ means } P \text{ is a probability measure on } \mathcal{B}(\Omega).$$

Corollary 4.2 Prohorov's Theorems. Suppose (Ω,τ) is a locally complete separable metric space and $\mathcal{P}$ a family of probability measures on $\mathcal{B}(\Omega)$. Then $\mathcal{P}$ is relatively precompact if and only if $\mathcal{P}$ is tight.

Proof: Let

$$\mathcal{D} := \{D : F \to [0,1] \mid D = P \text{ on } F,\ P \in \mathcal{P}\},$$

be the family of probability sc-measures on $F(\Omega)$. Now, $\mathcal{P}$ is tight if and only if $\mathcal{D}$ is tight, and in view of Theorem 4.1, the weak*-closure of $\mathcal{P}$ will consist of probability measures if and only if hypo-cl $\mathcal{D}$ consists of probability sc-measures on F. It now suffices to appeal to Corollary 2.3 and Theorem 3.6 to complete the argument. #

REFERENCES

[1] G.Salinetti and R.Wets. 'On the convergence in distribution of measurable multifunctions (random sets), normal integrands, stochastic processes, and stochastic infima'. *Mathematics of Operations Research*, to appear (CP - 82 - 87, IIASA, Laxenburg, Austria).

[2] E.De Giorgi. 'On a definition of Γ-convergence of measures'. In *Multifunctions and Integrands: Stochastic Analysis, Approximation and Optimization*, ed. G.Salinetti, Springer Verlag, Lecture Notes in Mathematics, Berlin (1984).

[3] S.Dolecki, G.Salinetti, and R.Wets. 'Convergence of functions: Equi-semicontinuity, *Trans. Amer. Mathematical Soc.* 276 (1983), 409-429.

[4] W.Vervaat. 'Random upper semicontinuous functions and extremal processes'. Manuscript Universiteit Nijmegen, (1982).

[5] G.Dal Maso and L.Modica. 'Nonlinear stochastic homogenization'. Technical Report, Università di Pisa, (1984).

[6] G.Choquet. 'Convergences'. *Ann. Inst. Fourier* (Grenoble) 23 (1947-1948), 55-112.

[7] G.Salinetti and R.Wets. 'Convergence of sequences of closed sets'. *Topology Proceedings*, 4 (1979), 149-158.

[8] Z.Frolik. 'Concerning topological convergence of sets'. *Czechoslovak Math. J.* 10 (1960), 168-180.

[9] R.Ash. '*Real Analysis and Probability*'. Academic Press, New York, (1972).

[10] C.Kuratowski. '*Topologie I*'. PNW, Warsaw, (1958).

[11] P.Billingsley. '*Convergence of Probability Measures*'. J. *Wiley*, New York, (1968).

Chapter 17

Stability Analysis in Optimization

T. Zolezzi

1. INTRODUCTION

This work is devoted to a survey of some recent results about the following topics:

1) convergence theorems for approximate and exact solutions to constrained minimum problems in an abstract setting;
2) applications to mathematical programming problems with inequality constraints; behaviour of global solutions and multipliers for non linear and non smooth problems under data perturbations;
3) applications to optimal deterministic control of ordinary differential equations: performance stability as related to relaxation stability.

Topics 1) and 2) have been studied from many viewpoints. See [4], [5],[10], [12],[13],[20] for general references and purposes. The relationships between performance stability and relaxation stability in optimal control have been firstly presented in [11]. Further results about this interesting problem are given here in the last section.

2. CONVERGENCE RESULTS FOR APPROXIMATE AND EXACT SOLUTIONS

We are given a convergence space X, a metric space T and a fixed element denoted by $0 \in T$, a point-to-set mapping

$$K: T \to \text{subsets of } X$$

and an extended real-valued function

$$f: T \times X \to (-\infty, +\infty],$$

which is never $-\infty$. We consider the following family of optimization problems depending on $p \in T$:

1) to minimize the objective function $f(p,x)$ subject to the constraints $x \in K(p)$.

The decision variable is $x \in X$ while p is the parameter. The *limit problem* is that corresponding to $p = 0$, while the *approximate problems* correspond to $p \neq 0$.

In most applications we shall consider, the space X is either the euclidean m-dimensional space R^m, or some subset of a fixed Banach space equipped with the strong or weak convergence.

We shall use the following notations.

$S(p)$ = argmin $f[p,K(p)]$= set of global solutions corresponding to p(if any):

$v(p)$ = inf $f[p,K(p)]$ = optimal value corresponding to p;

$S(p,\varepsilon) = \{x \in K(p): f(p,x) \leq v(p) + \varepsilon\}$ if $v(p) > -\infty$;

$S(p,\varepsilon) = \{x \in K(p): f(p,x) \leq -1/\varepsilon\}$ if $v(p) = -\infty$,

the set of ε-solutions corresponding to p, $\varepsilon > 0$.

The main objective of this section is to survey some basic results about the convergent behaviour of $S(p)$, $S(p,\varepsilon)$ and $v(p)$ as $p \to 0$. Since we assume that the limit problem (corresponding to $p = 0$) has been given in advance, we shall obtain closure results for (ε-) solutions and continuity theorems at $p = 0$ for the optimal value function v. The important issue about existence and representations of the limit problem (in some variational sense) will not be considered here. As far as parametric mathematical programming is involved, existence and explicit expressions of such limits are largely open problems.

The following basic definitions of *variational convergence* on the given convergence space X will be needed. Let us consider a sequence of extended real valued functions

$$q_n : X \to [-\infty, +\infty], \quad n = 0,1,2,\ldots$$

Definition 1: The sequence $q_n \xrightarrow{var} q_o$ iff

(2) $x_n \to x$ implies $\liminf q_n(x_n) \geq q_o(x)$;

(3) for every $u \in X$ and $a > 0$ there exists a sequence $u_n \in X$ such that

$$\limsup q_n(u_n) \leq a + q_o(u) .$$

Let us consider now

$$q : T \times X \to [-\infty, +\infty] .$$

Definition 2: $q(p,\cdot) \xrightarrow{var} q(0,\cdot)$ as $p \to 0$ iff for every sequence $p_n \to 0$ we have $q(p_n,\cdot) \xrightarrow{var} q(0,\cdot)$ according to definition 1.

In the next basic theorem we consider the constrained ninimum problems(1) defined above. We shall denote by

$$i(K,x) = \begin{cases} 0 & , \text{ if } x \in K \\ +\infty & , \text{ if } x \notin K \end{cases}$$

the indicator function of $K \subset X$.

Theorem 1

Assume

(4) $f(0,\cdot)$ proper on $K(0)$;

(5) $f(p,\cdot) + i\,[K(p),\cdot] \xrightarrow{var} f(0,\cdot) + i[K(0),\cdot]$ as $p \to 0$.

Then the following conclusions hold as $p \to 0$:

(6) $\limsup S(p,\varepsilon) \subset S(0,\varepsilon)$ <u>for every</u> $\varepsilon > 0$ <u>sufficiently small</u>;

(7) $$\limsup S(p) \subset S(0) ;$$

(8) $$\limsup v(p) \le v(0) .$$

Proof: Let us prove (6). Given $\varepsilon > 0$ and $p_n \to 0$, let $x_n \in S(p_n,\varepsilon)$ and $x_n \to x$ for some subsequence. Assume $v(p_n) > -\infty$ for every n large enough. Let $z \in K(0)$ such that $f(0,z) < +\infty$. Given $a > 0$, by (5) and (3) there exists a sequence $z_n \in X$ such that

(9) $$\limsup \{f(p_n,z_n) + i[K(p_n),z_n]\} \le f(0,z) + a.$$

Then by (9) $z_n \in K(p_n)$ for all n large enough. Therefore

$$f(0,z) + \varepsilon + a \ge \varepsilon + \limsup f(p_n,z_n) \ge \varepsilon +$$

(10) $$+ \limsup v(p_n) \ge \limsup f(p_n,x_n) \ge \liminf f(p_n,x_n) =$$

$$= \liminf \{f(p_n,x_n) + i[K(p_n),x_n]\} \ge f(0,x) + i[K(0),x]$$

by (5) and (2). Then (10) implies $x \in K(0)$ and $f(0,x) \le f(0,z) + \varepsilon$, thus giving $x \in S(0,\varepsilon)$. Moreover, by (10), $f(0,z) \ge \limsup v(p_n)$, yielding (8). A similar proof gives conclusion (7). Now assume $v(p_n) = -\infty$ for some subsequence. Analogous reasoning yields conclusions (6) and (8) if $\limsup v(p_n) > -\infty$ and $v(0) > -\infty$.

If $v(0) > -\infty$ but $v(p_n) \to -\infty$ with $v(p_n) > -\infty$ for a subsequence, then for some subsequence $f(p_n,x_n) \to -\infty$, therefore by (2)

$$f(0,x) + i[K(0),x] \le \limsup f(p_n,x_n) = -\infty$$

which contradicts properness of $f(0,\cdot)$. If $v(0) > -\infty$ and $v(p_n) = -\infty$ for n sufficiently large, then $f(p_n,x_n) \le -1/_\varepsilon \le v(0) + \varepsilon$ for ε small enough.

Then we get

$$f(0,x)+i\,[K(0),x] \le \liminf f(p_n,x_n) \le v(0) + \varepsilon$$

and (6) is proved in this case. If finally $v(0) = -\infty$, then $f(p_n,x_n) \le$ $\le -1/_{\varepsilon}$ for some subsequence, thus

$$-1/_{\varepsilon} \ge \liminf f(p_n,x_n) \ge f(0,x)+i[K(0),x],$$

hence $x \in S(0,\varepsilon)$. Given $a > 0$ and $z \in K(0)$ such that $f(0,z) \le -a$, we consider z_n fulfilling (9), thereby concluding $\limsup v(p_n) \le -a$. #

Remark Any condition implying lim sup S(p) non empty gives an existence theorem for optimal global solutions to the limit problem. This way has been largely ignored in trying to get existence in calculus of variations and optimal control theory, by firstly constructing suitable approximations to the given problem and then taking limits.

Example 1
Let $T = X = R^1$, $f(p,x) = p^2x^2 - 2p(1-p)x$. Here

$$S(p) = \{1/p-1\} \quad \text{and} \quad v(p) = -(1-p)^2 \text{ if } 0 < p < 1,$$

$$S(0) = [0,+\infty),\ v(p) = 0 \text{ if } p \le 0,$$

$$S(p) = \{0\} \text{ if } p < 0.$$

Therefore the sole assumptions (4), (5) do not imply neither continuity at 0 of the optimal value function, nor of S in any sense (including closure).

Some more (compactness-type) assumptions are needed to get continuity of v at 0.

Let $u : T \to X$.

Definition 3: u is called *pseudominimizing* (for problems (1)) iff $v(p) > -\infty$ for every p and

(11) $$f[p,u(p)] - v(p) \to 0 \quad \text{as} \quad p \to 0 .$$

u is called *asymptotically minimizing* iff it is pseudo-minimizing and $u(p) \in K(p)$ for every p.

Theorem 2

Assume hypothesis (4) and at least one of the following:

(12) assumption (5) holds and there exists u asymptotically minimizing and convergent as $p \to 0$;

(13) there exists u pseudominimizing and convergent to some point of K(0), moreover f is lower semicontinuous at $\{0\} \times X$; given $x \in K(0)$ there exists $w(p) \in K(p)$ such that

$$\limsup f[p,w(p)] \le f(0,x).$$

Then

(14) v is continuous at 0

and the limit point of u as $p \to 0$ is a global optimal solution to the limit problem.

Proof: Assume (12). Given $p_n \to 0$ and $z \in K(0)$ with $f(0,z) < +\infty$, by (10) we get

(15) $$f(0,z) \ge \limsup v(p_n)$$

Since u is asymptotically minimizing and $u(p_n) \to x$ say, then

$$f(0,z) \geq \limsup v(p_n) \geq \liminf v(p_n) =$$

$$= \liminf f\,[p_n, u(p_n)] \geq f(0,x) + i\,[K(0),x] \quad .$$

Then $x \in K(0)$ and $v(p_n) \to v(0)$ since z is arbitrary. Assume (13). Given $z \in K(0)$ with $f(0,z) < +\infty$, let $z_n = w(p_n)$ such that $\limsup f(p_n, z_n) \leq \leq f(0,z)$. Then (15) follows.
Remebering (11) and $u(p_n) \to x$, we conclude

$$f(0,z) \geq \limsup v(p_n) \geq \liminf v(p_n) =$$

$$= \liminf f\,[p_n, u(p_n)] \geq f(0,x) \geq v(0) \quad . \qquad \#$$

Theorem 3
Assume conditions (4) and (5), and suppose v(p) finite everywhere. Then v is continuous at 0 if

(16) $$\cap \{\liminf S(p,\varepsilon) \; ; \; \varepsilon > 0\} \neq \emptyset \; .$$

Proof: Let $p_n \to 0$ and $x \in \liminf S(p,\varepsilon)$ for every $\varepsilon > 0$. Then we find $x_n \in S(p_n, \varepsilon)$ such that $x_n \to x$. We obtain by (6) and (8)

$$v(0) \leq f(0,x) \leq \liminf f(p_n, x_n) \leq \varepsilon + \liminf v(p_n) \leq$$

$$\leq \varepsilon + \limsup v(p_n) \leq \varepsilon + v(0). \qquad \#$$

Some remarks about theorem 3 are in order. The intersection in (16) is included in S(0) by (6). Therefore by theorem 3, a further sufficient condition of continuity of v at 0 (assuming the variational convergence (5)) is that any optimal solution to the limit problem may be obtained as limit of ε-solutions to the approximating problems for every $\varepsilon > 0$. The following example shows that S(0) may be strictly larger than the intersection in (16), even if v is continuous.

Example 2

Let $f(p,x) = x^2$ if $p \neq 0$, $K(p) = R^1$, $f(0,x) = 0$ for every x. Then $S(0) = R^1$ while $v(0) = v(p) = 0$, and $S(p,\varepsilon) = [-\sqrt{\varepsilon}\ , \sqrt{\varepsilon}\ ]$, if $p \neq 0$.

Continuity of v at 0 is equivalent to equality between $S(0)$ and the intersection in (16) for unconstrained problems if variational convergence is strengthened to *epi-convergence*, according to the following

Definition 4: The sequence $q_n : X \to [-\infty, +\infty]$ is called epi-convergent to $q_o : X \to [-\infty, +\infty]$, written $q_n \xrightarrow{epi} q_o$ iff (2) is fulfilled, and moreover

(17) for every $u \in X$ there exists $u_n \to u$ such that

$$\limsup q_n(u_n) \le q_o(u).$$

Given $q: T \times X \to [-\infty, +\infty]$, we write

$$q(p,\cdot) \xrightarrow{epi} q(0,\cdot) \text{ as } p \to 0$$

iff $q(p_n,\cdot) \xrightarrow{epi} q(0,\cdot)$ for every sequence $p_n \to 0$ in T.

Theorem 4

Let $K(p) = X$ for every p, assume (16) and suppose $v(p)$ finite everywhere, $f(p,\cdot) \xrightarrow{epi} f(0,\cdot)$. Then the following are equivalent:

(i) v is continuous at 0;

(ii) $S(0) = \cap\{\liminf S(p,\varepsilon): \varepsilon > 0\}$.

The proof is given in [1]

While inclusions (6) and (7) hold under sufficiently general assumptions, the converse inclusion

(*) $$S(0) \subset \liminf S(p)$$

requires much more restrictive conditions.

See [9] for some related results. An approximate version of (*) is obtained in the following

Theorem 5

Let $K(p) = X$ for all p and assume

(i) for every $x \in X$ there exists $u(p) \to x$ such that

$$\limsup f[p, u(p)] \le f(0, x);$$

(ii) $\liminf v(p) \ge v(0)$.

Then for some $\varepsilon(p) > 0$ with $\varepsilon(p) \to 0$ we have

$$S(0) \subset \liminf S[p, \varepsilon(p)] .$$

The proof is given in [9] (theorem 2.7). A dual statement will be considered in the following section about the behaviour of Lagrange multipliers for convex programming problems (theorem 7).

Bibliographical remarks The definition of variational convergence was introduced in [19] (in a slightly more general way). The definition of epi-convergence (a particular case of the so called Γ-convergences) was introduced in [8]. Theorems 1 and 2 are extensions of some results in [21] . Similar results are discussed in [9]. Theorems 1 and 2 extend most of the known results about continuity of the value function and closedness (or upper semicontinuity) of the optimal solutions in the countable case (as referred to the parameter space T). See [21] for a detailed comparison. Further results are given in [14] .Theorem 3 is an extension of the sufficiency condition to optimal value continuity given by [1] . See [9] for related results. Example 1 is taken from [5] .

Relationship with a result of [6] Following the abstract framework of [6] , we consider two topological spaces U, Y and sequences of subsets and functionals

$$K_n \subset U \times Y \; ; \quad I_n : U \times Y \to [0,+\infty], \; n = 0,1,2,\dots \quad .$$

We assume the following sequential gamma-convergent behaviour:

(i) $u_n \to u$ in U, $y_n \to y$ in Y, $(u_n,y_n) \in K_n$ for some subsequence imply $(u,y) \in K_o$;

(ii) $(u,y) \in K_o$ and $u_n \to u$ in U imply the existence of $y_n \to y$ in Y such that $(u_n,y_n) \in K_n$ for all n large enough;

(iii) $u_n \to u$ in U, $y_n \to y$ in Y imply $\liminf I_n(u_n,y_n) \geq I_o(u,y)$;

(iv) for every $a > 0$ and $u \in U$ there exists a sequence $u_n \to u$ in U such that for every $y_n \to y$ in Y we have $\limsup I_n(u_n,y_n) \leq a + I_o(u,y)$.

We consider the sequence of problems P_n: to minimize $I_n(u,y)$ subject to $(u,y) \in K_n$.

The limit problem corresponds to n = 0. Let us denote by S_n the global optimal solution set for P_n. Then the main abstract result of [6] (theorem 2.1) shows that assumptions (i),...,(iv) are sufficient to get $\limsup S_n \subset S_o$. It is easy to check that (i),...,(iv) imply

$$I_n + i(A_n) \xrightarrow{var} I_o + i(A_o)$$

over U x Y equipped with the convergence inherited from the product topology. Therefore theorem 2.1 of [6] is a particular case of theorem 1. (The assertion made at p.386 of [6] about the generalization of results of reference [9] thereof is wrong since the results alluded to are both necessary and sufficient conditions for convergence).

Well-posedness and stability analysis A deeper (and far-reaching) point of view yields interesting relations between well-posedness and continuous dependence of optimal solutions on constraints for convex minimization problems whenever f is independent on the parameter. Roughly speaking, under uniqueness assumptions, the continuous dependence of the optimal solutions on the moving constraints (a form of well posedness in the classical sense of Hadamard) is equivalent to the convergence of any minimizing sequence to the optimal solution for any (convex) constraint set (a form of well-posedness firstly defined by Tyhonov, that is obviously relevant to the convergence of minimization algorithms).

We consider a real reflexive Banach space X, a continuous convex function $f: X \to (-\infty, +\infty)$.

Under uniqueness conditions we shall denote (as before) by arg min f(K) the (only) minimum point of f on K. Let us recall the following definitions. Given a sequence $K_n \subset X$ we write $K_n \xrightarrow{M} K_o$ iff K_o=seq-weak lim sup K_n = strong lim inf K_n (*convergence in the sense of Mosco*). Given $K \subset X$, we say that (K,f) is *Tyhonov well-posed* iff f has an unique minimum point x on K and

$$x_n \in K,\ f(x_n) \to \inf f(K) \ \text{imply}\ \ x_n \to x.$$

f will be called *Hadamard well-posed* with respect to a given convergence iff K_n bounded convex subsets of X, $K_n \to K$ in the given sense imply argmin $f(K_n) \to$ argmin f(K).

The basic result is the following

Theorem

(i) Assume f is locally bounded. Then (K,f) Tyhonov well-posed for every closed affine half-space K of X implies f Hadamard well-posed with respect to convergence in the sense of Mosco.

(ii) Assume f has exactly one minimum point on every closed convex subset of X. Then f Hadamard well-posed with respect to Hausdorff convergence implies (K,f) Tyhonov well-posed for every closed convex $K \subset X$.

The proof is given in [15] , which contains further results. Extensions to variational inequalities are presented in [16].

3. APPLICATIONS TO MATHEMATICAL PROGRAMMING

We consider the same framework as in section 1 but we assume some more structure. We are given a positive integer q, and functions

$$g_1, g_2, \ldots, g_q : T \times X \to (-\infty, +\infty).$$

We assume that for every $p \in T$

(18) $$x \in K(p) \ , \text{ iff } g_j(p,x) \leq 0 \ , \quad j = 1,\ldots,q.$$

We shall denote by <u>Q(p)</u> the mathematical programming problem (1) with K(p) defined by (18).

For a given $y \in R^q$, $y \geq 0$ means that $y_j \geq 0$ for every $j = 1,\ldots,q$.

The aim of this section is to survey some results about the behaviour as $p \to 0$ of the global solution set and of the multiplier set for the above mathematical programming problem. Due to the great (theoretical and practical) relevance of such problems, there exist many results in this area: for general reference see [5], [12], [18] and the references thereof.

We shall exploit some results from section 1. It is easily seen that conditions (4) and (5) imply

(19) for any sequence $p_n \to 0$, given $u \in K(0)$ such that $f(0,u) < +\infty$, there exists a sequence $u_n \in K(p_n)$ satisfying $\limsup f(p_n,u_n) \leq$ $\leq f(0,u)$.

As a consequence of theorems 1 and 2 we get

Corollary 1 <u>Assume conditions</u> (4) <u>and</u> (19) <u>and suppose</u> $f, g_1, \ldots, g_q$ <u>lower semicontinuous on</u> $\{0\} \times X$.

<u>Then the following conclusions hold:</u>

<u>for every</u> $\varepsilon > 0$, $\limsup S(p,\varepsilon) \subset S(0,\varepsilon)$;

$\limsup S(p) \subset S(0)$;

v <u>is upper semicontinuous at</u> 0.

<u>Moreover if</u>

(i) <u>there exists</u> u <u>asymptotically minimizing and convergent</u>

<u>or</u>

(ii) v <u>is finite and there exists</u> $u: T \to X$ <u>converging as</u> $p \to 0$ <u>such that</u>

$$\limsup g_j\,[p,u(p)] \le 0,\ j=1,..,q;$$

$$f(p,u(p)) - v(p) \to 0 \text{ as } p \to 0$$

<u>then</u> v <u>is continuous at</u> 0.

Throughout the remaining of this section we are concerned with the behaviour of the multipliers as $p \to 0$.

Convex programming Throughout this subsection we shall assume the following:

(20) X is a real Banach space equipped with either the strong or the weak convergence;

(21) $f(p,\cdot)$, $g_j(p,\cdot)$, $j = 1,\dots,q$ are proper convex functions on X, for any $p \in T$.

A role will be played by the following problem <u>Q(p,u)</u>:

(22) to minimize $f(p,x)$ subject to $g_j(p,x) \le u_j$, $j = 1,\dots,q$, for a given $p \in T$ and $u \in R^q$.

As well known, a (Kuhn-Tucker) *multiplier* for problem (22) may be defined as a vector $y \in R^q$, $y \ge 0$, such that the optimal value to (22)is finite and equal to

$$\inf \{f(p,x) + y'\,[g(p,x)-u]: x \in X\}.$$

Here a prime denote transpose, so that y'u denotes the usual scalar product between y and u. The following definition of *continuous convergence* will be used.

Definition 5 Let $g: T \times X \to R^1$. Then $g(p,\cdot) \xrightarrow{\text{cont}} g(p,0)$ iff for every $p_n \to 0$ and $x_n \to x$ we have $g(p_n, x_n) \to g(0,x)$.

Theorem 6

Assume that

(23) $$f(p,\cdot) \xrightarrow{\text{epi}} f(0,\cdot);$$

(24) $$g_j(p,\cdot) \xrightarrow{\text{cont}} g_j(0,\cdot), \quad j = 1,\dots,q.$$

Then

$x(p)$ optimal solutions and $y(0)$ a multiplier for $Q(p)$, $x(p) \to \bar{x}$ and $y(p) \to \bar{y}$ as $p \to 0$ imply $\bar{x}$ an optimal solution and $\bar{y}$ a multiplier to $Q(0)$, moreover $v(p) \to v(0)$.

Proof: Let $p_n \to 0$, $\bar{x}_n = x(p_n)$, $\bar{y}_n = y(p_n)$. Consider

$$L_n(x,y) = \begin{cases} f_n(x) + \sum_{i=1}^{q} y_i g_{ni}(x), & \text{if } y \geq 0, \\ -\infty & \text{, otherwise,} \end{cases}$$

where $f_n(x) = f(p_n, x)$, $g_{ni}(x) = g_i(p_n, x)$. Suppose $x_n \to x$ and let $y \geq 0$. Then $\liminf L_n(x_n, y) \geq L_o(x,y)$. Now given $y_n \to y$ in R^q, $y \geq 0$, let $x_n \to x$ such that $f_n(x_n) \to f_o(x)$. Then $\limsup L_n(x_n, y_n) \leq L_o(x,y)$. Thus we see that corollary 4.4 p.21 of [2] may be applied. Therefore $L_n \to L_o$ in the epi/hypo sense. This implies, thanks to theorem 3.10 p.18 of [3], that $(\bar{x},\bar{y})$ is a saddle point for L_o, moreover $L_n(\bar{x}_n, \bar{y}_n) \to L_o(\bar{x}_o, \bar{y}_o)$. This entails the conclusions of the theorem by well-known properties of convex mathematical programming problems. #

As a partial converse to theorem 6 we have

Theorem 7

Assume $X = R^m$, conditions (23) and (24) and suppose

(i) $\bigcup_{p \in T} \{x \in R^m: f(p,x) \le c, g_j(p,x) \le c, \quad j = 1,\dots,q\}$ is bounded for any real c;

(ii) there exists some $z \in R^m$ such that

$$g_j(0,z) < 0, \quad j=1,\dots,q.$$

Then for any y a multiplier for Q(0) there exists $u(p) \to 0$ as $p \to 0$ and multipliers y(p) for Q[p,u(p)] such that $y(p) \to y$.

The proof is given in [21] .

Bibliographical remarks Corollary 1, theorems 6 and 7 are taken from [21] .

Further results are given in [17] . The short proof of theorem 6 was suggested to me by R.Wets. The results described in theorem 7 may be extended to the infinite dimensional setting with equality constraints added: see [21] . Theorem 7 is a dual version of theorem 5.

Remarks about theorem 7 One may wonder whether the conclusions of theorem 7 follow (without assumptions (i) and (ii)) by using proposition 3.12 of [3] . There it is required that (notations in the proof of theorem 6)

(iii) for every x there exists $x_n \to x$ such that

$$\sup_y L_n(x_n,y) \to \sup_y L_o(x,y) .$$

Let q = 1, X = R, $f_n(x) = f_o(x) = x^2$, $g_n(x) = x/n$, $g_o(x) = 0$ for every x. The only optimal solution to the n-th problem is $\bar{x}_n = 0$, its optimal value is $v_n = 0$, and the only multiplier is $\bar{y}_n = 0$. The multiplier set for the limit problem is $[0,+\infty)$. Condition (iii) does not hold , since if $x > 0$ and $x_n \to x$, then $\sup_{y \geq 0} L_n(x_n, y) = +\infty$ for all large n, while $\sup_{y \geq 0} L_o(x,y) = x^2$.

In this example the conclusion of theorem 7 holds (but Slater's condition (ii) fails). This example shows that proposition 1.17 of [2] extends only conclusion (22) of theorem 4 in [21] , since conclusion (23) thereof cannot be deduced in such a way (as stated in [2] , p.11).

Mathematical programming with locally Lipschitz data In this subsection we shall consider the mathematical programming problems Q(p) (as defined by (22) with u = 0) and we shall suppose that $X = R^m$; $f(p,\cdot)$, $g_j(p,\cdot)$ are locally Lipschitz for every $p \in T$.

As shown by Clarke, if $x \in S(p)$ there exist numbers $y_o, y_1, \ldots, y_q$, not all zero, such that the following multiplier rule holds:

$$(25) \qquad \begin{cases} 0 \in y_o \, \partial f(p,x) + \sum_{j=1}^{q} y_j \, \partial g_j(p,x); \\ \\ y_j \geq 0 \quad \text{and} \quad y_j g_j(p,x) = 0, \; j=1,\ldots,q. \end{cases}$$

Here ∂ denotes the Clarke's generalized gradient (see [7]).

Given $x \in S(p)$ we denote by

M(p,x) the set of all vectors $y \in R^{q+1}$ satisfying (25) (F.John multipliers for Q(p) corresponding to the global optimal solution x).

Definition 6 (i) Let $h: R^m \to R^1$ and $x \in R^m$ be given. Then $u \in \partial^- h(x)$ iff for every z

$$h(z) \geq h(x) + u'(z-x) + o(z)$$

where $\frac{o(z)}{|x-z|} \to 0$ as $z \to x$.

(ii) Let $h: T \times R^m \to R^1$ be given. Then h is *equi-lower semidifferentiable* iff for every open ball $B \subset R^m$ there exists a continuous function k on $B \times B$ such that, for every $x \in B$, $\frac{k(x,y)}{|x-y|} \to 0$ as $y \to x$, and for every $p \in T$, x and y in B, any $u \in \partial^- h(p,x)$ we have

$$h(p,y) \geq h(p,x) + u'(y-x) + k(x,y) .$$

Theorem 8

Assume conditions (23), (24), (19) and the following hypothesis:

f, g_j are locally bounded and equi-lower semidifferentiable for all j.

Then

$$x(p) \in S(p) \quad \text{and} \quad x(p) \to x \quad \text{imply}$$

$$\emptyset \neq \limsup M\,[p,x(p)] \subset M(0,x)$$

and (of course) continuity of v at 0 together with $x \in S(0)$.

The proof is given in [22] . Extensions to equality constraints may be found therein.

Behaviour of multipliers for smooth data Assume that $f(p,\cdot)$ and $g_j(p,\cdot)$ belong to $C^2(R^m)$ and suppose that the Mangasarian-Fromovitz constraint qualification holds for Q(0) at a given x_o, which satisfies the classical second-order sufficient conditions for local optimality corresponding to p = 0. Then, roughly speaking, the portion of local solutions to Q(p) which are sufficiently near to x_o behaves in a (lower and upper semi) continuous way as $p \to 0$, while the set of corresponding multipliers is upper semicontinuous at p = 0.

This stability theorem, a particular case of the theory developed in [18] may be partially extended by the following corollary to theorem 8. We shall use the following terminology. The point-to-set mapping

$$V: T \to \text{subsets of } R^n$$

is called *closed* (the same as uppersemicontinuous by inclusion) iff

$$p_n \to p \text{ in } T\ ,\ w_n \in V(p_n) \text{ and } w_n \to w \text{ imply } w \in V(p).$$

V is called *lower semicontinuous by inclusion* iff

$$p_n \to p \text{ in } T \text{ implies } V(p) \subset \liminf V(p_n).$$

The multiplier point-to-set map M will be called here *closed* iff

$$p_n \to p \text{ in } T,\ x_n \in S(p_n),\ x_n \to x,\ y_n \in M(p_n, x_n) \text{ and}$$

$$y_n \to y \text{ imply } y \in M(p,x) \text{ (under conditions forcing } x \in S(p)).$$

Corollary 2 Let f and g_j be continuous and locally equibounded. Let $f(p,\cdot)$, $g_j(p,\cdot)$ belong to $C^2(R^m)$ with locally equibounded second partial derivatives, for all p and j. Assume that the feasible region K is lower semicontinuous by inclusion. Then both S and M are closed.

The proof is given in [22] , along with a comparison with known results in stability analysis of multipliers for smooth problems.

4.PERFORMANCE STABILITY AND RELAXATION STABILITY IN OPTIMAL CONTROL

In this section we consider the following standard problems in optimal control, depending on the parameter $p \in T$ (a metric space as before) through state constraints and dynamics: to minimize $f[x(b)]$ subject to

state equations $\dot{x} = g(t,x,u,p)$, $a \le t \le b$, $x(a) = x_o$;

state constraints $x(t) \in H(t,p)$, $a \le t \le b$;

control constraints $u(t) \in U$.

The time interval [a,b] is fixed throughout. *Admissible controls* are (as usual) measurable functions u on [a,b] such that $u(t) \in U$ a.e. The limit problem corresponds (as before) to p = 0. The *relaxed problem* of the limit one is defined as follows (in the standard fashion): to minimize $f[y(b)]$ subject to

state equations $\dot{y}(t) \in \text{co } g[t, y(t), U, 0]$, $a \leq t \leq b$; $y(a) = x_o$;

state and control constraints as before with $p = 0$.

We shall denote by $v(p)$ the optimal value corresponding to the parameter $p \in T$, and by v^* the optimal value of the relaxed problem.

Definition 7 For the above optimal control problems we have *performance stability* iff v is continuous at 0. For the limit problem we have *relaxation stability* iff $v(0) = v^*$.

In this section we wish to consider relationships between performance stability and relaxation stability, which may be useful on two counts. Firstly, if we are interested in performance stability, we shall find sufficient conditions to it (and necessary conditions too) by assuming relaxation stability. Conversely, if we are interested in relaxation stability, we shall get it by proving performance stability with respect to properly chosen perturbations of the given control problem.

To state the next theorem, given $x \in H(b,0)$, $\varepsilon > 0$ small enough, we need to define the following set:

$$Q(t,p) = \{y \in H(t,p) : f(y) \leq f(x) + \varepsilon\}.$$

Theorem 9

Relaxation stability implies performance stability under the following assumptions:

(i) g is continuous on $[a,b] \times R^m \times U \times \{0\}$ and bounded ; f is continuous; $H(t,\cdot)$ is upper semicontinuous at $p = 0$, closed valued with $H(t,0)$ compact; $H(\cdot,p)$ is closed; U is compact; given $x \in H(b,0)$ and $\varepsilon > 0$, for every p sufficiently small and for some admissible trajectory y we have

(ii) $y(b) \in Q(b,p)$

or (more particularly)

(iii) for some $t^* \in (a,b)$ $y(t^*) \in Q(t^*,p)$, $Q(t,p) \neq \emptyset$ if $t^* \leq t \leq b$ and for some admissible continuous control u, all $z \in Q(t,p)$, we have

$$\liminf_{h \to 0+} h^{-1} \operatorname{dist}(z + hg(t,z,u(t),p), Q(t+h,p)) = 0.$$

Sketch of the proof: The inequality $\liminf v(p) \geq v^*$ is obtained from (i) by standard means (which use theorem 3.1.7 of [7]). The inequality $v(0) \leq v^*$ comes from relaxation stability. The inequality $\limsup v(p) \leq$ $\leq v(0)$ comes from (8) of theorem 1 through assumptions (ii),(iii) which (by viability theory) give existence of feasible state trajectories y for the perturbed problems verifying the constraints $f([y(b)] \leq v(0) + \varepsilon$.

Example 3

We wish to minimize $-x(1)$ subject to

$$\dot{x} = u-p,\ x(0) = 0,\ 0 \leq t \leq 1,\ |u(t)| \leq 1,\ p > 0,$$

with state constraints $x(1) = 0$ or $|x(1) - 1| \leq p^\alpha$.

Theorem 9 may be applied if $\alpha \leq 1$, then we have performance stability. Such a conclusion is false if $\alpha > 1$. No variational convergence may be obtained in this case (since (19) fails).

In the next theorem we denote by $x(u,p)$ the state variable corresponding to the control u and parameter p (under uniqueness conditions).

Theorem 10

Performance stability implies relaxation stability under the following assumptions: hypothesis (i) of theorem 9, moreover $g(t,\cdot,u,0)$ is uniformly Lipschitz and of linear growth with $L^1(a,b)$ "constant"; there exists γ with $\gamma(p) \to 0$ as $p \to 0$ such that for every admissible control u

$$|x(u,p)(t) - x(u,0)(t)| \leq \gamma(p);$$

There exists an optimal state trajectory x^* to the relaxed problem such that for all t and $p \neq 0$

$$|y - x^*(t)| \leq \gamma(p) \text{ implies } y \in \text{int } H(t,p).$$

The proof is given in [11] (theorem 2).

By considering suitable perturbations to the original problem and assuming performance stability we get relaxation stability. This fact is intuitively clear (see e.g. [7], p.224). A simple statement may be obtained as follows.

Let us consider the following optimal control problem $\underline{Q}$: to minimize $f[x(b)]$ subject to

$$\dot{x} = g(t,x,u) \quad , \quad x(a) = x_o,$$

$$x(t) \in H(t) \quad , \quad u(t) \in U \ , \ a \leq t \leq b,$$

and the perturbations to it obtained by leaving unaffected the performance index, dynamics and control constraints, and by altering the state constraints as follows:

$$x(t) \in H(t) \ , \quad a \leq t \leq b;$$

$$x(b) \in H(b,p) \quad , \quad p > 0$$

where

$$H(b,p) = \{y \in R^m : \text{dist } [y,H(b)] \leq p\}.$$

Theorem 11

Relaxation stability holds for Q if we have performance stability (for the above defined problems) and

(i) solutions to $\dot{x} \in g(t,x,U)$, $x(a) = x_o$, are uniformly dense in those to $\dot{x} \in \text{cl co } g(t,x,U)$, $x(a) = x_o$;

(ii) $H(b)$ is closed, f is continuous.

The proof is standard.

Remarks Theorem 9 is a variant to theorem 1 of [11]. We use here more explicit assumptions involving problem data. The use of viability theory to get performance stability through variational convergence is new. Example 3 is taken from [11]. Further results about this topic will appear elsewhere (extending these results to an abstract setting akin to that considered by Ioffe-Tyhomirov and Ekeland-Temam in the study of relaxed problems in optimization and calculus of variations).

REFERENCES

[1] H.Attouch, R.Wets. 'Approximation and convergence in nonlinear optimization'. Nonlinear programming 4, edited by O.Mangasarian - R. Meyer - S.Robinson, *Academic Press*, (1981), 367-394.

[2] H.Attouch, R.Wets. 'A convergence for bivariate functions aimed at the convergence of saddle values'. *Lecture Notes in Mathematics*, 979, (1983), 1-42.

[3] H.Attouch, R.Wets. 'A convergence theory for saddle functions'. *Trans. Amer. Math. Soc.*, 280 (1983), 1-41.

[4] J.P.Aubin. 'Mathematical methods of game and economic theory'. *North Holland*, (1979).

[5] A.Bank, J.Guddat, D.Klatte, B.Kummer, K.Tammer. 'Non-linear parametric optimization'. Birkhäuser, (1983).

[6] G.Buttazzo, G.Dal Maso. ' Γ-convergence and optimal control problems'. *J.Optim. Theory Appl.*, 38 (1982), 385-407.

[7] F.Clarke. 'Optimization and nonsmooth analysis'. *Wiley - Interscience*, (1983).

[8] E.De Giorgi, T.Franzoni. 'Su un tipo di convergenza variazionale'. *Atti Accad. Naz. Lincei*, 58 (1975), 842-850.

[9] S.Dolecki. 'Convergence of global minima and infima'. *Seminaire d'Analyse Numerique*, Toulouse, (1982,1983).

[10] A.Dontchev. 'Perturbations, approximations and sensitivity analysis in optimal control systems'. *Lecture notes in Control and Informations Sciences*, 52, Springer, (1983).

[11] A.Dontchev, B.Morduhovic. 'Relaxation and well-posedness of nonlinear optimal processes'. *Systems and Control Letters* 3 (1983), 177-179.

[12] A.Fiacco. 'Introduction to sensitivity and stability analysis in nonlinear programming'. *Academic Press*, (1983).

[13] R.Lucchetti. 'On the continuity of the optimal value and of the optimal set in minimum problems'. Pubblicazioni *I.M.A.* 133, Genova, (1983).

[14] R.Lucchetti. 'On the continuity of the minima for a family of constrained optimization problems'. To appear in *Numer.Funct.Anal.Optim.*
[15] R.Lucchetti, F.Patrone. 'Hadamard and Tyhonov well-posedness of a certain class of convex functions'. *J.Math. Anal. Appl.* 88 (1982), 204-215.
[16] R.Lucchetti, F.Patrone. 'Some properties of "well-posed" variational inequalities governed by linear operators'. *Numer. Funct. Anal. Optim.* 5 (1982,1983), 349-361.
[17] R.Lucchetti, F.Patrone. 'Closure and upper semicontinuity results in mathematical programming, Nash and economic equilibria'. Submitted.
[18] S.Robinson. 'Generalized equations and their solutions, part 2:applications to nonlinear programming'. *Math. Programming Study* 19, (1982), 200-221.
[19] T.Zolezzi. 'On convergence of minima'. *Boll. Un. Matem. Ital.* , 8 (1973), 246-257.
[20] T.Zolezzi. 'Some convergence results in optimal control and mathematical programming'. Proccedings workshop in Differential Equations and their Control, Iasi, (1982), edited by V.Barbu-N.Pavel. University of Iasi and I.N.C.R.E.S.T., Iasi (1983).
[21] T.Zolezzi. 'On stability analysis in mathematical programming'.*Math . Programming Study,* 21 (1984), 227-242.
[22] T.Zolezzi. 'Continuity of generalized gradients and multipliers under perturbations'. To appear in *Math. Oper. Research.*